Romans, Rubbish, and Refuse

The archaeobotanical assemblage of Regione VI, insula I, Pompeii

Charlene Alexandria Murphy

Archaeopress Publishing
Gordon House
276 Banbury Road
Oxford OX2 7ED

www.archaeopress.com

ISBN 978 1 78491 115 7
ISBN 978 1 78491 116 4 (e-Pdf)

© Archaeopress and C A Murphy 2015

Cover image: Photomosaic of Insula VI.I from the Via Consolare
(Photomosaic © 2007 Jennifer F. Stephens · Arthur E. Stephens)

All rights reserved. No part of this book may be reproduced, stored in retrieval system,
or transmitted, in any form or by any means, electronic, mechanical, photocopying or otherwise,
without the prior written permission of the copyright owners.

Printed in England by Holywell Press, Oxford
This book is available direct from Archaeopress or from our website www.archaeopress.com

For

Dr. S. E. Stephenson

Contents

List of Figures v
List of Tables..viii
Acknowledgements ix
Glossary xi

Chapter 1: Introduction 1
    1.1 Pompeii: historical background 1
    1.2 Pompeii: a case study in urban archaeobotany 3
    1.3 Research aims and objectives... 3
    1.4 Research questions 3
    1.5 Status, ethnicity and Romanisation 3
    1.6 City and hinterland 4
    1.7 Pompeii as a type site and economic models 4
    1.8 Situate environmental data 5
    1.9 Domestic space... 5

Chapter 2: Historical Background 6
    2.1 Introduction to the Mediterranean Environment 6
    2.2 Palynological evidence.. 6
    2.3 The modern environment 7
    2.4 Historical overview of the city of Pompeii 8
    2.5 Roman literary sources... 10
    2.6 Ancient literary biases 10
    2.7 Viticulture and villas 11
    2.8 Latifundia 11
    2.9 The hinterland... 12
    2.10 Economy.. 13
    2.11 Ceramic evidence from Pompeii 13
    2.12 Regione VI, Insula I: brief historical overview 15
    2.13 The House of the Vestals (Casa delle Vestali) (VI.IVII) 17
    2.14 The House of the Surgeon (*Casa del Chirurgo*) (VI.I.x) 21

Chapter 3: Methodology...25
    3.1 Research methods 25
    3.2 Recovery methods 25
    3.3 Sampling.. 26
    3.4 Distribution of sample number by type 26
    3.5 Identification 27
    3.6 Quantification 27
    3.7 Fruit 27
        Grape 28
        Olives 28
        Fig.. 29
        Nutshells 29
    3.8 Cereals 29
    3.9 Millet 30
    3.10 Pulses... 30
    3.11 Limitations of the data 31
        Taphonomy 31
        Modern taphonomic factors... 31
    3.12 Historical factors 32
    3.13 Current research 33
    3.14 The field school 33
    3.15 Contexts 33
    3.16 Preservation.. 34

Carbonisation	34
Mineralisation	34

Chapter 4: Results 37
    4.1 Contexts examined from Insula VI.I 37
    4.2 Contexts by time period 37
    4.3 Total archaeobotanical remains 37
    4.4 Preservation 39
    4.5 Carbonised archaeobotanical remains from Insula VI.I 39
    4.6 Mineralised archaeobotanical remains from Insula VI.I 42
    4.7 Residue volume 44
    4.8 Unknowns/indeterminate archaeobotanical material from Insula VI.I 45
        Carbonised unknowns/indeterminate archaeobotanical material from Insula VI.I 45
        Mineralised unknowns/indeterminate archaeobotanical material from Insula VI.I 46
    4.9 Recovery method flot/residue/*in situ*/dry sieved 47
    House of the Surgeon (VI.I.x) (*Casa della Chiurugo*) 48
        House of the Surgeon (VI.I.x) 1st century BC 48
        House of the surgeon (VI.I.x) 1st century AD 48
    House of the Vestals (VI.I.vii) (*Casa delle vestali*) 50
        House of the Vestals (VI.I.vii) 1st century BC 51
        House of the Vestals (VI.I.vii) 1st century AD 51
    Shrine (VI.I.xiii) 52
        Shrine (VI.I.xiii) 1st century BC 52
        Shrine (VI.I.xiii) 1st century AD 53
    Soap factory (VI.I.xiv) 54
        Soap factory 1st century AD 55
    Triclinium 56
        Triclinium 1st century AD 56
    Via Consolare 57
        Via Consolare 1st century BC 57
        Via Consolare 1st century AD 57
    Vicolo di Narciso 58
        Vicolo di Narciso 1st century BC 58
        Vicolo di Narciso 1st century AD 58
    Comparison of archaeobotanical assemblage from roads 59
    Well 59
        Well 1st century BC 59
        Well 1st century AD 60
    4.10 Unique deposits from insula VI.I 60
        Ritual deposit 60
        Latrine 60
        Drain 61
    4.11 Unique deposits from the House of the Vestals (VI.I.vii) 63
        Mineralised archaeobotaanical remains 63
        Carbonised taxa from the House of the Vestals (VI.I.vii) 64
    4.12 Summary 67

Chapter 5: Comparison of Archaeological Evidence 69
    5.1 Initial observations on plants at Pompeii 69
    5.2 Archaeological research on Pompeii 69
        Earliest evidence from outside of Pompeii 69
        Shops 69
        Bars 69
        Domestic houses within Pompeii 69
        The House of the Vestals (VI.I.vii) 69
        The House of Amarantus (I.IX.xii) 70
        The House of Hercules' Wedding (VII.IX.xlvii) 71
    5.3 Villae 71
        Villae rusticae 71
    5.4 Archaeobotanical evidence from outside Pompeii 72
        Herculaneum 72

  Oplontis 72
 5.5 Garden evidence 72
  Garden evidence at Pompeii 72
  Pollen evidence from gardens 73
  Garden evidence from the House of the Vestals (VI.I.vii) 74
  Garden evidence from outside of Campania 75
  Limitations 75
 5.6 Ritual botanical evidence 75
  Pompeii 75
 5.7 Medicinal drugs 76
 5.8 Charcoal analysis from Pompeii 76
 5.9 Trade 77
 5.10 Broad trends in archaeobotanical evidence from the Roman world 77
  Rome 77
 5.11 Summary 78

Chapter 6 : Food and Roman Culture 80
 6.1 Roman agriculture 80
 6.2 Food, ideology and Roman culture 80
 6.3 Cuisine 80
 6.4 Symbolism 81
 6.5 Cultural identity 81
  Wheats 82
 6.6 Late Republican to Early Imperial Period Dietary shifts in meat preference 82
 6.7 Dietary shifts in cereal preferences 84
  Dining 85
  Luxury 86
 6.8 Historical perspective on food 87
  Cookery 88
  Gardens 88
  Cereals 90
  Einkorn 90
  Emmer 90
  Barley 91
  Free threshing/soft bread wheat 92
  Oats 92
  Millet 93
 6.9 Spices 94
  Opium 94
  Pepper 94
  Sesame 95
 6.10 Fruit 95
  Olive 95
  Olive oil 97
  Grapes 97
  Peaches 97
  Fig 97
  Dates 98
  Pomegranates 98
  Apples 98
  Citron 98
 6.11 Nuts 99
  Hazelnut 99
  Walnut 99
  Almond 99
 6.12 Pules 100
  Peas 100
  Lentils 100
  Chickpeas 100
  Broad bean 100

  Bitter vetch 101
 6.13 Trees 102
  cypress 102
  Stone pine 102
 6.14 Summary 102

Chapter 7: Discussion: Rubbish, Floors and Food 104
 7.1 Interpretation of reults from Insula VI.I 104
 The roads 104
  Via Consolare 104
  Vicolo di Narciso 104
  Commercial triangle 105
  Bar of Pheobus (VI.I.xviii) 105
  Bar of Acisculus (VI.I.xvii) 105
  Soap Factory/Metal workshop (VI.I.xiv) 106
  North 107
  Triclinium 107
  INN 107
  Inn Bar 108
  Commercial properties 108
  Domestic properties 108
  House of the Vestals (VI.I.vii) 108
  Vestals bar 109
  House of the Surgeon (VI.I.x) 110
  Shrine (VI.I.xiii) 110
  Well 111
 7.2 Sampling recovery 112
 7.3 Indeterminates/unknowns 112
 7.4 Disposal 113
 7.5 Household changes 114
 7.6 Changes over time in Insula VI.I. 115
  Pre-Roman to Roman 115
  Cereal by-products 115
  Evidence of cereal by-products 116
  Insula VI.I 116
 7.7 Evidence of Urbanisation 116
 7.8 City and country 117
 7.9 Preservation 117
 7.10 Lack of archaeobotanical remains 118
 7.11 Summary 118

Chapter 8: Conclusions 119
 8.1 Integration of results 119
 8.2 Preservation 119
 8.3 Evidence of urbanisation 119
 8.4 Shifts in food preferences 120
 8.5 Spatial distribution and patterns 121
 8.6 Future research 121
 8.7 Summary 121
 8.8 Final remark 121

Bibilography 122

Appendix 136

List of Figures

Figure 1: Pompeii and the surrounding region 1
Figure 2: Photomosaic of Insula VI.I from the Via Consolare 15
Figure 3: Map of the city of Pompeii 15
Figure 4: Insula VI.I (AAPP 2005 Resource Book p. 26) 16
Figure 5: House of the Vestals (VI.I.vii) frontage (Photomosaic © 2007 Jennifer F. Stephens · Arthur E. Stephens)...................17
Figure 6: Map of excavation of Pompeii 17
Figure 7: The southern end of Insula VI.I 19
Figure 8: House of the Surgeon (VI.I.x) frontage 21
Figure 9: House of the Surgeon (VI.I.x) 22
Figure 10: House of the Surgeon (VI.I.x) entranceway with view into the *hortus* 23
Figure 11: Timeline for Regione VI, insula I 24
Figure 12: Total volume of residue (L) examined per property from Insula VI.I 26
Figure 13: Number of samples examined from different properties within Insula VI.I 27
Figure 14: Number of Samples examined per property type 27
Figure 15: Mineralised *Malus domestica* (Apple) seed from Insula VI.I 27
Figure 16: a. Carbonised *Vitis vinifera* (Grape) pips and b. Mineralised *Vitis vinifera* (Grape) pips 28
Figure 18: SEM of carbonised *Vitis vinifera* (Grape) pip from Insula VI.I 28
Figure 19: a. Carbonised half fragment of *Vitis vinifera* (grape) pip and b. Carbonised grape pip 28
Figure 20: SEM of carbonised *Vitis vinifera* (grape) petiole from Insula VI.I 28
Figure 21: Whole carbonised *Olea europea* (olive) 29
Figure 22: *Ficus carica* (fig) carbonised mesocarp from Insula VI.I 29
Figure 23: a. and b: SEMs of carbonised *Ficus carica* (fig) mesocarp from Insula VI.I 29
Figure 24 a. and b: Mineralised *Ficus carica* (fig) from Insula VI.I 29
Figure 25: SEM of one mineralised *Ficus carica* (fig) from Insula VI.I 29
Figure 26: a. Carbonised *Pinus pinea* (pine nut) shell fragment a. inside view b. outer shell view from Insula VI.I....................29
Figure 27: a. Spikelet fork of *T. dicoccum* (Emmer) b. Spikelet fork of *T. dicoccum* (Emmer) from Insula VI.I 30
Figure 28: a. *Hordeum vulgare* (Barley) (dorsal view) and b. *Hordeum vulgare* (Barley) (ventral view) from Insula VI.I 30
Figure 29: SEM of carbonised *Hordeum vulgare* (Barley) (dorsal view) from Insula VI.I 30
Figure 30: Badly damaged carbonised *Triticum* sp. from Insula VI.I..30
Figure 31: a. Carbonised *Panicum* b. Carbonised Setaria both from Insula VI.I 30
Figure 32: Carbonised *Vicia* sp. from Insula VI.I 30
Figure 33: a. Carbonised *Vicia sativa* and b. *Vicia faba* from Insula VI.I..31
Figure 34: Total number of contexts examined grouped by time period 37
Figure 35: Number of contexts examined by time period shown by property from Insula VI.I 37
Figure 36: Total count of archaeobotanical remains (carbonised and mineralised remains) per property from Insula VI.I 37
Figure 37: Percentage of total count of archaeobotanical remains recovered from Insula VI.I by property...............................37
Figure 38: Total number of taxa (carbonised and mineralised) per property from Insula VI.I 37
Figure 39: Number of mineralised and carbonised taxa from published sources on Vesuvian sites..38
Figure 40: Total number of archaeobotanical (carbonised and mineralised) remains per context 38
Figure 41: Total volume of residue examined plotted 38
Figure 42: Archaeobotanical remains (carbonised and mineralised) 38
Figure 43: Percentage of archaeobotanical remains recovered from Insula VI.I by preservation 39
Figure 44: Total count of charred botanical remains per property from Insula VI.I 39
Figure 45: Total count of carbonised archaeobotanical remains per property from Insula VI.I 39
Figure 46: Total count of charred botanical remains per property from Insula VI.I by time period...39
Figure 47: Total count of carbonised botanical remains and total volume of residue examined per property from Insula VI.I 40
Figure 48: Charred botanical remains per L of residue examined per property from Insula VI.I 40
Figure 49: Number of carbonised taxa per property from Insula VI.I 40
Figure 50: Histogram of carbonised taxa from Insula VI.I 40
Figure 51: Number of carbonised taxa from Vesuvian Sites 41
Figure 52: Histogram of carbonised taxa from Vesuvian sites 42
Figure 53: Carbonised archaeobotanical remains per context examined per property from Insula VI.I....................................42
Figure 54: Histogram of recovered carbonised botanical remains per context from Insula VI.I 42
Figure 55: Total count of mineralised archaeobotanical remains per property from Insula VI.I 42
Figure 56: Total count of mineralised archaeobotanical remains by property from Insula VI.I..42
Figure 57: Total count of mineralised botanical remains recovered per property from Insula VI.I by time period....................42
Figure 58: Total count of mineralised botanical remains recovered...43
Figure 59: Mineralised botanical remains recovered per L of residue examined per property from Insula VI.I 43
Figure 60: Number of mineralised taxa per property from Insula VI.I...43
Figure 61: Histogram of mineralised taxa from Insula VI.I 43
Figure 62: Number of mineralised taxa from Vesuvian sites 44
Figure 63: Histogram of mineralised archaeobotanical remains from published sources on Vesuvian sites 44
Figure 64: Mineralised botanical remains per context per property from Insula VI.I..44
Figure 65: Histogram of recovered mineralised botanical remains per context from Insula VI.I 44
Figure 66: Total volume (L) of residue examined per property from Insula VI.I 44
Figure 67: Total residue volume (L) per property from Insula VI.I 44

Figure 68: Breakdown of Residue Volume (L) by time period 45
Figure 69: Percentage of total carbonised archaeobotanical unknowns/indeterminates from Insula VI.I by time period 45
Figure 70: Count of carbonised indeterminates/unknowns by property and time period from Insula VI.I 45
Figure 71: Percent of carbonised archaeobotanical remains per property from Insula VI.I that were unknown/indeterminate 45
Figure 72: Percent of total count of mineralised unknowns/indeterminates from Insula VI.I 46
Figure 73: Total count of mineralised indeterminate by property by time period from Insula VI.I 46
Figure 74: Percentage of unknown mineralised remains per property from Insula VI.I 46
Figure 75: Total count of carbonised and mineralised unknowns from Insula VI.I by time period 46
Figure 76: Excavation areas for the House of the Surgeon (VI.I.x) (AAPP05 Resource Handbook p. 27) 47
Figure 77: 1st century BC archaeobotanical remains from the House of the Surgeon 48
Figure 78: Count of archaeobotanical remains from the House of the Surgeon (VI.I.x) by category from the 1st century BC 48
Figure 79: Histogram of count of archaeobotanical remains from the 1st century BC from the House of Surgeon (VI.I.x) 48
Figure 80: Count of archaeobotanical remains from the House of the Surgeon (VI.I.x) from the 1st century AD 48
Figure 81: House of the Surgeon (VI.I.x) areas where no archaeobotanical remains were recovered 49
Figure 82: Ubiquity of archaeobotanical remains from House of the Surgeon (VI.I.x) from flotation 49
Figure 83: shows the distribution of the total count of archaeobotanical (carbonised and mineralised) remains per room. 49
Figure 84: Excavation areas for the House of the Vestals (VI.I.vii) 50
Figure 85: House of the Vestals (VI.I.vii) Room numbers 50
Figure 87: 1st century BC histogram of archaeobotanical remains from the House of Vestals (VI.I.vii) 51
Figure 86: 1st century BC archaeobotanical remains from the House of the Vestals (VI.I.vii) 51
Figure 89: 1st century AD histogram of archaeobotanical remains from the House of the Vestals (VI.I.vii) 51
Figure 88: Total count of archaeobotanical remains (mineralised and carbonised) 51
Figure 90: Excavation areas from the Shrine (VI.I.xiii) (AAPP05 Resource Handbook p. 27) 52
Figure 91: Archaeobotanical remains recovered from the Shrine (VI.I.xiii) from the 1st century BC or earlier 52
Figure 92: Histogram of archaeobotanical remains from the Shrine (VI.I.xiii) from the 1st century BC or earlier 52
Figure 93: 1st century AD archaeobotanical remains from the Shrine (VI.I.xiii) 53
Figure 94: Histogram of archaeobotanical remains recovered from the Shrine (VI.I.xiii) from the 1st century AD. 53
Figure 95: Excavation areas from the Soap Factory (VI.I.xiv) (AAPP05 Resource Handbook p. 27) 54
Figure 96: Archaeobotanical remains from 1st century BC from the Soap Factory (VI.I.xiv) 54
Figure 97: Histogram of 1st century BC archaeobotanical remains from the Soap Factory (VI.I.xiv) 54
Figure 98: Archaeobotanical remains from 1st century AD from the Soap Factory (VI.I.xiv) 55
Figure 99: Histogram of count of archaeobotanical remains from the Soap Factory (VI.I.xiv) from the 1st century AD 55
Figure 100: Excavation areas from the Triclinium located within the Inn (AAPP05 Resource Handbook p. 27) 56
Figure 101: Archaeobotanical remains recovered from the Triclinium by preservation and general category from Insula VI.I 56
Figure 102: Carbonised archaeobotanical remains from the Triclinium from Insula VI.I 56
Figure 103: Excavation areas from Via Consolare 57
Figure 104: Archaeobotanical remains from 1st century BC from Via Consolare 57
Figure 105: Archaeobotanical remains from 1st century AD from Via Consolare 57
Figure 106: Excavation areas from Vicolo di Narciso (AAPP05 Resource Handbook p. 27) 58
Figure 107: 1st century BC archaeobotanical remains from Vicolo di Narciso 58
Figure 108: 1st century AD count of archaeobotanical remains from Vicolo di Narciso 58
Figure 109: Excavation areas from the Well 59
Figure 110: 1st century BC archaeobotanical remains from the Well from Insula VI.I 59
Figure 111: Histogram of count of archaeobotanical remains from the Well from the 1st century BC 59
Figure 112: 1st century AD archaeobotanical remains from the Well from Insula VI.I 60
Figure 113: Mineralised *Agrostemma githago* L. (corn cockle) from Insula VI.I (not from the latrine deposit) 61
Figure 114: Total count of mineralised botanical remains recovered from the late 1st century BC toilet 61
Figure 115: Mineralised lentil recovered from Insula VI.I 61
Figure 116: Total number of mineralised and carbonised taxa from the House of the Vestals (VI.I.vii) 62
Figure 117: Archaeobotanical material recovered from Vicolo di Narciso from the 1st century BC 62
Figure 118: Archaeobotanical material by category recovered from Vicolo di Narciso from the 1st century AD 62
Figure 119: Mineralised Remains from the 1st century AD from Vicolo di Narciso 62
Figure 120: Carbonised *Ornithopus perpusillus/sativus* seed from Insula VI.I 63
Figure 121: Count of mineralised taxa from the House of the Vestals (VI.I.vii) 63
Figure 122: Count of Mineralised Remains of Fruits and Nuts from the House of the Vestals (VI.I.vii) 63
Figure 123: Mineralised taxa from specific contexts within the House of the Vestals (VI.I.vii) 64
Figure 124: Mineralised weed seeds from specific deposits from the House of the Vestals (VI.I.vii) 64
Figure 125: Carbonised taxa from the House of the Vestals (VI.I.vii) 64
Figure 126: Carbonised taxa from specific deposits within the House of the Vestals (VI.I.vii) 64
Figure 127: Count of carbonised taxa recovered from the House of the Vestals (VI.I.vii) 65
Figure 128: Carbonised botanical remains per L from specific contexts from the House of the Vestals (VI.I.vii) 65
Figure 129: Count of *T. dicoccum* from Insula VI.I 91
Figure 130: Count of Hordeum from Insula VI.I 92
Figure 131: Count of *T.aestivum/T.durum* from Insula VI.I 92
Figure 132: Count of *Triticum* sp. from Insula VI.I 92
Figure 133: Total count of Millet from Insula VI.I 93
Figure 134: Total count of Millet from Insula VI.I by time period 93
Figure135: Total count of Millet per property from Insula VI.I 93
Figure 136: Percentage of Millet from Insula VI.I by century 93
Figure 137: Total count of Millet per Property from Insula VI.I 94
Figure 138: Commercial triangle 105

Figure 139: Bar of Pheobus 105
Figure 140: Bar of Acisculus 105
Figure 141: Soap Factory/Metal workshop 106
Figure 142: North end of Insula VI.I 107
Figure 143: Inn 107
Figure 144: Inn Bar 108
Figure 145: House of the Vestals 108
Figure 146: Vestals Bar 109
Figure 147: House of the Surgeon 110
Figure 148: Shrine 110
Figure 150: House of the Vestals (VI.I.vii) presence of mosaic floors 113

List of Tables

Table 1: Results of Amphorae Study from the House of the Vestals (VI.I.7) 14
Table 2: Number of samples examined from properties within Insula VI.I 27
Table 3: Summary of results from Insula VI.I by property by time period 47
Table 4: Presence and Absence table for Roads from Insula VI.I 65
Table 5: Presence and Absence table for Domestic properties within Insula VI.I 66
Table 6: Presence and Absence table for Commercial properties within Insula VI.I 67
Table 7: Presence and Absence table for Ritual contexts from Insula 68
Table 8: Insula VI.I Presence and Absence 89
Table 9: Pompeii sites Presence and Absence 89
Table 10: Pompeii sites continued Presence and Absence 90
Table 11: Via Consolare 104
Table 12: Vicolo di Narciso 105
Table 13: Bar of *Pheobus* (VI.I.xviii) 105
Table 14: Bar of Acisculus (VI.I.xvii) 106
Table 15: Soap Factory (VI.I.xiv) 106
Table 16: Triclinium 107
Table 17: Inn Presence and Absence 107
Table 18: Inn bar Presence and Absence 108
Table 19: House of the Vestals (VI.I.vii) 109
Table 20: Vestals Bar 109
Table 21: House of the Surgeon 110
Table 22: Shrine (VI.I.xiii) 111
Table 23: Well 111
Table 24: Presence and Absence of common taxa from properties within Insula VI.I 112
Table 25: Presence of spikelet forks from Insula VI.I 116

Acknowledgements

I wish to first thank the directors of the Anglo-American Project in Pompeii (AAPP) Dr. Rick Jones and Dr. Damian Robinson for the wonderful opportunity to pursue this research. I would like to extend a special thanks to the environmental specialists with the AAPP: Dr. Andrew Jones, Dr. Jill Thompson, Dr. Jane Richardson, and Dr. Robyn Veal. In particular, I wish to thanks Dr. Andrew Jones and Dr. Jill Thompson for their always kind and helpful assistance at Bradford University and permission to access the environmental storeroom.

I would like to thank all the staff working on the AAPP including project manager, Dr. Hillary Cool, for her assistance with access to data, particularly dating information and advice regarding matters relating to Insula VI.I. Thanks goes to coin specialist, Dr. Richard Hobbs for access to his analysis of the AAPP coin data for dating purposes. I would like to extend my thanks to all the past and present AAPP Pottery team especially Dr. Jaye Pont and the late Dr John Dore. Thanks goes to Dr. Michael Anderson and Briece Edwards for their practical help as co-field directors during the 2005 and 2006 field seasons with the AAPP and especially to Dr. Anderson for his assistance with stratigraphy and dating issues regarding the House of the Surgeon. To my lovely neighbours Arthur and Jennifer Stephens for their support and friendship during my field seasons in Pompeii and for use of their photomosaic image of Insula VI.I. I would like to issue a general thanks and acknowledgement to all the hard work contributed by previous staff and students over the course of the AAPP which made this research possible.

Sincere thanks go to my primary supervisor Dr. Dorian Q. Fuller, particularly for his assistance with archaeobotanical identifications and my secondary supervisor Dr. Kris Lockyear. I would like to thank Dr. Michéle Wollstonecroft for her excellent editing skills and helpful comments on the early drafts of my PhD thesis. Thanks to Dr. Sue Colledge, for her guidance, counsel and support during my time as a PhD student at UCL. Another mention must be made to my colleague Dr. Robyn Veal for her helpful editing and comments on early drafts of my PhD thesis and much appreciated emotional support, friendship and practical assistance during our field seasons together in Pompeii. A special thank you goes to my friend and colleague Dr. Louise Parkinson for her extremely generous practical help, guidance and assistance with editing. A warm thanks to Dr. Alison Weisskopf and Alan Michels for their generous hospitality during the final stages of my PhD.

I wish to thank Dott. Pietro Giovanni Guzzo (*Soprintendente*), Dr. Antonio D'Ambrosio, the helpful *custodi* and laboratory staff of the *Soprintendenza Archeologica di Pompei*, particularly the late *Dott.ssa* Annamaria Ciarallo for the kind use of the laboratory facilities over the course of the AAPP field seasons in Pompeii.

I would like to thanks to my PhD examiners Dr. Mark Robinson, Oxford University and Dr. Andrew Gardner, UCL, whose comments and suggestions have improved this work.

This book was completed under the European Research Council grant (ComPAg, no. 323842). And finally, much thanks to my family for all their support.

Glossary

Atrium	A formal anteroom for the reception of clients within the Roman house.
Cauponae	A commercial shop which only served drinks.
Cella vinaria	A fermentation room for grapes for the purposes of wine-making.
Cenacula	Dining room, usually located on the upper storey, within the Roman house.
Cubiculum	A small room traditionally defined as a bedroom within the Roman house.
Culina	Kitchen
Dolia (plural) *Dolium* (singular)	An underground large pottery vessel used for the storage of food or the fermentation of wine.
Exedrae	A small open side room within the traditional Roman atrium house.
Fauces	The main narrow entrance hallway into the traditional Roman atrium house.
Hortus	Roman garden.
Latifundia	A large agricultural estate growing cash crops harvested by slave labour and owned by wealthy Roman elite.
Oecil	Large open room within traditional Roman house.
Oecus	*Main hall in Roman house.*
Olea perforata	Portable ceramic planting pots used to transport trees or other plants and used as a potting vessel in gardens.
Opus quadratum	A type of Roman construction style in which square blocks of stones of the same height where laid in parallel courses without mortar.
Opus signinum	A Roman building material which utilised broken fragments of tiles and pottery mixed with mortar.
Otium	Roman concept of total relaxation.
Penariae	Where food was kept in Roman house.
Peristyle	Garden and ambulatories within Roman house.
Taberna	Bar
Tablinium	Room in the Roman house off to one side of the atrium.
Thermopolia	A public food and drink establishment.
Torcularium	A wine press for crushing grapes for wine.
Triclinium	The label given to the dining room within a traditional Roman atrium house.
Villae rusticae (plural) *Villa rustica* (singular)	Large agricultural estates outside the city gates.
Villa suburbana	Agricultural Roman luxury residence located just outside the city gates.

Chapter 1

Introduction

Due to its catastrophic destruction and subsequent outstanding preservation, Pompeii has entered into the annals of history and archaeological lore, a relatively unimportant Mediterranean port city in southern Italy during its brief life span, it now rivals the fame of Rome itself. It has had a widespread and profound influence upon philosophy, art and culture in Western society and has sparked innumerable heated debates on a plethora of academic and popular subjects since its rediscovery, 250 years ago. Arising from its humble status as an early Italic village on a volcanic outcropping it is now regarded as the most famous and well-preserved UNESCO world heritage site due to the rare glimpse it offers to visitors and scholars alike, into a past way of life that was captured so precisely on the 24th of August 79 AD with the eruption of Mount Vesuvius. Despite the fact that Pompeii is the earliest continuously excavated Roman site and one of the most visited and well-studied archaeological sites there is still much that is not understood about it, especially in terms of its long-term development from pre-Roman times (Laidlaw 2007, 620).

Despite its world renown as an archaeological site, the past twelve years of archaeological excavations (1995-2006) by the Anglo-American Project in Pompeii (AAPP) provides one of the few examples of chronological depth in Pompeii. The analysis of all the recovered archaeobotanical material from the only triangularly shaped insula, Regione VI, insula I from Pompeii, adjacent to the Herculaneum Gate, provided a unique research opportunity of a highly documented case study of urban environmental analysis. This research is able to provide a diachronic analysis of wider patterns of food consumption and natural resource use though the approximate three hundred years of occupation of this insula and examine the cultural impacts that subsequent waves of immigration may have had on this multiethnic city with the invasion of the Samnites, the influence of the nearby Greek colonies of southern Italy and finally the colonization by the Romans after the Social Wars in 89 BC.

The ultimate goal of the Anglo-American Project in Pompeii is to reconstruct a holistic view of everyday life in the ancient city of Pompeii and to examine, in detail, issues of intensification, urbanisation and inequality using Regione VI, insula I as a lens through which to understand the complexity and changing nature of this provincial city through its entire period of occupation. The analysis of the archaeobotanical remains will assist in all these spheres of enquiry. As a highly documented case study of urban archaeobotany it has the potential to examine issues of economic and social differentiation as reflected in access to and processing of agricultural produce and food consumption.

1.1 Pompeii: historical background

The Sarno River supplied the entire fertile plain south-east of Mount Vesuvius and supported the agricultural villages which were situated along the upper reaches of its valley and near the mouth of the river around 700 BC. The original Oscan or Greek settlement at Pompeii dates back to the 6th century BC and was built on an isolated volcanic ridge about 39 metres above sea-level at the time (Jashemski 1979a, 4). Pompeii was beholden to and became the port city of *Nucceria Alfaterna*, or *Noceria*, a city situated at an important road conjunction and which dominated the southern region of Campania (Grant 2005, 17).

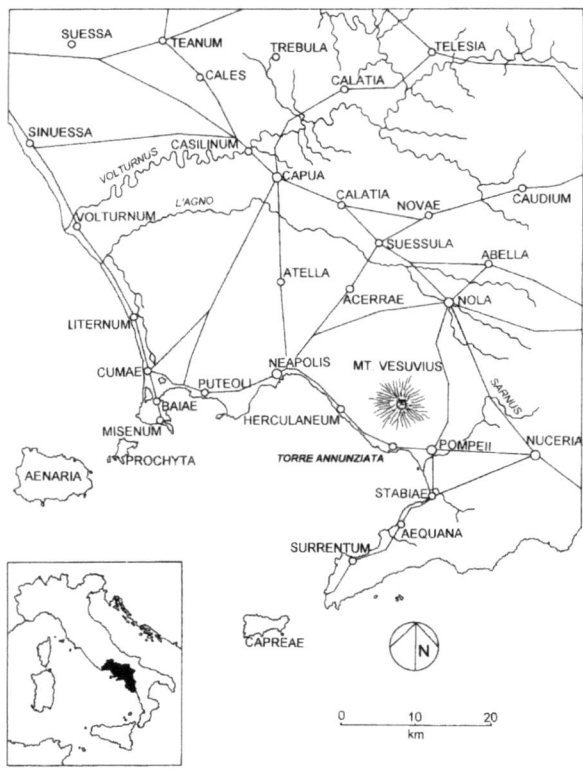

FIGURE 1: POMPEII AND THE SURROUNDING REGION (AAPP 2005 RESOURCE BOOK, P. 37)

During ancient times, the region of Campania (highlighted in black in the small map in the left-hand corner of Figure 1) has been argued to have been the chief granary of the peninsula. In some districts as many as three grain crops were produced each year, in addition to a crop of vegetables (Jashemski 1979a, 4). Ancient sources exclaim over the wonderful variety of crops grown throughout the year with harvests postulated to have been six times greater than the average crop yield for the rest of the Italian peninsula (Grant 2005, 15).

Recent archaeological evidence uncovered at the site of Capua has supplied evidence that Etruscan intrusions into the region of Campania begun *circa* 650 BC. It appears that towards the end of the next century the Etruscans extended their control over larger areas of the Campanian plain. Etruscans came into contact with the Greeks at the city of Cumae, thought to be the earliest Greek settlement in Italy dating to 750 BC, which subsequently became the Etruscans' main centre of Greek commerce (Grüger et al. 2002, 240). The Etruscans attacked the city of Cumae in 524 and 474 BC and were rebuffed each time. At this time it seems likely that Etruscans controlled the city of Pompeii, as inferred from known Etruscan trading contacts at Pompeii, Etruscan bucchero pottery which has been found throughout Pompeii and fragments of Etruscan inscriptions found on black vases excavated from beneath the Temple of Apollo (Grant 2005, 20).

By the mid- 5th century BC the Samnites, renowned as tough indigenous mountain people, swept down from their original homeland in the Apennines of central and southern Italy towards the coast. The Etruscans were defeated by the Samnites in 474 BC. In 423 BC, while mainland Greece was engaged in the Peloponnesian War and unable to send reinforcements, the Samnites attacked the Greek colony of Capua, which at the time was the most important Etruscan centre in Campania, and other Greek colonies along the coast. In 421-420 BC the Samnites took over the Greek colony of Cumae and subsequently moved into the Campanian plain and resettled Pompeii and Herculaneum.

The Samnite people spoke Oscan, one of the three main dialects, along with Latin and Umbrian, of the Italic branch from the Indo-European family of languages (Descœudres 2007, 15). This new 'Oscan' Campania, although probably still retaining traces of past Etruscan influences, formed itself into a separate federation of towns in 445 BC under the leadership of the city of Capua. Neapolis (Naples) located further north became a bicultural city with elements of both Greek and Samnite influence (Grant 2005, 20). Thus, by the 4th century BC Pompeii may have followed Stabiae, Herculaneum and the villages of the Sarno Valley as part of a loose confederation with controlling influence upon it from *Nuceria*.

Around 358-354 BC the Samnites entered into a defensive alliance with Rome, possibly fearing an invasion by the Gauls from the north. Roman sources claimed that non-Samnite Campanian cities made an appeal to Rome fearing further Samnite invasion and thus triggered the First Samnite War, 343-341 BC, and the entrance of the Romans into Campania. During the Second Samnite War, 326-304 BC, in which the Romans' legendary defeat in 321 BC at *Caudine Forks* took place in the interior of the region, the Romans captured Neapolis in 327-326 BC. The Romans sacked *Nuceria*, and took Pompeii and possibly Herculaneum in 310-302 BC. By 300 BC the region of Campania was now wholly under the control of the Romans (Grant 2005, 22).

The 2nd century BC saw a period of construction in Pompeii in response to a major expansion in the local population and by the end of the 2nd century BC was an urban community ((Jashemski 1979a, 4). By the 2nd century BC, Pompeii is known to have had intense commercial contacts with several areas throughout the Mediterranean region (Ciaraldi 2001, 32). It was the expansion of trade in the Mediterranean in the later part of the 1st century BC and the subsequent expansion of the empire from Augustus onwards which allowed the 1st century AD Pompeii to have become a densely occupied urban centre (Parker 1990, 330).

Towards the end of the Republican period much of the land surrounding the city was farmed by *villae rusticae*, large farming estates in the country, or cash crops grown from market gardens. Along the coast of the Gulf of Baiae, the western section of the Gulf of Pozzuoli and the Bay of Naples were numerous luxury villas, owned by the upper classes, and this was an especially popular area when the Emperor Tiberius owned the entire island of Capri during the Early Imperial period (Grüger et al. 2002, 240; Jashemski 1979a, 289). However, recent archaeological discoveries suggest that small landowners continued to exist and farm in this area in the 1st century AD (Frayn 1979, 20; Lomas 1993, 118). In 91 BC, during the Social War in which the Italians attempted to gain equal rights and privileges granted to Roman citizens, the Vesuvian cities of Pompeii, Herculaneum and Stabiae defected from their previous allegiance with Rome (Descœudres 2007, 15). In 89 BC Pompeii was attacked by General Sulla and his army and archaeological evidence testifies to the damage caused by the siege-artillery. Ballista and sling bullet indentations on city walls have been found outside the vulnerable Herculaneum Gate (Jones and Robinson 2007, 395). At the end of the war Pompeii received Roman citizenship. However, a colony of Sulla's retired soldiers, known as *Colonia Cornelia Veneria Pompeianorum*, was imposed in 80 BC upon Pompeii (Grant 2005, 23). Cicero writes of the violent ill-will of the citizens of Pompeii against the newly imposed Roman colonists. However, other Classicists have argued that the land for the Roman soldiers was taken from the people of nearby ruined Stabiae rather than Pompeii (Grant and Forman 1976, 23).

As evident from the extensive building projects outside the city walls during the late 1st century BC with the *Pax Romana*, the inhabitants of Pompeii were no longer worried about invasions as the Romans had conquered the Italian peninsula by this time (Cooley 2003, 111; MacKendrick 1960, 200). Indeed, the early years of the Roman colony provided evidence of building and refurbishment of villas in the countryside surrounding Pompeii (Ling 2005, 63). At the time of its destruction, the city of Pompeii was approximately 167 acres (63.5 ha); elliptical in shape with a circumference of 2 miles (3km), (an average size for an Italian town at the time of the early Roman Empire), with a population of approximately ten to twenty thousand people (Cooley 2003, 33; Grant 2005, 33) with a more conservative estimate of population of 8,000 to 12,000 (Jongman 2007, 513; Ling 2005, 98; Westfall 2007, 129).

1.2 Pompeii: a case study in urban archaeobotany

Although there was interest expressed in the later part of the 19th and the early 20th century in exploring Pompeii's pre-Roman roots, this was undertaken by studying the surviving standing masonry structures and changes in building techniques and materials. Paradoxically, despite Pompeii's parallel development with the discipline of archaeology, its potential as a case study for refining analytical methodologies of wider archaeological utility has been underexploited. Hence, Pompeii was regarded as a static single-layer site. This synchronic perspective, focusing mainly on the Roman period leading up to the destruction of the city in AD 79, is reflected in the limited archaeological investigations carried out in the recent past beyond the AD 79 destruction layer (Robinson 2002, 93). It was not until the early Pompeian archaeologist Maiuri, during the 1930s and 40s, began excavations beneath the AD 79 layer that its pre-Roman roots began to come to light (Cooley 2003, 113-114).

There has been a recent movement to allow full-scale major stratigraphic excavations below the AD 79 destruction layer, under the former leadership of *Soprintendente* Prof P. Guzzo, the Archaeological Superintendent of Pompeii, in houses and areas that have been disturbed by excavations in the recent past. During the course of the AAPP, excavations have taken place below the AD 79 destruction layer, where no substantial intact floors have survived, and in which a dated construction sequence based upon evidence from sealed deposits has been created. Along with other recent excavations (Wallace-Hadrill and Fulford 1999 with the British School at Rome, Carafa 1997, the Pompeii Forum Project directed by J. J. Dobbins 1994 and the Temple of Apollo by De Caro 1986) the overall result has been a growing trend towards understanding the complexity of the chronological sequence for urbanisation and the Roman influence upon the city of Pompeii (Robinson 2002, 94).

Practically this research is significant as an entire insula block within the city of Pompeii has never been excavated in its entirety to date. This research represents a unique opportunity to undertake a diachronic study of urban archaeobotany from a variety of different contexts. As a case study Pompeii has the potential to contribute to questions of economic and social differentiation as reflected in access to and processing of agricultural and trade products, on-site processing and preparation, the use of domestic and public space based upon the preserved archaeobotanical assemblage, differential food consumption in a complex urban society and the role that archaeobotanical analyses can contribute to these topics.

1.3 Research aims and objectives

The main objective of this research is the analysis of the archaeobotanical assemblage, the preserved seeds and macro-botanical remains, with respect to the chronological and stratigraphic record, recovered over the course of the past twelve years from the excavation of Regione VI, insula I by the AAPP. The study of this one city block, Regione VI, insula I, is a unique opportunity to examine the preserved plant material both across contemporaneous households, from a variety of domestic and commercial contexts, and diachronically, over the three hundred years for which suitable data has been excavated. This research will offer a new view of Roman and pre-Roman households and local economies and how they developed over time. It will also provide a test ground for urban archaeobotanical methodology, an area which is largely underdeveloped since most environmental research has tended to focus on larger scale issues such as societal development and the origins of farming.

1.4 Research questions

Pompeii is well-situated as a firmly documented archaeological context from which to explore issues of food distribution and consumption in a complex urban society and the role that archaeobotanical analysis can contribute to studies of social differentiation through both food and agricultural economy. In the recent past, literary and art historical sources from frescos and mosaics from Pompeii and Herculaneum, have been used to study Roman cuisine and diet (Jashemski et al. 2002, 80-180). However, these are potentially elite-biased idealized representations and contrasting these constructed images against preserved food refuse (cereal grains and chaff, fruit pips, agricultural weeds, etc.) provides a more holistic view of the Roman diet from all levels of society.

1.5 Status, ethnicity and Romanisation

One of the aims of this research is to examine the archaeobotanical assemblage through time to see if there are changes in the socio-economic status of the inhabitants of Insula VI.I and explore any economic, political, ethnic and tra*de facto*rs which may have influenced dietary changes over the occupation of the site.

Food in Greco-Roman society was a powerful and constant visible marker of ethnic and cultural differences. It repeatedly reinforced the social and economic distinctions between rich and poor. It served as a vehicle not only for communicating status, but also for maintaining familial, social and political relationships and served to maintain the rigid social hierarchy present within Roman society. Conspicuous consumption and displays of food in Roman society were powerful signifiers of wealth, status and power, especially when exotic or imported items were on the menu (Garnsey 1999, xi; van der Veen 2003, 405).

Roman-type ingredients and similar methods of procurement and preparation of food items have been employed as indicators of a Roman cuisine or lifestyle in previous studies and will be one aspect considered in this research (Jones 1991 cited in Meadows 1999, 105).

Although it has to be acknowledged that Roman veterans and citizens often came from disperate areas of the Roman Empire and were increasingly, during the Roman Imperial and Early Empire, exposed to new influences and cultures through warfare, trade and migrations of new and subjected peoples. This study will use Pompeii to examine aspects of ethnicity and the process of the incorporation of a Roman colony within Pompeii through the evidence of cuisine in the archaeobotanical record from Insula VI.I.

Previous studies have utilised the faunal evidence to examine the process of Romanisation. During the late Republican period in Italy, a high pork consumption pattern emerges as a feature of the Roman diet. Present faunal data suggests that a similar pig-dominated pattern emerged earlier, in the 3rd to 2nd century BC, in the region of Latium, north of Campania. This chronological distinction in meat consumption is most clearly seen in the Bay of Naples area. Thus, the cattle-dominant diet of the 1st century BC was replaced by the high pig pattern present in the Imperial period (King 1999, 169). Indeed, preliminary research on the faunal assemblage from Regione VI, insula I at Pompeii by Dr. J. Richardson (2006) revealed a similar high pig pattern indicating that dietary preferences appear to have shifted over time and class distinctions in differential meat consumption are noticeable in the archaeological record.

Historic sources discussing Roman agriculture such as Cato, Columella and Varro write of the general trend from mixed farming towards larger estates with specialized vine and olive cultivation, yielding higher profits, between the 2nd century BC and the 1st century AD. Archaeological evidence compiled by Lomas (1993, 118-119) also illustrates this shift from smaller farms in the early 2nd century BC towards larger specialized estates by the 1st century AD; with the caveat that significant numbers of smaller mixed farmsteads survive into this period (Rosenstein 2008, 3). Preliminary charcoal evidence by Veal and Thompson (2008, 10-11) found a decline in beech from the 1st century BC onwards and a corresponding increase in orchard and vineyard cuttings. This correlates with the above mentioned known archaeological evidence and suggests that the overall charcoal patterns reveal 'increasing penetration of Roman influences into the area over time' (Veal and Thompson 2008, 11). Following along these multiple lines of evidence this research will investigate if the process of the incorporation of previous Samnite and other native Italic peoples into an imposed Roman colony and the general trends in agricultural production can be observed through changes in vegetation, seasonality, exploitation of the natural environment, and the consumption and trade of different plant materials in the archaeological record.

1.6 City and hinterland

Another area of research to be investigated is the pattern of interconnections and economic changes over time between Pompeii and its rural hinterland and its participation in the trade networks within the wider Mediterranean basin using the archaeobotanical assemblage from Insula VI.I. The ancient city in general was known to have encompassed both the rural hinterland and an urban centre, where the administration and its public cults were situated. The difficulty lies in integrating previous paleoenvironmental research on the Vesuvian cities of Pompeii, Herculaneum and the surrounding areas with the complexity of the archaeology and history of the region and the daily interaction between town and country (Ciaraldi 2007, 19). A town the size and complexity of Pompeii must have relied upon its surrounding hinterland for the majority of its food, fuels and other natural resources to maintain the population. Pompeii possessed excellent river connections with the towns of its economic hinterland. With the construction of the Roman harbour in 194 BC at Puteoli, Pompeii was well-situated to take full advantage of its position within maritime trade (De Caro 2007, 76).

Recent research into archaeobotany within the vicinity of Mount Vesuvius (Borgongino 1999, 2006; Ciaraldi 2001, 2007; Jashemski et al. 2002; Meyer 1980, 1988, 1994; Robinson 1999, 2002) has revealed a change in the number and diversity of plants recovered and this represents a significant change in the economic history of Pompeii and pre-Roman regional trade. Exotic goods discovered at Pompeii by Ciaraldi, such as black pepper and *citrus* fruit, suggest that some of the items may have passed through the main Italian port city of Puteoli, on the Campanian coastline, implying that Pompeii may have been engaged to some degree with the trade networks throughout the Mediterranean basin (Ciaraldi 2001, 32).

1.7 Pompeii as a type site and economic models

This study will attempt to use the archaeobotanical evidence from Insula VI.I to address some of the issues surrounding trade and the economy at Pompeii. For decades there has been an ongoing debate amongst Classicists and Romanists regarding the appropriateness of theoretical models used to study the ancient economy of the Roman city. Pompeii, due to its chance preservation, has erroneously in the past, been labelled and perceived as a 'type-site' of Roman urbanism and economy. However, Pompeii's economy was unique in the sense that it was closely integrated into the luxury villas on the Bay of Naples and within the wider Campanian economy of trade centred upon Puteoli (Laurence 1994, 53). Rostovtzeff (1957) argued for the use of modern economic models, using the cities of 20th century Western Europe and North America in which the increase in traded goods resulted in the rise of a middle class, to understand the Roman economy in general and the ancient economy of Pompeii in particular (Laurence 2007, 9).

The initial academic debate began with Finley's adaptation of the Greek polis as the ideal ancient city as a 'consumer city' economy model based upon Werner Sombart and Max Weber's description of the evolution of the ideal Western capitalist city (Jongman 2007, 502; Lomas

1995, 1), which was never intended for application to the ancient city. As opposed to the 'producer city', in which agricultural produce from the surrounding countryside was purchased in part through the manufacture and export of luxury goods to the countryside and external markets, the 'consumer city' is defined as a city which is economically parasitic upon the agricultural production of its hinterland (Finley 1973, 131; Jongman 2007, 502; Laurence 1994, 51). This model was used in the recent past as the accepted theoretical framework for understanding the economy of the Roman city. Within the Weberian model no economic growth or development occurred (Jongman 2007, 502; Wallace-Hadrill 1992, 241).

Environmental data is another area of study being used to examine the economy of the city of Pompeii. Veal and Thompson (2008, 10) argue that the majority of the taxa identified from the charcoal evidence from the House of the Vestals (VI.I.vii) was believed to have grown in the lower level managed montane deciduous forests or the higher altitude natural forests surrounding the city of Pompeii; with the majority of the surrounding plain given over to cereal cultivation (Jongman 2007, 502; Lomas 1993, 118-119; Veal 2009, 229; Veal and Thompson 2008, 10). The low diversity of wood recovered from an environment with a diversity of ecosystems and different woods to choose from suggests that strong human selection was occurring. Six different taxa are observed in the 2nd century BC, nine in the 1st century BC and ten in the 1st century AD (Veal and Thompson 2008, 6). The trend by the 1st century AD revealed that Fagaceae (Beech) diminished in importance in comparison with other woods, such as fruits and vines, which while relatively scare in the 2nd and 1st centuries BC appears in greatest numbers during the 1st century AD, increasing to 20% of the assemblage.

Veal and Thompson (2008, 7) argue that significant cultural and economic implications can be inferred from the continued dominance of beech in the analysed charcoal assemblage through time. With its preferred ecological niche, some 800 metres above sea level, the nearest source of beech would have been at least 15 km from the city. Transport of raw timber or charcoal down the Sarno River, which originates in the Apennines, was a known major trade route linking Pompeii to the inland cities. However, lower montane managed taxa such as hazel and oak increased along with orchard and vineyard cuttings through time towards the 1st century BC. This increase in taxonomic diversity within the charcoal assemblage over time, with the use of highly managed woods and an increase in orchard and vine cuttings, Veal and Thompson (2008, 1) argue (that this selection of wood) represents a 'developed market structure for wood supply to the city'. This evidence correlates well with archaeological evidence from Pompeii which suggest increasingly sophisticated consumption and greater efficiency of production by the 1st century AD.

1.8 Situate environmental data

Another research goal is to situate and compare the known environmental data from the Vesuvian region, including recent wood charcoal analysis from Pompeii, with the current archaeobotanical results from Insula VI.I. Interestingly, no *Olea* sp. (olive) charcoal was recovered for the entire occupation of the House of the Vestals (VI.I.vii) (Veal and Thompson 2008, 7). Yet, the archaeobotanical background noise of the House of the Vestals (VI.I.vii) is largely composed of charred olive stones which increase through time towards the 1st century AD. Similarly, Walnuts and Peaches, which are eastern imports, have been recovered from the 2nd century BC onwards (Ciaraldi 2007, 149). Thus, one may ask if the trends elucidated from the charcoal data are visible in the archaeobotanical assemblage and question if the developing and increasingly sophisticated market system was influencing the food items being brought into and consumed within the city of Pompeii.

1.9 Domestic space

This study hopes to examine the spatial distribution of the archaeobotanical material recovered from Insula VI.I in order to contribute to the understanding of the use of space within an urban setting. Spatial analysis of Roman domestic space by Wallace-Hadrill (1990, 1995), Zanker (1998) and Laurence (2007) and recent work by Allison (2007) on the material cultural remains from the Insula of Menander, Regione I, insula IX, have challenged traditional concepts of Roman domestic space in Pompeii. Allison (1999, 2004, 2007) highlights her concerns regarding the unquestioned assumptions about Roman domestic household behaviour. Based upon her assessment of the Insula of Menander (I.X.iv) she argues that food preparation and storage within the Roman house appears to have been flexible in terms of location. Therefore, by examining all phases from every room within the different properties within Insula VI.I this study hopes to further examine the question of Roman domestic behaviour and attempts to test Allison's more flexible concept of food preparation, consumption, storage and disposal within the Roman household through comparison of the different assemblages both spatially and chronologically from Insula VI.I.

Chapter 2

Historical Background

2.1 Introduction to the Mediterranean Environment

The Mediterranean region comprises a single 'climatic region' characterized by strong seasonality with heavy winter rains and long summer droughts (Dormoy et al. 2009, 737; Finley 1975, 31; White 1970b, 54). The region of Campania today has a modern annual precipitation of 830 mm (Foss 1988, 131). It has a richer flora than the rest of continental Europe (Grove and Rackman 2001, 45; Sallares 2007, 17). Due to the climate it is an area of relative ease for its inhabitants and in the past as today, involves much outdoor living (Jashemski 1979b, 89). The most fertile soils are located on the coastal plains and the large inland plateaus. Staple cereal grasses, vegetables and fruits, especially grapes and olives, have been cultivated in the area for centuries (Ciaraldi 2007, 198). Pasture land is suitable for small animals, such as sheep, pigs and goats but not cattle rearing.

The ecology of the Mediterranean Basin has been influenced over the last several thousand years by numerous civilizations: 'complex 'coevolution' has been claimed to shape the interactions between ecosystem components and human societies' (Blondel 2006, 713). The two prevailing folk myths of the Mediterranean environment which continue to persist include: the 'ruined landscape' which solely attributes human actions to the desertification and destruction of the pristine natural landscape and the 'lost Eden' in which humans have been responsible for the ecological diversity since the last glacial period (Blondel 2006, 713; Hughes 1994, 5).

Jashemski et al. (2002, 476) has argued that there is evidence for 184 plants based upon art historical and literary evidence known at the time of the eruption in AD 79, with an additional 95 species known from the pollen record totaling 279 plants. However, it should also be kept in mind that the plant remains recovered from Pompeii all have potential economic value and thus could have been imported to the city (Ciaraldi 2001, 42). Therefore, like all archaeobotanical assemblages, it is a biased sample of the total environmental material present from the city of Pompeii (Fuller and Stevens 2009, 37).

2.2 Palynological evidence

The Mediterranean climate has only been in existence for the past few thousand years (Grove and Rackham 2001, 45). Only a few unevenly distributed pollen studies have been undertaken in the Mediterranean area, which are unable to capture the full complexities of the Mediterranean vegetation for a variety of reasons, including the predominance of insect-pollinated species as opposed to wind-pollinated species. Roman farmers likely tolerated trees and bushes in their fields and vineyards, as seen today with hedges and wild trees amongst modern farmland, and their presence amongst crops often obscures the evidence of cultivated plants within palynological studies (Ciarallo 2000, 57).

Pollen diagrams have been created for Italy from the ancient maar lake sediments of *Valle di Castiglione*, *Lagaccione* and *Stracciacappa* (Follieri et al. 1998, 3). For the region of Campania the closest source of palynological data analyzed to date is from the maar lake of Monticchio in south central Italy, east of Naples (Allen et al.1999, 740; Watts 1985, 491; Ciaraldi 2001, 25). Pollen data suggests a steppe-like environment during the Late Pleistocene, with semi-desert vegetation, such as *Artemisia* (sage-brush) and *Chenopodium*, and drier and colder temperatures than in modern times.

According to the pollen record the earliest human impact upon the environment was at Lago dell'Accesa, Tuscany around 8000 cal. BP (6050 BC). The Etruscan period witnessed the appearance of *Castanea* (Chestnut) and *Juglans* (Walnut) ca. 2800 cal. BP (850 BC). The impact of Etruscan settlement near the lakeshore is evident from palynological values of arable crops, species of secondary forest canopy including *Ericacea, Pinus, Pistacia, Mytrus* and anthropogenic indicators such as *Chenopodiaceae, Plantago lanceolata,* and *Rumex* (Drescher-Schneider et al. 2007, 279).

Pollen of *Castanea, Juglans* and *Secale* appeared in northern Italy with the Romans around 200 BC with the conquest of the floodplain of the Po. The *Castanea* curve increased around the 5th and 6th century AD (Drescher-Schneider 1994, 47). In central western Italy, the crater lake Lago di Vico showed a similar pattern with the sudden appearance of *Castanea* and *Olea* pollen around 3710±50 BP (uncalibrated) (Frank 1969, 67). The expansion of these two species was followed by the simultaneous appearance of *Vitis* and *Juglans*, cultivars dated to 2630±95 BP (uncalibrated) (Ciaraldi 2001, 25-26; Magri 1999, 171; Magri and Sadori 1999, 258). Samples taken from Lake Avernus dating to the Roman period from the region of Campania contained pollen of *Castanea, Platanus* and *Juglans* (Grüger et al. 2002, 264).

Hemp grew wild throughout Italy. The highest and earliest peak of hemp in the pollen record occurs in the 1st century AD (AD 50-190). This sudden increase in hemp pollen is followed by the rise of cultivated tree pollen including *Castanea, Juglans* and *Olea* and is associated with the increase in cereals and ruderal plants and is therefore considered an indicator of Roman cultivation in central

Italy (Mercuri et al. 2002, 263). At Lago di Monterosi there is a gradual decline in oak since the Roman period with tree pollen continuing into most recent layers. At Lago di Baccano, oak and pine continued to dominate until deposition was terminated by Roman drainage activity (Grove and Rackham 2001, 172).

2.3 The modern environment

The native vegetation of the Vesuvian area was probably evergreen oaks. Human's have extensive modified the vegetation with intensive replanting programs of conifer forests in the Campanian region (Ciaraldi 2001, 28-29; Ciarallo 2005, 169; Veal 2009). Vegetation in non-agricultural areas consists mainly of thickets with species such as laurel, myrtle, broom, oleander, and brambles (Foss 1988, 128). Several different types of environments surround Pompeii, including coastal areas, fertile coastline plains and the medium-height Lattari Mountains. Modern day lowland landscapes are the result of intensive human occupation and agriculture, largely based upon fruits, such as tomatoes, tobacco, potatoes, vegetables, corn, onions, cauliflower and flowers (Ciaraldi 2001, 26). Although plentiful in the area today, oranges and lemons were not present in the region in ancient times. Indigenous to the region are the olive, vine and fig, which became plentiful in Campania (Costantini and Giorgi 2001, 245; Grant 2005, 15).

The modern Vesuvian flora was described by Agostini (1952) who separated it into two archaeological coenoses on the basis of their exposure. *Macchia mediterranea*, occurring at an altitude of 600-800m, is composed of a xerothermic community with a prevalence of herbaceous, thermophilic and xeromorphic shrubs, e.g. 'Mediterranean evergreens such as tree heath (*Erica arborea* L.), sage-leaved cistus (*Cistus salvifolius* L.), narrow-leaved cistus (*Cistus monspeliensis* L.), and strawberry tree (*Arbutus unedo* L.) and myrtle (*Myrtus communis* L)' (Ciaraldi 2001, 27).

The other coenosis is composed of mesophilic-mesothermic species including 'sweet chestnut (*Castanea sativa* L.), hop hornbeam (*Ostrya carpinifolia* Scop.), manna ash (*Fraxinus ornus* L.), common hazelnut (*Corylus avellana* L.), small-leaved elm (*Ulmus minor* Miller), common privet (*Ligustrum vulgare* L.), field maple (*Acer campestre* L.), hybrids of lime (*Tilia* x *euroaea*) and English oak (*Quercus robur* L.)' (Ciaraldi 2001, 28). The mesotherm-igrophile component includes species such as 'pussy willow (*Salix caprea* L.), eared willow (*Salix aurita* L.), Italian alder (*Alnus cordata* Loisel), *Betula alba* L. and aspen (*Populus tremula* L.)' (Ciaraldi 2001, 28).

Pliny (NH XV. 1) quotes the historian Fenestella that the olive (*Olea europea* L.) was introduced to Rome in 581 BC during the reign of Tarquinius Priscus (Ciarallo 2000, 57). Archaeobotanical and palynological data refute this ancestral evidence and have shown that the olive tree has been part of the ecosystem of the Mediterranean region during most of the Quaternary (Liphschitz et al. 1991, 441; Zohary and Speigel-Roy 1975, 319). Surviving the Glacial Quaternary the olive tree became a major component of Holocene thermophilous Mediterranean vegetation (Terral 2000, 127-128). The olive tree is indigenous to and confined to the Mediterranean Basin by its intolerance of frost as a subtropical evergreen plant (Grove and Rackham 2001, 45; Terral et al. 2004, 64). Today, the olive tree is ubiquitous throughout the Mediterranean, one of the essential crops of agriculture in the region (Costantini and Giorgi 2001, 246). Charcoal analysis has shown that the spread of the olive tree, associated with species characteristic of the Oleo-lenticetum, has increased since the Neolithic period and this corresponds with the regression of mesophilous and semi-deciduous oak forests. Later, the olive tree was exposed to both climate and intense anthropogenic constraints and manipulation (Terral 2000, 127-128).

Meschenelli, an Italian botanist, observed beech (*Fagus sylvatica* L.) among the ancient arboreal vegetation. However, Agostini claims that it is absent in modern times (Ciaraldi 2001, 28). Beech forests are present near the Lattari Mountains, particular *Mte Pizzone*, within a national park (Veal 2009). On this basis Agostini interpreted the presence of Betula in the assemblage as evidence of relic vegetation belonging to the association *Fagetum*, which is typical of more mountainous vegetation. The presence of beech in archaeological deposits from AD 79 has been confirmed by the waterlogged buds recovered from *Villa Vesuvio*. In addition, *Fagus sylvatica* (beech) dominates the analyzed charcoal assemblage from the 3rd century BC to AD 79 from Regio VI, insula I and Veal and Thompson (2008, 11) suggest that this selection and use of beech represents a fairly sophisticated form of wood economy and management.

Ricciardi et al.'s (1986) botanical survey lists 906 modern taxa, comprising both native and introduced plant species. Of these species, 115 were among the Vesuvian flora at the time of the eruption. Ricciardi et al.'s (1986) survey also reported very low percentages of Boreal and Atlantic species (respectively 4.5% and 3%). This contrasts with De Fiore's (1915) results in which a number of these species were identified among the fossil flora. The present study suggests that the current vegetation composition of Mount Vesuvius is of recent origin. Intense changes in geological substrata have resulted from the continuous volcanic activity in the region (Ciaraldi 2001, 28-29).

The exceptional fertility of the land is largely due to its spongy volcanic earth. The palaeosols of Pompeii are rich in phosphorus and potash deposited over centuries of repeated volcanic eruptions and includes pumice, ash, volcanic glass, cindery grains and some clastic material. The soils possess a high content of extractable nutrients including Nitrogen, Phosphorus and Potassium and have an alkaline pH throughout their profiles, likely due to Ca^{++} and Mg^{++} ions leaching into overlying materials and recharging the surface horizons of the palaeosols (Foss 1988, 131-133; Foss et al. 2002, 77-78).

The soil composition also allows for the retention of the abundant seasonal rainfall and thus suffers no ill effects from the annual long summer droughts. All these factors contribute to the agricultural potential of these soils (Foss 1988, 133). The favourable climate can be largely attributed to the moist south-westerly winds which cross the plains and the mild, short winters. These factors make the region well suited for viticulture and the growing of fruit and cereals (Grant 2005, 2). However, it was the access to maritime trade and the subsequent building of two main Roman roads, the Via Appia and the Via Latina, which helped make this region so prosperous and attractive to commercial exploitation in the later Republican and the Imperial periods (Frayn 1979, 19-20; Guzzo 2007, 7).

2.4 Historical overview of the city of Pompeii

Proto-historic finds suggest that people were living in the area that was to become the future site of the city by the second millennium BC. An Eneolithic polished stone axe was found near what is now the Nola Gate and a Bronze Age deposit of material was recovered from an area near the *Nuceria* Gate. The first city fortifications, constructed from soft local volcanic tuff called pappamonte tufa, are the earliest known features in Pompeii dating from *circa* 600-550 BC (Carafa 2007, 63-65; Jones and Robinson 2007, 389; Ling 2005, 20, 29). Archaeological research suggests that standing structures at Pompeii (excluding the cult sanctuaries of the Temple of Apollo and the Greek Doric Temple in the Triangular Forum which date to 6th century BC) date from the latter years of 3rd century BC.

The earliest settlement was built on a lava plateau surrounded by a flat marshy area, which was subjected to the periodic flooding of the River Sarno. With its advantageous viewpoint over the coast of the Bay of Naples it was the only defendable position in the area (Carafa 2007, 65). The nearby harbours of Puteoli (*Pozzuili*), the Greek colony of *Dicaearchia* (*Dikaiarchia*) and Naples (*Neopolis*) acted as entrepôts for trade with Africa, Greece and the East (Grüger et al. 2002, 266; Lomas 1993, 121). Until the end of the 1st century AD, much of the grain supply to Rome passed through the port city of Puteoli, adding to the economic importance of the Bay of Naples within the Roman world. Situated south of the mouth of the River Sarno, according to the geographer Strabo (V.4.8), Pompeii was the entrepôt for the Sarno river valley (Guzzo 2007, 3-4; Laurence 1994, 52) and acted as the port for *Nuceria*, Nola and even Acerrae, to the north (Ling 2005, 19). Hence, Pompeii was likely a distribution centre for agricultural products to the region of Campania where it would have been sent on to external markets. The latter implies that the surrounding countryside was farmed intensively at this time.

As a maritime outlet Pompeii played an important role in trade and communication with other communities along the river valley and ultimately the rest of the Roman world (Cooley 2003, 17). Although the initial settlement at Pompeii was less than half a kilometre from the sea today, due to successive volcanic eruptions and changes to sea level, the current coastline is two kilometres west of the ancient city (Jashemski 1979a, 5). The River Sarno has been diverted further south than its previous course in ancient times (Ling 2005, 19).

The free port at Delos which was established in 166 BC may have stimulated trade and thus be partially responsible for the influx of coins and ceramics from different parts of the Mediterranean region. Based upon the earliest presence of rat bones in Italy in late 2nd century BC contexts at Pompeii Arthur (1986, 41) argues that there is evidence of trade from Egypt or the Near East. He further argues that it indicates direct maritime contact with Pompeii, as opposed to commercial redistribution of goods via the larger port city of Puteoli.

For the majority of its lifespan Pompeii was not Roman but occupied by non-Roman Italic peoples yet the ethnic origins of Pompeii's original inhabitants have been difficult to ascertain (Cooley 2003, 18; Lomas 1993, 172). According to Strabo (V.4.8):

> 'The Oscans used to occupy both Herculaneum and Pompeii next to it, past which the River Sarno flows. Then the Etruscans and the Pelasgians, and after that, the Samnites; these peoples were also thrown out of these places'.

A similar passage from Pliny the Elder (NH III. 60-2) reinforces the theory of an Oscan settlement. A number of arguments have been put forth against a Greek foundation of Pompeii. These include the lack of Greek inscriptions dating to the early period and the layout of the city on a grid rather than following the standard Greek city plan (Ling 2005, 34). The Etruscans were known to have widespread political and military hegemony in northern and central Italy during the 7th and 6th centuries BC, extending into Campania. Pottery fragments dating to the archaic period of Etruscan bucchero vessels, with inscriptions written in Etruscan and some with Greek letters, suggest that Pompeii was Etruscan, at least in terms of its cultural habits (Descœudres 2007, 14). These multiple cultural influences may be reflected in food-stuffs used or imported within the city.

The Greeks are known to have founded the town of Stabiae, located below the hills towards the southern end of the Bay of Naples. To the north of Pompeii the Greeks founded the smaller coastal settlement of *Heracleion*, later known as Herculaneum (Figure 1). Bounded by the sea to the west and Mount Vesuvius to the north, the proximity of these Greek settlements meant that Pompeii was geographically constricted to a few kilometres of land (Grant 2005, 4). It is thought that the Vesuvius gate permitted access to the farmland near the base of Mount Vesuvius. Pompeii's territorium is thought to have extended to the mid-point between those of its neighbours, Herculaneum, Nola and *Nuceria*. It may have also extended south to the River Sarno beyond the territorium of Stabia, which was believed to have been subsumed with that of *Nuceria* after the Social War. Dividing the Campanian plain in half has yielded an estimated economic territory for Pompeii of between 130 to 200 square kilometres (Jongman 1988, 106-107; Ling 2005, 99). Agricultural preferences may have been heavily influenced by these spatial constraints (Ciaraldi 2001, 224).

The final circuit of the street grid was constructed shortly after 600 BC (Ling 2005, 30). During the late 4th to early 3rd centuries BC Pompeii was still largely an agricultural settlement. However, by end of the 4th century BC urban buildings covered the majority of the fortified area of Pompeii's 165 acres (Guzzo 2007, 4). Roman influence, with the 3rd century Roman victory over Hannibal after the Second Punic War (218-202 BC), was now firmly entrenched in Italy (Bispham 2007, 113). By the close of the 3rd century BC Pompeii began to flourish economically. By the beginning of 2nd century BC the town was completely built up and within the next 50 years it was a Hellenized urban centre (Descœudres 2007, 15).

Literary, epigraphic and archaeological evidence indicates a growing interest in Hellenistic culture in the 1st and 2nd centuries BC among the wealthy Roman elite in the Bay of Naples area. Aristocratic villas were built in the Bay of Naples area by Romans as luxury vacation houses. These residences were used for the practice of *otium*, the Roman ideal of total relaxation (Mattusch 2008, 18). More than a hundred villas, including rural *villae rusticae* and coastal holiday villas, are now known from archaeological and literary evidence from the Vesuvian area (Lomas 1993, 92, 95, 122; Moorman 2007, 440). However, many of these villas are currently located on private property and have never been properly excavated (Jashemski 1979a, 289).

Pompeii, like other Italic cities, held a unique position within peninsular Italy. From the end of the 4th century BC up until 90 BC Pompeii had been a *socius* or ally of Rome and remained largely politically autonomous and self-governing (Descœudres 2007, 15). This status ended with the Social War, 91-89 BC, when Pompeii was captured by Roman forces after a lengthy siege in 89 BC. Stone-throwing ballistae directed at Regione VI caused known devastation in the House of the Labyrinth (VI.II.ix-x), the House of the Vestals (VI.I.vii), smaller properties within the northern end of Regione VI.I and the city walls themselves (Jones and Robinson 2005b, 696). The Samnite period ended with the defeat and subjugation of Pompeii to Rome (Cooley 2003, 18). Like other Italic communities south of the River Po, at the conclusion of the Social War, Pompeii was granted full Roman citizenship in 89 BC and admitted into the Roman commonwealth (Guzzo 2007, 16; Ling 2005, 51; Wallace-Hadrill 1990, 192).

In 80 BC a colony, *Colonia Cornelia Veneria Pompeianorum*, was imposed upon Pompeii by the Roman General *Lucius Cornelius Sulla Felix* as a penalty for the city's earlier disobedience and resistence during the Social War (Cooley and Cooley 2004, 17; Descœudres 2007, 16; Nappo 1998, 164). In practice, the city still largely governed itself but was pulled into a closer relationship than before with Rome (Laurence 1994, 37; Lomas 1993, 89; Lomas and Cornell 2003, 1). Pompeii, rarely mentioned in ancient literary sources before the 1st century BC, appears in Cicero's speeches in 62 BC in which the Pompeians were nearly implicated in the 'conspiracy' of Catiline (Cooley 2003, 19). The Sullan settlement is thought to have numbered as high as 4000-5000 to the more accepted 2000 Roman soldiers and their families (Guzzo 2007, 16). Latin, as opposed to the Italic dialect of Oscan, was adopted as the official language, although Oscan still persisted into 1st century BC.

During this period southern Italy and the region of Campania were defined by issues of land confiscation, redistribution and Roman colonization (Lomas 1993, 85-87). The cityscape of Pompeii was transformed by the restructuring of the monumental centre of the city. Public amenities were built for the Sullan colonists in a Roman style including the Forum, Macellum, Basilica and Temple of Jupiter (Laurence 1994, 36). In the countryside, during this early colonial period the building or reburbishment of many villas was taking place. There appears to have been an upsurge during the Sullan period and under Augustus with numerous buildings dedicated to leisure and entertainment. This trend was followed by decline in the middle and later years of the 1st century AD (Lomas 2003, 37). Pompeii was thus transformed from a 2nd century BC Hellenised Samnite town into a Roman colony by the early Imperial period (Ling 2005, 57).

Pompeii was now a Roman *colonia*. Its polyglot cultural identity was drawn from disparate elements of southern Italian culture based on strong cultural connections with Rome (Laurence 1994, 37; Zanker 1998, 3): 'The cities of southern Italy were societies in a state of cultural transition, caught between their native Greek culture, the Oscan elements which had been absorbed to a greater or lesser degree, and the increasing cultural dominance of Rome' (Lomas 1993, 125). Pompeii's Hellenistic culture gave the city a unique and privileged position within Roman society due to the Romans' simultaneous fascination and ambivalence towards this foreign culture. In addition, the high levels of immigration and colonization in *Magna Graecia* during the Late Republic and Early Empire probably influenced the cultural heritage of this region (Lomas 1993, 4).

Archaeological evidence shows that Pompeii was the epicentre of a severe earthquake on the 5th February 62 AD from which it never fully recovered (Jones and Robinson 2005b, 702; Ling 2005, 88). However, there is new evidence for a long pre-eruptive period of intense ground deformation or a series of earth tremors prior to the Plinian eruption of Mount Vesuvius on August 24th, AD 79, which ultimately destroyed the city (Cooley 2003, 23; Marturano and Varone 2005, 957). Allison (1991, 54) proposed that the complex processes of abandonment, damage, relocation and repair that occurred during the final seventeen years of Pompeii cannot simply be attributed to the known AD 62 earthquake or the hypothetical one of AD 70 proposed by Zanker (1988). Patterns of hoarding and multiple activity areas are common in a number of Pompeian houses which suggests that abandonment, damage and repair occurred in a more piece-meal manner than previously thought. 'Maiuri (1942:113) observed that at the time of the eruption, many houses in Region VI Insulae 15-16 were still in ruins resulting from the AD 62 earthquake. In contrast, Gioacchino Francesco La Torre (1988: 86) observed that the houses best restored after this earthquake were in Regions VI and VII' (cited in Allison 2004, 196). Recent evidence from Regione I in Pompeii suggest that at least two serious earthquakes happened during the period AD 63 to 79, the last one occurring only a few months before the final eruption of Mount Vesuvius. This would explain the extensive building and restoration work undertaken throughout the city during the final years.

Pliny the Younger, *Plinius Caecilius Secundus*, wrote two detailed epistles at the insistence of the historian Tacitus based on his eye witness account from Misenum of the eruption of AD 79. Like his great uncle, Pliny the Elder, who also made observations of natural phenomena, Pliny the Younger's description of this event were so vivid and precise that his descriptions of the umbrella pine shaped cloud which arose from Mount Vesuvius have been named by vulcanologists as a Plinian eruption after him (Ling 2005, 13-15).

Modern vulcanologists studying Mount Vesuvius believe that before the Plinian eruption in AD 79 there had been a period of quiescence for over 700 years (Sigurdsson 2007, 46). Hence, its volcanic nature would likely have passed from living memory of the inhabitants of the city. Andrew Wallace-Hadrill (1990) recently estimated that a total of only 1,150 or 1,300 bodies recovered from the ruins were victims of the volcanic eruption (Ling 2005, 98. Lazer's (2009) recent re-examination of the skeletal evidence from Pompeii found it to be a normally distributed population. From what is believed to have been a population of 8,000 to 20,000 this suggests that the majority of individuals escaped (Ling 2005, 98; Westfall 2007, 129). For over 18 hours about 2.6 km3 of pumice rained down on the city (Sigurdsson et al.1982, 39). Victims of the disaster probably returned to attempt to reclaim some of their possessions. After AD 79 the volcanic debris would have impeded people's movement near the coast and the city of Pompeii was abandoned (Guzzo 2007, 7).

2.5 Roman literary sources

No contemporary or near contemporary written records on agricultural development exist for the first six centuries of Roman history. The earliest classical Roman agronomist, Marcus Porcius Cato, better known as Cato the Elder (BC 234-149) wrote *De Agri Cultura* in the middle of the 2nd century BC. It was the first agricultural work written in Latin and was largely concerned with issues involving 'investment farming'. *Marcus Terentius Varro* (BC 116-28) wrote De Re Rustica which was published in 37 BC.

Lucius Junius Moderatus Columella (fl. BC 4- AD 65) wrote his twelve volume work *De Re Rustica*, which is arguably the best preserved ancient source on Roman agriculture (White 1970b, 15). *Gaius Plinius Secundus*, or Pliny the Elder (AD 23-79), cited by many as the father of the discipline of botany, is considered the best Roman source on ancient plants (Meyer 1980, 403). The above mentioned ancient texts were typically biased towards large-scale agricultural farms, the majority of which were found in the Po valley, the plains surrounding Rome, and the fertile regions of Campania (MacKinnon 2004, 16).

Gaps in the literary sources exist from the 2nd century BC and during the 2nd and 3rd centuries AD, for which no agricultural manuals or writings have been found. Unfortunately, part of this gap occurs during the period of Pompeii of interest here. Because ancient agronomists often borrowed material and ideas from earlier authors, it is uncertain whether what is described in the ancient agrarian texts reflects current Roman practices in agriculture or established traditions from earlier times (MacKinnon 2004, 16). However, it is likely that these agricultural writings were based to some extent on practical farming experience or observations and thus provide invaluable insight into Roman values and attitudes towards agriculture (Meyer 1980, 403; White 1970b, 15).

Based upon his study of the ancient agrarian texts, Kron (2000, 277) suggests that Roman farming was more sophisticated and productive than is generally thought by Romanists. Specifically, Kron (2000, 277) argues that ley-farming (also known as up-and-down husbandry), which is the practice of leaving a parcel of land fallow for a period of time to prevent soil exhaustion, was known to Roman agronomists. Ley-farming appears to have been the standard Roman method of managing arable cultivation on mixed farms combining livestock-rearing with tillage. Although none of the sources explicitly describe this practice of alternate husbandry, Kron (2000, 277) contends that its widespread use is implicit in the technical terminology and that Roman agronomists would have been familiar with the principles behind its application.

2.6 Ancient literary biases

Archaeological evidence from Campania has revealed disparities with ancient literary sources about agrarian issues and the villa culture. These literary sources tend to focus on specialist crops such as vines, olives and fruits and rarely mention cereal or pulse cultivation (Lomas 1993, 121; 1995, 5). Lomas (1993, 121) attributes this literary bias to several factors: the importance placed upon viticulture, oeloculture and other fruit cultivation which were regarded as elite activities for the production of 'luxury' items; and a disinterest in holistic accounts of agricultural production. During the Republican period Cato cultivated vines and olives on his estates near Monte Cassino and Venafro in northern Campania (White 1970b, 27). Thus, Lomas (1993, 199) cautions against using these texts as evidence of economic specialization. Modern example of this bias are found in White's (1970, 27) seminal work, *Roman Farming*, in which he dismisses 'the treatment of the cereals and legumes in Book II is a straightforward resume, reflecting the decreased importance of this type of husbandry in central Italy in Columella's day'.

2.7 Viticulture and villas

Archaeobotanical seeds of both *Vitis sylvestris* (*Vitis vinifera* ssp. *sylvestris*) and *V. vinifera* have been recovered from prehistoric sites throughout Europe (Renfrew 1973, 125). Grapes are thought have been widely cultivated in Italy and the Western Mediterranean by the 3rd century BC (Borgongino 1999, 89; Ciarallo 2000, 57). The Romans are thought to have brought viticulture to the rest of Europe *circa* 100 BC (Ciarallo 2000, 57; Jashemski et al. 2002b, 7). Significantly, grapes are the most commonly depicted plant on wall paintings and frescoes in Pompeii (Jashemski et al. 2002b, 7). Guzzo (2007, 4) called the grape vine the principal crop of Pompeii.

Romans claimed that the famous Campanian wines were the oldest of the Italian peninsula (Purcell 1985, 7). Pliny the Elder (NH XIV. 35) wrote of the Murgentine vine, also known as the Pompeian vine, which although from Sicily, was confusingly attributed to the region of Campania (Cooley and Cooley 2004, 160). Columella (RR 3.2.27 cited in Cooley and Cooley 2004, 160) argued that the region's other famous vine, the '*Horconian*', was in fact a variant of '*Holconian*', which may have been named after the Pompeian *Holconii* family. Pliny the Elder famously commented unkindly about Pompeian wines, stating that they gave one a headache which lasted until noon the next day (Jongman 2007, 505). Combining his own research with Jashemski, Ling (1996, 346-7) calculates that the diversity in terms of vines recovered 'greatly exceeds the quota considered the maximum possible by Columella, and comes close to the levels achieved in modern viticulture'.

Viticulture is a risky agricultural enterprise compared with other non-staple crops which are suitable for the Mediterranean climate. However, fermented grapes were the commonest source of alcohol in the ancient world and alcohol was readily available, and therefore probably the cheapest, intoxicant available (Purcell 1985, 2). Sherratt (1987, 1999) proposed that due to its intoxicating, mind-altering properties wine could have been viewed as a special drink used in particular social contexts of consumption including feasting (Dietler and Hayden 2001, 1). Thus, wine may have been served and used in the negotiation of power and the maintenance of social cohesion in the Aegean region (Andreou 2003 cited in Valamoti et al. 2007, 54).

Purcell (1995, 173) attempted to resolve this disparity by suggesting that the villa was used by elite Romans to symbolise urban elite values, therefore those who lived the villa lifestyle prioritized wine and olive oil production over more functional crops such as cereals and pulses (Lomas 1995, 5). Olive trees can live for hundreds of years and significantly, the olive came to be a symbol of a sedentary existence and longevity. Olive cultivation, which is less labour-intensive than other Mediterranean crops, separated the agrarian Romans from their more nomadic neighbours (Finley 1973, 31; Lomas 1993, 117-118).

Archaeologically, the greater visibility of villas and wine presses has reinforced this bias of viticulture and oeloculture dominating cultivation in the region of Campania created by the ancient authors. This has resulted in the dismissal or overlooking of cereal agriculture and small farms, which most likely had a greater role in the economy than the villas (Lomas 1993, 117-118). One confounding factor is the false identification and labeling of small estates based on mixed farming as villas (Lomas 1993, 121; Moorman 2007, 440), creating an unrepresentative picture of the economic life of a large part of the Pompeian territory (Jongman 2007, 504). Wine processing plants have been identified from *Villa della Pisanella* and another farmhouse both at *Boscoreale* (Rossiter 1981, 345). At Pompeii itself, in the southwest corner of Insula II.V.i, a well-preserved winepress (*torcularium*) and fermentation room (*cella vinaria*) were discovered (Rossiter 1998, 282).

Jongman (1988, 2007, 505) thoroughly debunks the idea that Pompeian agriculture was monopolized by villas dedicated to viticulture. Using estimates of land fertility, territory size and the particularly high population density of early Imperial period Italy, Jongman (2007, 505) argues that the majority of agricultural input, largely cereals, must have been intended for the subsistence needs of Pompeii's population. The Bay of Naples was so heavily populated that Strabo (V.4.8) described its appearance as a single city (Jashemski 1979b, 4; White 1970b, 73). Indeed, during the 18th and 19th centuries, the Pompeian Plain was predominantly given over to cereal agriculture to feed a smaller population than in ancient times (Jongman 2007, 505). It is likely that in previous time cereals were grown on the level ground areas within the Campanian plains and that the hill slopes were planted with vineyards and orchards, while the nearby mountains used for timber and summer pasture land of sheep (Veal 2009, 229; White 1970b, 73).

Both past and recent palaeobotanical research refute Dupont's (2000, 126-7) unfounded argument that Romans living in the countryside ate mostly vegetables and little wheat. In fact, in Pompeii substantial cereal cultivation must have been undertaken to feed the local population (Jongman 2007, 505). However, there is little evidence for agricultural specialisation at Pompeii (Ciaraldi 2007, 149). The scholarly misunderstanding of the importance of cereals for Roman Pompeii may be due to a misreading of the pollen record: wheat pollen, which is poorly defined, does not disperse far in the environment; whereas vine, olive, walnut and chestnut pollen are easily detected by palynologists (Grove and Rackham 2001, 48).

2.8 Latifundia

After the Hannibalic invasion, at the end of the 3rd century BC, there was widespread economic damage throughout southern Italy due to the 'scorched earth' policy adopted by both the Roman and Carthagian armies. It is likely that cereal crop cultivation would have readily recovered. However, arboreal cultivars would have been severely impaired e.g. olive trees, which can take generations to recover and do not bear fruit for the first ten to twelve years (Finley 1973, 31; Lomas 1993, 122; Palamarev 1989, 93). As a result, the post-Hannibalic period (after 200 BC) is characterised as a period of widespread population decline, failure and abandonment of small farms, migration to the cities and the creation and monopoly of land by the large estates, *latifundia*. *Latifundias* were manned by slave labour for the production of cash crops as opposed to subsistence farming (Carter and Costantini 1994, 101; Lomas 1993, 117-119).

The concept of the *latifundia* is debated amongst Classicists. Few ancient sources mention this specific term. These large estates were likely facilitated by an increase in *ager publicus*, which could not be owned by an individual unless given out by the Senate as *ager privates*. However, this land could be leased in maximum 500 *iugera* per person units. The creation of these large estates by the Roman elite, and probably the Italian nobility, suggests that the Roman Senate was unable to uphold the restrictions on the use of *ager publicus* at this time (Lomas 1993, 119). Barker (1994, 100) argues that it is not possible to correlate changing sequences of sedimentation or vegetation during the Roman Empire with corresponding changes in settlement patterns in any region of Italy, particularly with the appearance of large farming estates.

Recent archaeological evidence suggests that the agrarian changes of the 2nd century BC involved shifts towards a villa economy rather than a fundamental crisis of the entire economic system including the disappearance of small tenant farmers (Rosenstein 2008, 3). The villa economy was based upon small-to-medium sized estates that continued to practice a mixed economy but which now included a larger role for slave labour (Lomas 1993, 119). Barker (1994, 100) argues that during the Imperial period there is a consistent trend towards the growth of villa estates in tandem with a decrease in smaller farms. These large agricultural estates offered greater profits through more efficient production; they would have produced surplus foods intended to meet growing needs in Rome and other urban centers (Garnsey 1999, 10).

The nature of and changes to land ownership are difficult to establish from literary and epigraphic evidence, yet a rise in estate ownership amongst the Roman elite in Campania is inferred to have occurred between the late 2nd to 1st century BC onwards. However, agricultural patterns from the 2nd century BC remained fairly stable until the major upheavals of the 3rd century AD when the widespread appearance of vast latifundia occurred (Lomas 1993, 117-119).

2.9 The hinterland

New archaeological research at Pompeii has raised questions about the relationship between the city and the surrounding countryside (Cooley 2003, 112). A complex symbiotic relationship had developed between the Roman urban centres and their surrounding rural communities (Lomas 1995, 5; Purcell 1995, 171). Pompeii was known to have encompassed the *chora*, an amalgamation of the rural hinterland and an urban centre, where the administration and its public cults were situated (Finley 1973, 139). Therefore, for the Romans, the city was included within the natural landscape; in other words, unlike the later medieval or early modern world view, the town did not end at the city gate but rather the surrounding countryside was integrated politically with the city (Jongman 2007, 513). Thus the correct way to understand the Roman perception of the urban and its surrounding rural environment is to view it as a continuum (Hales 2003, 5; Laurence 1995, 72).

Pompeii's role as a consumer city with its' close links with its agricultural hinterland have dominated discussions of Pompeii's economy (Finley 1973, 131; Jongman 2007, 508). Pompeii was effectively connected to other urban centres by an extensive road system. Farms and villas were situated within rural settings between the towns and cities. The concentration of suburban villas to the north of Pompeii, are situated along the same roads that radiate out from the city. Thus farms and suburban villas appear to have been built on this early established grid, although their locations were probably also influenced by topography, soil fertility and in some cases, stunning views over the Bay of Naples (Adams 2006, 6; Guzzo 2007, 3; Jongman 1988, 106-107).

The land just outside the city walls was converted into *villae rusticae* by the end of the Republic (Lomas 1993, 119, 122). Strabo (V.4.8) writes that, aside from the summit, Mt. Vesuvius was covered in beautiful farmland. Orchard fruits, nuts and vines may have been cultivated on plots of land just outside the city walls, particularly on the lower fertile slopes of Mount Vesuvius. They were probably also grown in garden plots within and immediately outside the city walls of Pompeii, suggested by the archaeological evidence for small farms and market gardens within the city (Ciarallo 2000, 38; Ciarallo and De Carolis, 1998; Jashemski, 1979, 1984, 1993; Jashemski et al., 2002a, 4; Veal 2009, 229). This would have allowed for the supply of fresh produce with little difficulty of transport to the local surrounding towns (Lomas 1995, 118; MacKinnon 2004, 649; Moorman 2007, 440).

Election posters from AD 79 in Pompeii mention the *aliarii*, *pomarii*, and *lupinari* (garlic sellers, market gardeners and greengrocers respectively) (Descœudres 2007, 15). Also, inscriptions describe Pompeii's weekly markets (Cooley and Cooley 2004, 157). Some commodities and staples could be exported and imported from greater distances through marine trade via the River Sarno to the Bay of Naples (Jongman 2007, 506; Richardson, Thompson and Genovese 1997, 88; Veal 2009, 236). Certainly during the 2nd century BC food shortages in Rome, and its surrounding rural communities, necessitated the import of foods (Grove and Rackham 2001, 88). The expansion of urban markets probably encouraged small-scale agricultural specialisation in some locally-grown markets crops, e.g. peaches and walnuts (Jongman 2007, 508).

Understanding town-country relations are integral to understanding the ancient economy and social structure of ancient Pompeii. 'To study Pompeii is to study Roman urbanism under a magnifying glass' (Jongman 1988, 56). Laurence (1995, 72) argues that the social relationship between the inhabitant and visitor to Pompeii was expressed in the spatial layout of the city. He noted that the highest number of doorways and street messages were along the routes that guided movement from the city gates to the centre of the city, the Forum, the centre of Pompeii's economy, social and religious space (Robinson 1996, 370). Ellis (2004, 381) argues that it is the spatial distribution of food and drink outlets within the city that highlights the relationship between the city of Pompeii and its hinterland. Based upon the spatial distribution and orientation of the bars themselves, Ellis (2004, 381) posits that these commercial shops were intended to attract the business of those entering rather than those leaving the town.

2.10 Economy

Agriculture probably formed the base of the Pompeian economy, with the majority of small farmers or tenant farmers cultivating primarily cereals (Jongman 2007, 505-506, 513) (although previous debates about the Pompeian economy were typically dominated by discussion about trade e.g. pottery and amphorae studies). Like Finley (1973, 131), Jongman (2007, 502) agrees that the Pompeian economy did not show signs of economic growth or development. The archaeological remnants of workshops are evidence of a variety of occupations throughout the town. Production of these crafted items locally is thought to have supplied the needs of the residents of the town rather than being intended for export and trade. Jongman (1988, 60, 158) uses this lack of evidence of large scale production to support his theory that there was no middle class. It is therefore likely that Pompeii's urban elite secured the revenues of a rural economy rather than making a profit in trade.

Pompeii evidently experienced economic urban prosperity during the late Republic and early Imperial periods (Wallace-Hadrill 1990, 192). During the Republican period commercial ventures were primarily small-scale, of limited duration, and aimed at local consumption, involving less expensive commodities like olive oil or wine (Woolf 2001 cited in De Sena and Ikäheimo 2003, 301). The expansion of traffic on the Mediterranean Sea in the late 1st century BC, and the greater economic, political and social stability established under the more than forty year reign of the Emperor Augustus and the enforced *Pax Romana* created a prosperous trade environment. Hence, sea journeys were now shorter than in the past and allowed for the trading of more perishable food items than during the Republican period (Parker 1990, 330). However, this period of economic stability and increase in imports is not immediately archaeologically visible at Pompeii (De Sena and Ikäheimo 2003, 312).

2.11 Ceramic evidence from Pompeii

De Sena and Ikäheimo (2003) made a preliminary analysis of the pottery assemblage from the House of the Vestals (VI.I.vii). Their study focused upon the overall supply of amphora-borne agricultural goods and domestic pottery to the city of Pompeii from *circa* 150 BC to AD 79. These authors observed a clear shift from regional self-sufficiency, largely from around the Bay of Naples, towards a heavier reliance upon extra-regional goods. In particular, goods were imported from the Roman provinces within Africa and Spain, especially during the 30 year period prior to the destruction of the city (Table 1).

Many of the deposits from the House of the Vestals (VI.I.vii) were of secondary fill, containing various pottery types including domestic wares, transport amphorae, lamps, glass, bone and other building debris, which had been dumped into the house to raise or level floor surfaces during construction or repair of this property. The archaeological record shows that by the 2nd century BC (150-100BC) a sharp increase in building activity had occurred. It is believed that Pompeii prospered at this time, continuing to be a self-sufficient town relying upon its own agricultural production and craft manufactured items. Pompeian regionally-produced containers, represented exclusively by Black Sand amphorae, account for nearly two-thirds of all amphora-borne goods found here. Another 21% of the amphora came from the nearby coastal areas of Etruria. 13% came from North Africa, and approximately 1/10 are of unknown provenience (De Sena and Ikäheimo 203, 307).

Ceramic evidence suggests that strong trade contacts with the northern rim of the Bay of Naples, the Phlegrean Fields and nearby regions in peninsular Italy including northern Campania, Latium and Etruria were in place by the late 2nd century BC. Also, there appears to have been limited trade with North Africa, Asia Minor and possibly Greece at this time. The newly acquired Roman provinces, Iberia, Gaul, Sardinia and Sicily, were not fully exploited for their agricultural wealth at this time (Woolf 2001). In 123 BC *Gaius Gracchus* established the first grain dole for the city of Rome; shipments of grain from the provinces were unreliable. It was not until around 60 BC that a steady supply line was established with North Africa (Garnsey 1988, 182 cited in De Sena and Ikäheimo 2003, 310).

The majority of wine consumed in Pompeii during the later Republican and much of the Julio-Claudian period appears to have been produced in the Campanian region with lesser amounts imported from other Italian regions. Italy was a very strong producer of wine for both the domestic and foreign markets, particularly the region along the Tyrrhenian coast between Etruria and the Bay of Naples (Tchernia 1983, 87). There is little evidence that wine was routinely imported into Italy at this time. Based upon the amphorae evidence, the wine drunk by the inhabitants of Pompeii appears to have been exclusively Italian in origin, and about 90% from the Vesuvian region and 10%

from west-central Italy. Around AD 50 a sharp decline in regional amphorae made of black sand fabric occurred in tandem with an increase in central Italian amphorae and the introduction in limited quantities of Gaulish, Aegean, Anatolian and Hispanic wine amphorae (De Sena and Ikäheimo 2003, 315). Indeed, Pompeian wine amphorae have been recovered over a wide geographic area including, *Ostia* (Italy), *Ampurias* (Spain), *Alesia* (Gaul), *Vindonissa* and *Augst* (Switzerland), *Trier* (Germany) and as far afield as Stanmore, Middlesex in Britain.

In contrast, olive oil and fish sauce are only represented by North African amphorae during the late 1st century BC. Iberian amphorae, also 1st century, were found within other contexts at Pompeii, although none were recovered from the House of the Vestals. From ancient literary texts and archaeological evidence olive oil is known to have been produced intensively in many regions of Italy, especially northern Campania, Latium and Etruria. All these areas had communication and trade ties with the Bay of Naples by sea and overland routes (Mattingly 1988, 49-50). Olive oil was probably transported in oil skins or wooden casks from farms throughout the region. Organic containers are less likely to survive archaeologically than pottery and therefore this may explain the absence of evidence for olive oil. Indeed, while it is thought that a fair amount of central Italian olive oil and fish products consumed in Pompeii were probably produced in Italy, no trace of such commodities exist in the ceramic record of the city (De Sena and Ikäheimo 2003, 309).

The ceramic evidence suggests that African oil and possibly fish sauce were first imported into Italy after the sack of Carthage in 146 BC and possibly no earlier than 111 BC (Rostovtzeff 1957, 279). Garum, a Roman fermented fish sauce, was produced in Italy, including in Pompeii (Curtis 1991, 89-96 cited in De Sena and Ikäheimo 2003, 309). It should be noted that amphorae were not required for the local distribution of garum and therefore the counts of amphorae of African oil and fish sauce may represent a fraction of these specific items consumed in Pompeii at this time. Of the provincial producers of oil and fish products, North Africa appears to have been the sole source between the mid 2nd century and the end of the 1st century BC, when Iberian products were introduced to Italy (De Sena and Ikäheimo 2003, 315).

By the 1st century BC Pompeii obtained many types of agricultural and craft goods locally as well as importing them from other places within the Mediterranean basin (Laurence 1994 cited in De Sena and Ikäheimo 2003, 304). It was not until the founding of the Sullan colony, in 80 BC that Pompeii's position within the Roman Mediterranean trade network was solidified and Roman tastes and ideals were firmly installed at this new colony (De Sena and Ikäheimo 2003, 311). The period following the colony's founding was one of increased wealth and visible displays of Romanisation through both public and private building programs.

During the period between AD 1-50 fairly dramatic changes in the supply of amphora-borne commodities occurred despite the stability of the economic structure of the Roman Empire throughout the transition from the reign of Augustus to that of Claudius. The period was characterised by a steady continuation of North African production, estimated at over 3/5 of the imported oil and fish sauce amphorae, and a significant increase in the import of trade items from Iberia into Pompeii. Primarily Italian-made wine continued to be imported, being transported both in Black Sand containers and west-central Italian amphorae (De Sena and Ikäheimo 2003, 311).

From AD 50 to 79 there is a noticeable decline in regional amphora-borne products at Pompeii, accounting for 1/5 of all goods at this time. Imports increased, including goods from other Italian centres as well as the Roman provinces. Indeed, from the 1st century AD onwards Rome was increasingly supplied with wine from central Italy with the development of an important center of viticulture in the region of Umbria (Tchernia 1983, 253-259); particularly after the eruption of Mount Vesuvius destroyed the Campanian vineyards in AD 79 (Parker 1990, 325-326). In terms of olive oil and fish sauce Iberian amphorae accounts for 40% and North African 60%. Interestingly, during this final phase of the city new provincially manufactured domestic wares were introduced at Pompeii (De Sena and Ikäheimo 2003, 314).

Time Period	Wine	Olive Oil	Garum (Fish sauce)
150-100 BC	90% Vesuvian region 10% west-central Italy	North African	North African
100-50 BC	Majority from Campanian region with an increase in central Italian amphorae	North African	North African
50-1 BC	Majority from Campanian region with rest coming from other regions of Italy	North African	North African
AD 1-50	Increase in Italian amphorae with corresponding drop in regional Vesuvian amphorae	60% African	60% African Introduction of Iberian amphorae
AD 50-79	20% Regional amphorae Majoirty Italian amphorae	40% Iberia 60 African	40% Iberian 60% African

TABLE 1: RESULTS OF AMPHORAE STUDY FROM THE HOUSE OF THE VESTALS (VI.I.7) (DE SENA AND IKÄHEIMO 2003)

Despite the fact that the populous of Pompeii could have maintained itself from the fertile countryside of Campania during the Roman period there appears to be ever increasing wealth in circulation that thrived on novel and exotic items. As *Puteoli* (modern Pozzuoli), rather than *Ostia*, served as the principal port of Rome until the time of Claudius, with its close proximity to Pompeii (30 km up the coastline) it would be expected that a fraction of these imported foodstuffs and luxury items would have been distributed within the region of Campania and the city of Pompeii itself. This heavier reliance upon long-distance goods corresponded to a net decrease in the exportation of supplies, such as wine, to distant markets. Other ceramic studies have also noted the decline of Vesuvian products in other areas of Italy and the western Mediterranean. This suggests that by the 1st century AD Pompeii was exporting far less and importing more goods to meet its growing demands (De Sena and Ikäheimo 2003, 315).

The fine chronology used by De Sena and Ikäheimo (2003), of 50 year (or less) intervals, should be regarded with caution. Such a precise delineation of time within such a small span of archaeological time (around 200 years), with few obvious archaeological horizons, seems unlikely. Although the amphorae recovered from the elite House of the Vestals (VI.I.vii) provide an interesting case study and offer broad generalizations regarding trade patterns, it is uncertain how widely applicable the results are for the entire city of Pompeii.

2.12 Regione VI, Insula I: brief historical overview

FIGURE 2: PHOTOMOSAIC OF INSULA VI.I FROM THE VIA CONSOLARE (PHOTOMOSAIC © 2007 JENNIFER F. STEPHENS · ARTHUR E. STEPHENS)

The standard convention of naming used in this work follows the modern convention designed for the archaeological site of Pompeii in the late 19th century to unambiguously identify individual properties within the site. There are nine regions within Pompeii. Within these regions are insulae, which are separate blocks of buildings, numbered in systematic order. Within the insulae the doorways of each property are numbered in sequence, either clockwise or anti-clockwise and this will be the standard convention used in this volume (Ling 2005, 25) (Figure 2).

Regione VI, insula I is the only triangularly shaped insula within Pompeii. It is situated just inside the *Porta Ercolano* (Herculanum Gate) and is bounded on the north by the main street, Via Consolare and the back street, Vicolo di Narciso, which appears to follow the alignment to the axis of the Via di Mercurio (Descœudres 2007, 12) (Figure 4), which is thought to have been in existence

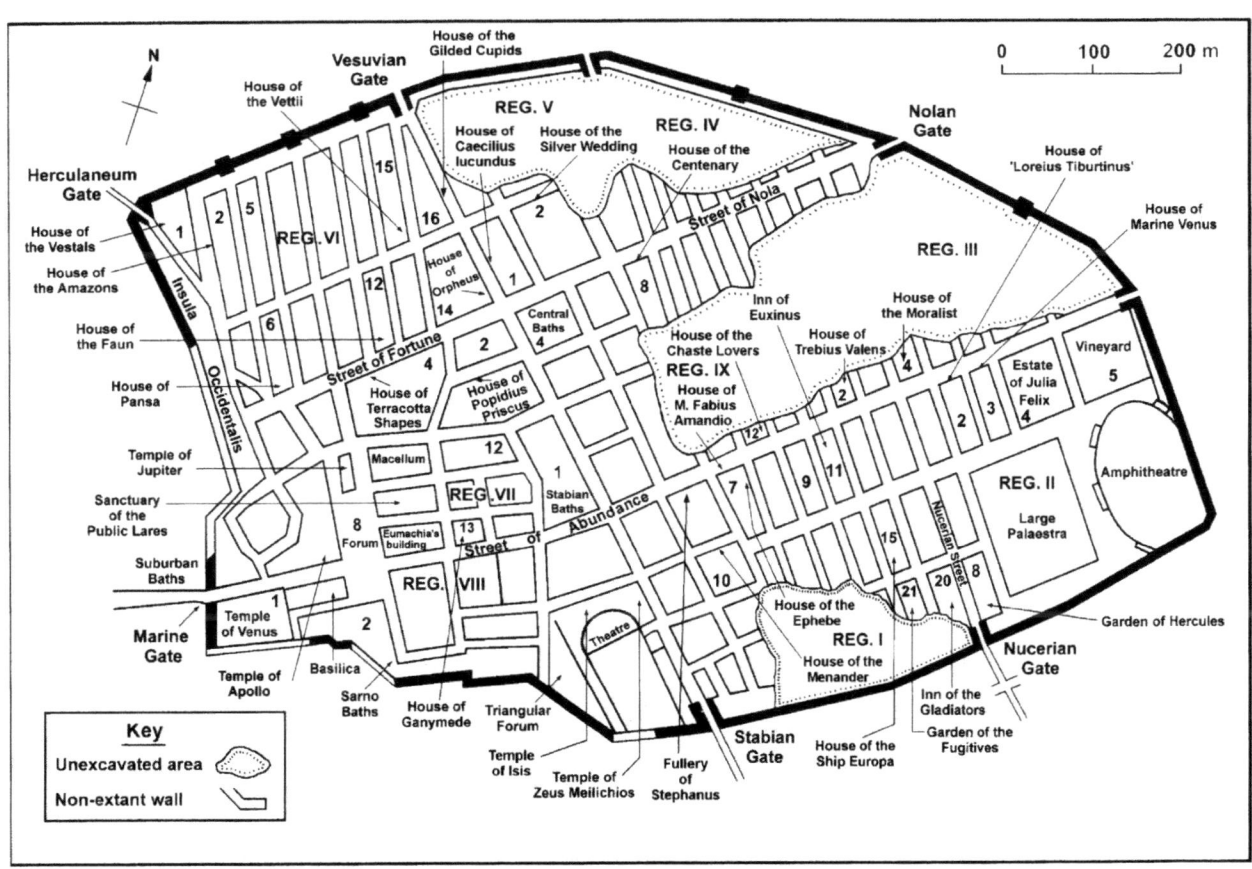

FIGURE 3: MAP OF THE CITY OF POMPEII (AAPP 2005 RESOURCE BOOK, P. 39)

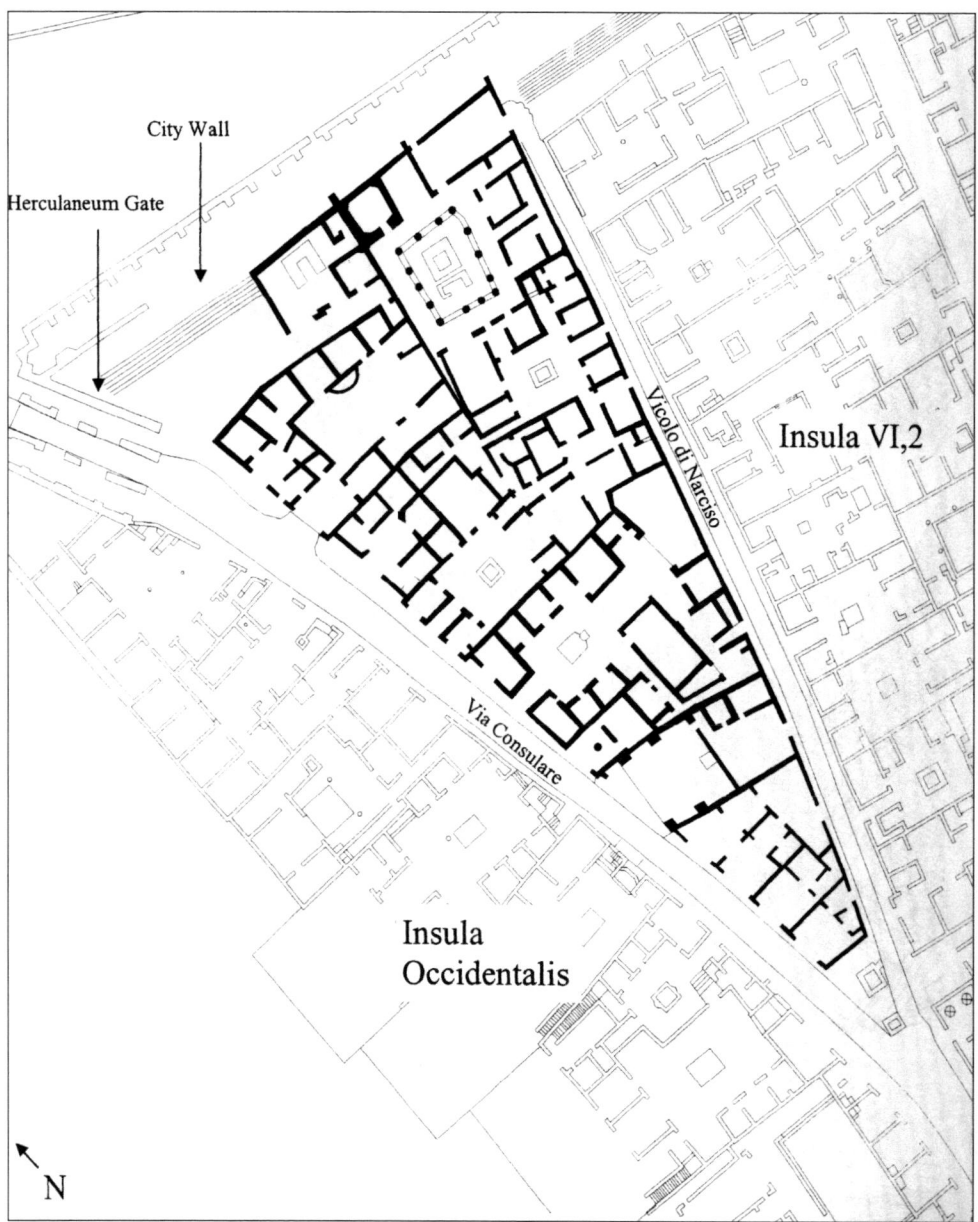

FIGURE 4: INSULA VI.I (AAPP 2005 RESOURCE BOOK P. 26)

in some form by at least the 2nd century BC (Jones and Robinson 2006b, 42, 44). It is believed that the majority of the insulae within Regione VI were laid during the 4th century BC with the slightly later establishment of Insula VI.I during the 3rd century BC (Robinson et al. 2008, 10). Traces of the earliest structural elements discovered within this insula include a small fragment of an earth-built wall which aligns with the Via Consulare. The latter suggests that this insula pre-dates the street grid of Regione VI to the east which has a regular orthogonal pattern (Jones and Robinson 2006b, 42; Jones and Robinson 2007, 390).

The original anthropogenic layer underneath Regione VI, insula I was identified during excavations carried out by the AAPP. It was directly above the natural soil and consisted of a series of rammed earth surfaces, which were found across most of the insula, which were occasionally cut by stake holes and postholes (Jones and Robinson 2006b, 42). Little 'archaic' material was recovered from the area of Insula VI.I (Robinson et al. 2008, 4). A sherd of black-gloss ware from one of the early rammed earth layers was dated to the second half of the 4th century BC (Jones and Robinson 2006b, 42, Jones and Robinson 2007, 389-390).

The northern section of Insula VI.I was an open area during the archaic period with human activity only first appearing at the beginning of the 4th century BC (Jones and Robinson 2006b, 51). Early features on the pitted earth surfaces in this section of the insula include part of an un-mortared wall beneath the Inn (VI.I.iv), a 'toilet' feature beneath the House of the Vestals (VI.I.vii) and compacted earth or mud brick features under the Bar (VI.I.v).

The first two properties built within the insula were the House of the Surgeon (VI.I.x) and the House of the Vestals (VI.I.vii) (Jones and Robinson 2007, 393). The earliest occupation layers were removed by subsequent terracing events for both the House of the Vestals and the House of the Surgeon. Each property now sits on a plot of seventy Oscan feet wide, which suggests that by end of the 3rd century BC or early in the 2nd century BC the terracing and plot divisions were completed. Subsequently, stone-built architecture within the rest of the insula appears after this period (Jones and Robinson 2007, 391).

2.13 The House of the Vestals (*Casa delle Vestali*) (VI.IVII)

FIGURE 5: HOUSE OF THE VESTALS (VI.I.VII) FRONTAGE (PHOTOMOSAIC © 2007 JENNIFER F. STEPHENS · ARTHUR E. STEPHENS)

In 1748 Pompeii was 'rediscovered' (Cooley 2003, 13). Preliminary clearance of the city of Pompeii began in the second half of the 18th century under the guidance of archaeologist F. La Vega. Clearance began outside the Herculaneum Gate and the Street of the Tombs. From 1764 onwards it proceeded along the Via Consolare. The House of the Vestals (VI.I.vii) was among the first houses of the town to be cleared of volcanic debris and much of its decoration was subsequently removed (Figure 6) (Jones and Robinson 2004, 107). It also has the dubious honour of being one of the first Pompeian houses to be disturbed by early Bourbon looters. Therefore, unfortunately, the majority of archaeological material from the AD 79 destruction layer was lost along with other contextual information including the majority of the wall paintings and floor mosaics (Jones and Robinson 2006b, 42, 51).

Despite these early historic taphonomic problems, the House of the Vestals (VI.I.vii) has several major archaeological horizons. Among these is a clear destruction layer from the General Sulla's attack on the city in 89 BC. Another horizon is the piped water and mosaics in the house in the 20s AD, introduced after the town's aqueduct had been built. Damage and rebuilding activity after the AD 62 earthquake is also preserved. The chronological and stratigraphic sequences are fragmentary for the earliest periods of this house. From the 2nd century BC onwards the chronology becomes clearer for the development of the insula (Jones and Robinson 2004, 123; Jones and Robinson 2006b, 44).

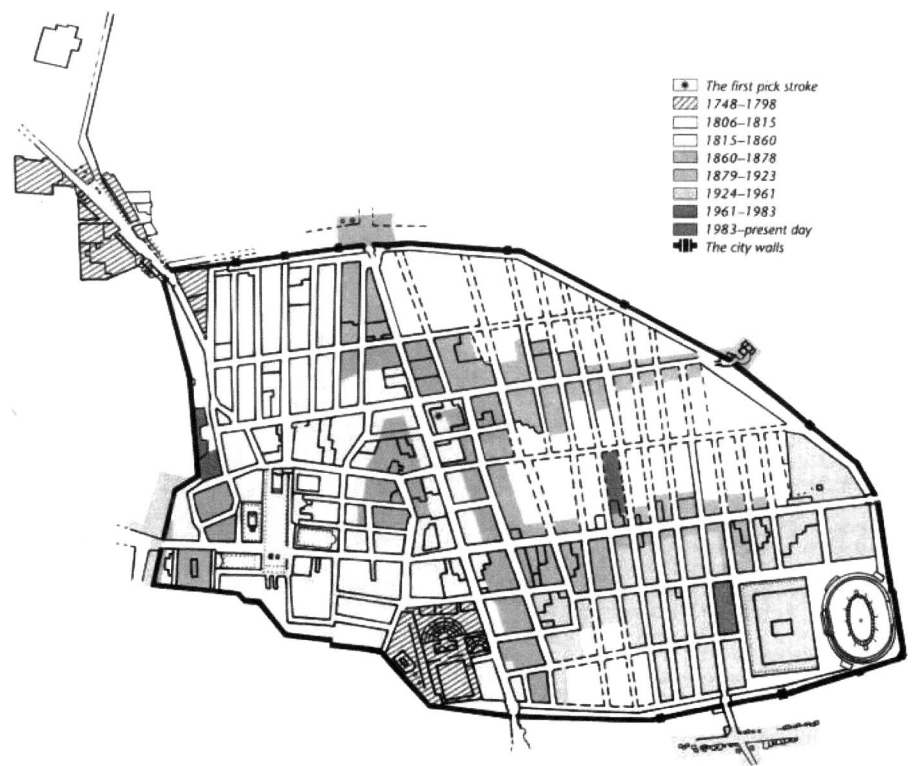

FIGURE 6: MAP OF EXCAVATION OF POMPEII (AAPP 2005 RESOURCE BOOK, P. 40)

The name, the House of the Vestals or *Casa delle Vestali*, was most likely the creation of the 18th century tourist industry at Pompeii and has no firm connection to any ancient reference (Jones and Robinson 2004, 127). During the 18th century, Pompeii became a compulsory site of interest on the increasingly popular European and subsequently North American elite's Grand Tours (Cooley 2003, 14). The House of the Vestals was listed in Grand Tour guide books of the 18th and 19th centuries among the elite residential houses to visit (Allison 2004, 119; Ciaraldi 2001, 139). The House of the Vestals was the most significant property in Regione VI in antiquity and is still one of the most important Pompeian grand residences excavated to date within the city walls (Jones and Robinson 2004, 128; Jones and Robinson 2006b, 42). The House of the Vestals was built shortly after the House of the Surgeon. Initially it was a structurally simple property that developed around the first masonry structure that formed the core of this house (Jones and Robinson 2007, 393).

Little is known of the original house plan (Jones and Robinson 2006b, 43). Most of the street front area is unexcavated due to the presence of mosaic floors from the 1st century AD. Destruction from a World War II bomb exposed the foundations in three rooms. These included a central room flanked by two smaller rooms at the rear of what is believed to have been a central court. The foundations contained orthostats with rubble infill, the superstructure of the walls being composed of yellow earth and occasional stones. The central room had a large pit believed to have been used for ritual purposes during the initial construction of this house. Fragments of black-gloss ware, neonatal pig bones, offering of bread, fruits and nuts and a few votive cups, were recovered from this deposit. The pottery dates from the 4th to 2nd centuries BC (Bon et al. 1997, 44). Little datable evidence was recovered from the rest of the construction phase and no datable links were established with the neighbouring House of the Surgeon (VI.I.x) (Jones and Robinson 2006b, 43).

The previously undeveloped open landscapes to the north and south of the House of the Surgeon (VI.I.x) and the House of the Vestals (VI.I.vii) within the insula started to become filled in on either side with a variety of small courtyard houses and commercial properties with a corresponding rise in the town's population. During the 2nd century BC the town became more densely packed as it took on urban characteristics, with a population largely engaged in non-agricultural commercial pursuits (Jones and Robinson 2007, 393, 395). At the northern section of the insula industrial structures were built on the earth ramparts, on what was later to become the backstreet to Insula VI.I, Vicolo di Narciso, beside the city walls (Figure 4).

During the 2nd century BC, the section to the east of the House of the Vestals (VI.I.vii) was being developed for the first time with substantial foundations, which appear to respect the approximate line of the back street, Vicolo di Narciso (Jones and Robinson 2006b, 44). Excavation only recovered a stretch of wall which survived beneath the later walls from the House of the Vestals (VI.I.vii) on the Vicolo di Narciso side and extended southwards under the rear garden wall of the House of the Surgeon. Interestingly, it appears that the domestic sections of the properties were constructed along the axis of the Vicolo di Narciso (Jones and Robinson 2004, 499). The southern section of Insula VI.I probably had earlier structures that were removed with the terracing event. This event entailed the removal of the natural ground surface to create a stepped sequence of building platforms and regularly laid out plots. Thousands of tonnes of earth were removed over an area stretching south to the House of Sallust (VI.II.iv) (Jones and Robinson 2007, 390-391).

Near the end of the 2nd century BC the Vicolo di Narciso area was evidently extensively re-organized. Buildings were torn down and the ground level was raised through the addition of large amounts of earth. A number of distinct buildings were added to the original house providing a rear range to this property. Around the same time, or slightly earlier, rooms were added to the north side of the Via Consolare house. During this construction event a wall was erected. It remained a constant feature of the spatial division of the insula until its destruction in AD 79 (Jones and Robinson 2006b, 44).

A series of tanks lined with plaster that contained abundant fish remains was recovered from the north end of the insula from workshops VI.I.ii and VI.I.v. These water-proof tanks appear to have been intended to hold some kind of liquid (Jones and Robinson 2006b, 45). Similar plaster-lined tanks were also discovered in the Soap Factory (VI.I.xiv), Bar of Acisculus (VI.I.xvii) and from the Shrine (VI.I.xiii) indicating that there may have been a phase of tank-centred industrial activity of fish processing, possibly for the purposes of salting fish, taking place throughout this insula during the late 2nd and early 1st centuries BC (Jones and Robinson 2007, 394). Dating from the end of the 2nd century BC or early years of the 1st century BC small workshops VI.I.xiv-xv, xvii-xviii were constructed, possibly as a single event (Jones and Robinson 2006b, 54). Each possessed a wide 'shop' doorway fronting onto the busy Via Consolare, a large front room, and one or two smaller rooms at the back of the property.

Whether General Sulla took the town of Pompeii by military force during his siege in 89 BC is unclear. However, the northern sector of Insula VI.I was devastated by the attack. Ironically, this section of Pompeii was again attacked militarily by the allied forces during the 20th century. A bomb explosion from the Second World War destroyed floors from all periods with the tablinum area and the cistern beneath it from the House of the Vestals (Bon et al. 1997, 155; Jones and Robinson 2004, 110), debunking Binford's Pompeii premise in a spectacular manner. The former attack created a clear dating horizon for the northern end of Insula VI.I with the presence of missiles, including stone balls and lead sling shot, the destruction of buildings to foundation level and subsequent clearing deposits.

Impact marks are still visible on the exterior city wall beside the Herculaneum Gate where General Sulla ordered stone balls fired into the city. Many of the missiles that cleared the city walls landed on the roofs of the northern end of Insula VI.I devastating commercial buildings along the Via Consolare, including the houses along the Vicolo di Narciso and part of the Via Consolare house (Jones and Robinson 2004, 498; Jones and Robinson 2006b, 45).

Rebuilding took place over the next few decades, retaining the previous divisions between residential and commercial spaces. All the buildings along the Via Consolare were rebuilt upon their original foundations. One visible difference is that some of the commercial properties seem to have changed functions. From the south end of the insula the series of fish-processing tanks are thought to have been abandoned as the plaster-lined tanks were found in-filled with rubble (Jones and Robinson 2007, 398). In place of fish-processing activities, food and drink shops and a smithy were constructed (Jones and Robinson 2007, 397-398).

In the poorly named Soap Factory (VI.I.xiv) (*Fabbrica del Sapone*) (Figure 7), (as there is no evidence of soap making), archaeological evidence suggests that it took over the neighbouring workshop (VI.I.xv) and again significant quantities of hammerscale, slag and hearth bottoms were recovered from the final occupation layer above a hard mortar floor. Thus it is thought to have operated as a metal smithy between the mid 1st century BC and mid 1st century AD. At the same time the back room of the Bar of Acisculus (VI.I.xvii) was converted into space for large-scale food processing and equipped with a cooking facility. It is likely that food products were being sold out of the front room which faced onto Via Consolare (Jones and Robinson 2006b, 49, 52, 56; Jones and Robinson 2007, 398, 402). The cooking platform in the Bar of Acisculus (VI.I.xvii) was removed during the construction of an upper story over the commercial properties in the south end. A surface of *opus signinum* was installed throughout this property as well as a marble-topped bar counter facing Via Consolare, which is still visible today.

FIGURE 7: THE SOUTHERN END OF INSULA VI.I, INCLUDING FROM LEFT TO RIGHT THE SOAP FACTORY (FABBRICA DEL SAPONE), THE BAR OF ACISCULUS (TABERNA DI ACISCULUS), THE BAR OF PHOEBUS (TABERNA DI PHOEBUS) AND THE WELL/FOUNTAIN (PHOTOMOSAIC © 2007 JENNIFER F. STEPHENS · ARTHUR E. STEPHENS)

At the end of the 2nd century BC the insula had a row of commercial properties fronting the street to target travellers leaving the city rather than local people (Figure 7). People travelled along the two-way street Via Consolare leading out of the *Porta Ecolano* or Herculaneum Gate along this coastal road which passed through the area later known as *Oplontis* (today *Torre Annunziata*) through the city of Herculaneum and further on turning towards *Neapolis* and *Cumae* (Guzzo 2007, 4; Mau 1902, 31; Poehler 2006, 72). The Herculaneum Gate may have also been known as *Porta Salis* or *Saliniensis*, from the salt works carried out along the coast to the west of the city (Ling 2005, 24).

With the creation of the Sullan colony, there was an apparent change in demand within the internal market of the city for the supply of everyday goods and foodstuffs along with the growing importance of the city as a regional market. The production of olives, fruit, beans, perfume, cloth, fish sauce production and wine making was particularly important to the economy (Mattusch 2008, 13). The evidence suggests that there was a transformation and fundamental shift in the urban economy within Insula VI.I from fish-processing, possibly for the export of salted fish, to the manufacture of metal goods and supplying food to travellers. With the exception of the House of the Surgeon (VI.I.x) and the House of the Vestals (VI.I.vii) all of the properties along the street front of Via Consolare were actively engaged and developed for participation in this urban economy (Jones and Robinson 2007, 398, 400, 403).

At the end of the 2nd century BC the House of the Vestals (VI.I.vii) doubled in size, thus extending to the Vicolo di Narciso by incorporating another property. A rear entrance was created in 100 BC leading into Vicolo di Narciso. A newly constructed atrium and separate service area were added. The large garden space in the middle of the house was retained (Jones and Robinson 2007, 393). Two small adjacent courtyard houses were built on the Vicolo di Narciso. One was constructed as part of the development at the rear of the House of the Vestals which suggests it was being used as a rental property and that the owners of the House of the Vestals (VI.I.vii) had begun a foray into investment property speculation (Jones and Robinson 2004, 128, 2007, 394). Also at this time the House of the Vestals was heavily investing in the construction of a series of new workshops along the Via Consolare (Jones and Robinson 2004, 498; Jones and Robinson 2007, 394-395).

During the mid-1st century BC following the Social War, the House of the Vestals incorporated two small courtyard houses, both badly damaged from the Sullan siege. With the addition of these acquired properties the House of the Vestals expanded to cover the entire northern section of Insula VI.I (Jones and Robinson 2005b, 696). Architectural evidence shows the integration of earlier properties into a single house by the mid 1st century BC. Along the Via Consolare frontage the commercial property was separated off from the residential house (Jones and Robinson 2007, 398). Interestingly these commercial properties were not included in the grand redevelopment of the House of the Vestals (VI.I.vii) despite also being damaged during Sulla's attack in 89 AD (Jones and Robinson 2004, 499).

An elaborate series of reception rooms surrounded by a peristyle were created in the House of the Vestals. By the second half of the 1st century BC the House of the Vestals was an opulent aristocratic property with a traditional axial ground plan (Jones and Robinson 2007, 397). It was among the twenty largest and most imposing residences in the city of Pompeii (Jones and Robinson 2005b, 696). It began to rival its neighbours, particularly the House of the Surgeon, in size and architectural complexity, financed in part by its new commercial centres.

The merging of the two separate houses into a larger elite house highlights the increasing economic and social inequalities within Pompeii. At this point it appears that the wealthier inhabitants began to monopolise urban space (Jones and Robinson 2004, 499; Jones and Robinson 2007, 397). In the mid 1st century AD an upper storey was constructed across the entire range of these separate commercial properties which strongly suggests a single ownership. A staircase opening onto Via Consolare (VI.I.xvi) was added. An apartment, probably for rental, was added to the new upper storey for additional income purposes. In its final form before the eruption in AD 79 the commercial triangle in the south end of the insula was composed of two bar counters, the so-called Bar of Pheobus (VI.I.xviii) and the Bar of Acisculus (VI.I.xvii), modern names inspired by epigraphic evidence (Figure 7) (Jones and Robinson 2006b, 49, 53, 56).

The north end of the insula was also involved in commercial property development with the building of a bar, large coach inn and upstairs rental apartment. These commercial properties were now structurally separate from the House of the Vestals. One theory is that this development was independent of the House of the Vestals (Jones and Robinson 2004, 499). Significant spatial division of the insula occurred with the creation of an inn that opened onto the Via Consolare (as the yard in the middle of the insula was no longer accessed through the House of the Vestal's peristyle). The bar, coaching inn and upstairs rental apartment continued to be owned by the House of the Vestals but was likely managed by slaves or freed men attached to this household (Jones and Robinson 2007, 402). This incorporation of tabernae into private houses demonstrates intensive exploitation of urban properties within the town during this period (Pirson 2007, 457).

Archaeologist Giuseppe Fiorelli suggested that the space in the southern section of insula VI.I.xiii was a religious Shrine. Architectural similarities exist between this building and the *Lares Augusti* shrine excavated near the *Porticus Aemilia* in Rome, which include the four large brick piers constructed to support a roof. To create room for this religious space within this densely populated urban area it may have been necessary to destroy commercial buildings. Excavation has revealed that the original floor level was raised by a metre and paved with white mosaic. With the addition of a lower step, it was paved in white signinum and decorated in diamond pattern by black tesserae. Perhaps this costly religious shrine was financed by wealthy Pompeian/s as a benefaction to the community, which would have been highly visible to passers-by along the busy Via Consolare (Jones and Robinson 2007, 400-401).

Piped water was made available to the city by the construction of the Serino branch of the Campanian aqueduct by Agrippa *circa* 27 BC, with the original aqueduct built by Augustus for the naval base at *Misenum* (Descœudres 2007, 18; Guzzo 2007, 18; Jansen 2001, 51). The *castellum aquae*, located near the Vesuvian gate, was built to regulate the distribution of water to the rest of the city. Water pipes were embedded in a newly constructed pavement built at the same time as the paving of this street in the last quarter of the 1st century BC (Jones and Robinson 2005b, 697). They were installed on the west side of Vicolo di Narciso at the fountain, at the southern most point of Insula VI.I (Jones and Robinson 2006b, 47). These new street arrangements respected the original access points into the House of the Vestals. Nevertheless, within two decades the house was re-organized to take advantage of the new aqueduct-fed water supply and water display features were installed inside the house. Domestic water continued to be sourced from extant underground cisterns.

Among the improvements in the House of the Vestals (VI.I.vii) were the installation of a larger private bath-suite and a decorative marble-lined impluvium (fountain) in and around the peristyle and in the two atria. A swimming pool was constructed in the centre of the large peristyle garden (Jones and Robinson 2006b, 47). The elaborate nature of this house in the 1st century BC was renewed in the 20s AD with increasingly lavish displays of luxury, including redecoration in the latest Third Pompeian wall painting style. Although the decoration left *in situ* from this house has been badly damaged from over two centuries of exposure to the elements, based upon past archaeological records (Fiorelli 1860) and those published by antiquaries (Gell and Gandy 1817-19) it must have necessitated tremendous financial resources, as luxury items such as precious stones and pigments were used in its construction and embellishment (Jones and Robinson 2004, 501; De Sena and Ikäheimo 2003, 305). Floors were ripped up, wall plaster stripped away and the resulting rubble dumped into pits. New floors were decorated in black mosaic with a white double-banded border. A second atrium off Vicolo di Narciso was also lavishly decorated (Jones and Robinson 2007, 398). Another room adjacent to the large peristyle was adorned with marble veneers and blue-dominated wall paintings.

During the early Empire additional architectural changes to the House were made, including the creation of a more imposing entrance from the Via Consolare with four brick columns leading directly into the atrium. Brick replaced tufa on both the peristyles. The previous axial design of the house was transformed with an enlarged service area in the centre of the property which now formed part of a service zone with upstairs rooms (Jones and Robinson 2007, 398). Interestingly, little space was allocated to service areas within this, now grand, residence. Service areas within the house were removed from view of the elite family and their guests (Jones and Robinson 2006b, 47, 50). Social differentiation through architectural spatial divisions became more pronounced over time, in particular from the middle of the 1st century BC.

During the early Empire the House of the Vestals had exclusive access to pressurised water. The owners of the House built a series of ornate fountains that allowed excess water to overflow into the street. This display of conspicuous consumption was a powerful signifier of wealth. It would have been visible to passers-by looking in through the front entrance, to guests, visitors and clients as they were welcomed into the entrance during the early Imperial period (Jones and Robinson 2005b, 696, 707; Jones and Robinson 2007, 398).

After the earthquake of AD 62 the public fountain at the southern apex of the insula was quickly repaired and the water supply reinstated, although it no longer supplied private houses (Jones and Robinson 2007, 401). No direct evidence of structural damage to the House of the Vestals was found during excavations. Yet the piped municipal water supply was not reinstated to the house, possibly because the piped water was unreliable as the house was on higher ground (Jones and Robinson 2006b, 49).

Decorative fountains were the focal point of the previous design so the inhabitants of the House of the Vestals were forced to begin another phase of construction and redecoration to ensure their social status by maintaining their residence in the current style during the brief post-earthquake period (Jones and Robinson 2004, 501). Water pipes were removed and discarded, and the overlying mosaics were repaired. The bath-suite was removed and a coin dated to AD 72 found in the deposit provided a *terminus post quem* (Jones and Robinson 2006b, 57). A new upper storey was built to provide a suite of reception rooms over the north side of the peristyle. Large windows were added with breathtaking views over the city walls towards Mount Vesuvius and the Bay of Naples. Major public rooms were redone in the latest wall painting style (Jones and Robinson 2007, 401).

Despite a lack of access to a reliable source of piped water the inhabitants of the House of the Vestals appear to have wanted water features. Attempts to decrease their water usage included reducing the volume of the swimming pool and changes to the bath-suite. The fountains were replaced by standing pools of water (Jones and Robinson 2007, 401). There is even some evidence that a substitute water system was put in during this post-earthquake period. A new above-ground cistern was added by converting a room on the north side of the large peristyle. Reception rooms, with raised roof levels for the collection of rainwater, were added unto the north and east sides of the large peristyle with walls reinforced and sealed with waterproof plaster (Jones and Robinson 2006b, 49, 57).

These costly redecoration phases of this property demonstrate the intense social competition of the upper classes of Pompeian society during the early years of the Imperial period (Jones and Robinson 2005b, 696, 708; Jones and Robinson 2007, 401). The luxury of space within the urban landscape becomes apparent with the uneven distribution of it within this insula. Small households were pushed out of the insula. The wealthy elite gained space and lived in comparative luxury: 'Indeed so close is the junction between the development of economic space and of aristocratic domestic space that it is tempting to suggest that these two phenomena must be linked' (Jones and Robinson 2004, 501). Based upon the distribution of space during the 1st century AD the majority of the populous within the insula were working to support the owners of the House of the Vestals in their luxurious lifestyle. The residents of this house continued to live in luxury until its final destruction (Jones and Robinson 2006b, 50).

Within the House of the Vestals a diverse cross-section of Roman society can be studied within this small urban area. Although the phases of change taking place over the occupation of Insula VI.I do not differ significantly from the overall scheme proposed for the city by Zanker (1998). It does reveal the complexity and continuous nature of change which took place over time which cannot be simplistically attributed to a few known historical events (Jones and Robinson 2006b, 50).

2.14 The House of the Surgeon (*Casa del Chirurgo*) (VI.I.x)

FIGURE 8: HOUSE OF THE SURGEON (VI.I.x) FRONTAGE (PHOTOMOSAIC © 2007 JENNIFER F. STEPHENS · ARTHUR E. STEPHENS)

The House of the Surgeon (*Casa del Chirurgo*) (VI.I.x) (Figure 8, Figure 9) was named after the bronze and iron surgical instruments discovered from one room during its initial clearance in 1770-1 (Nappo 1998, 148). This famous residential house, adjacent to the House of the Vestals (VI.I.vii), is well-known as one of the oldest and best preserved Pompeian houses to maintain the traditional simple atrium style plan of an early Italic house with few modifications to the original plan and appearance. Early Pompeian archaeologist August Mau (1902, 280) placed the House of the Surgeon to the so-called period of limestone atriums based on the austere limestone façade of *opus quadratum* facing the Via Consolare (Adam 2007, 99). It is made of Sarno stone, a calcareous sedimentary rock quarried in the Sarno valley, with ashlar work and its inner walls made of limestone framework (Ling 2005, 35; Robinson et al. 2008, 2). The House of the Surgeon dates from *circa* 200 BC based upon a coin dated to 214-212 BC in a foundation trench of the atrium of this property (Jones and Robinson 2007, 392; Veal 2009, 58).

FIGURE 9: HOUSE OF THE SURGEON (VI.I.x) (AAPP 2005 RESOURCE BOOK, P. 27)

It was the Italian archaeologist A. Maiuri (1930) who first pioneered at Pompeii a methodology of 'wise destruction' (Carrington 1933) in which small sondages or trenches were dug to address specific questions regarding the historical development of the House of the Surgeon (Robinson et al. 2008, 2). Maiuri believed it to be an example of a Samnite house (Nappo 1998, 148). This limited view of the archaeology 'impeded his full appreciation of the stratigraphy and the underdeveloped state of knowledge of the ceramic assemblage only allowed him to suggest a date between the early 4th to 3rd centuries for the construction of this property' (Robinson et al. 2008, 2). The earliest 3rd century BC date of the House of the Surgeon was obtained from the ceramic evidence collected in Maiuri's excavation. Maiuri also reasoned that the Sarno stone blocks of the foundations of the house were recycled as they were covered with plaster from the 6th century city wall. He further proposed that the Samnites dismantled the wall when they built a second and larger fortification after their invasion at the end of the 5th century BC. Thus, the House of the Surgeon could not be dated to before the invasion of the Samnites (Robinson et al. 2008, 3).

Spatial arrangement of this property has been long held up academically as a classic example of the inward looking early atrium house, which has not been largely altered since Mau's work. R.C. Carrington (1933) argued that this property was the best preserved example of the 'second 'old Tuscan atrium-house' developmental stage of Roman domestic architecture' (cited in Robinson et al. 2008, 3) as it appears to possess an architectural form which corresponds roughly with known classical references.

Renewed excavations at Pompeii on the House of the Surgeon (VI.I.x) were conducted between 2000 and 2006 by the AAPP based at the University of Bradford and Oxford University. New archaeological research has revealed that what was thought to have been a well-known example of the classic inward looking atrium style house deviates from this model with several important rooms that open outwards onto the colonnade which surrounded the house on both its eastern and southern sides. Thus, it appears that the house was not centered upon and entirely focused upon the central courtyard. Indeed, the southern wall of room 6C allowed access through a wide doorway between the atrium and the southern colonnade and doorways from rooms 6A and 6B also opened upon the garden space, beyond which was an important area (Robinson et al. 2008, 10, 14, 17).

The proto-Surgeon's House dating to the 3rd century BC had a different owner and this property was situated on a different axis. This early house was lost with the removal of over one metre of natural ground surface to create a level building platform for the construction of the new Surgeon's House. The earliest traces of occupation are from the original ground surface dating later than the north section of the Insula VI.I. Underneath the *tablinum* in the House of the Surgeon, there is a small section of wall composed of yellow earth mortared *opus incertum*, and a lamp sherd of black-gloss ware from a construction trench dating from the 3rd and 2nd centuries BC and a coin *circa* 214 BC. This wall was later buried by another levelling episode during an *opus quadratum* phase. Finally, part of an impluvium-like feature was discovered associated with a beaten earth floor made of *opus signinum* and surrounded by low masonry in what was later to become the 'service quarters' of the property (Jones and Robinson 2007, 390).

Room 23 of the House of the Surgeon contains the earliest signs of human activity. The initial terracing period protected the subsequent terracing events beneath this room. In contrast, two terracing events, the first in the 3rd century BC to create the so-called 'Shrine' and the commercial 'triangle' building plots, and the second at the end of the 1st century BC which was employed to create rooms 3 and 4 of the House of the Surgeon, both were stripped of any material from this early period (Robinson et al. 2008, 4).

Perhaps due to increasing prosperity in the community, the House of the Surgeon underwent substantial construction events around 100 BC. The construction of an enlarged triclinium and a service area, now separate from the main core of the property, hints at the increasing wealth of the residents of the House of the Surgeon. Another extensive period of construction took place during the final quarter of the 1st century BC with the redecoration of the main reception rooms and the construction of a new *oecus* (room 19) which had a view of the formally planted garden (Robinson et al. 2007, 21). The first permanent latrine within the House of the Surgeon was probably located in room 13. An external cesspit drains into the Vicolo di Narciso sidewalk. Room 13 may have been used as a kitchen due to the close connection between latrines and kitchen areas within Pompeii (Hobson 2009; Robinson et al. 2008, 20-21).

Commercial shops appear relatively late on this property compared with the insula in general and the House of the Vestals (VI.I.vii) in particular, in the 1st century BC. The House of the Surgeon retained ownership of the southern shop unit (of unknown economic functions) (Robinson et al. 2008, 25, 37). Commercial properties adjacent to elite residences were common in the ancient city and indeed continued into the medieval and early modern period in Europe (Ling 2005, 24).

The property was redeveloped during the Augustan period to create a grand axial view from fauces (entranceway) to *hortus* (garden) (Figure 9, Figure 10), which according to architectural classical sources should have been in place in the original house plan. The *hortus* was created from large quantities of topsoil. During excavations ten planting pots (*olea perforata*) were discovered buried at regular intervals of 0.6-0.7m along the eastern wall. Seven more were recovered along the northern wall of the garden, spot dated to the last quarter of the 1st century BC. It has been posited that planting pots were put in simultaneously to create a pleasing designer garden space (Robinson et al 2008, 20-22).

FIGURE 10: HOUSE OF THE SURGEON (VI.I.x) ENTRANCEWAY WITH VIEW INTO THE *HORTUS* (PHOTO BY AUTHOR 2007)

The house was largely in a state of disrepair during its final 17 years. Post-holes were discovered for timbers that are thought to have propped up the roof. Some rooms were in a poor state of repair. The large room to the south had been converted into a lime pit, possibly for on-going repairs to the house (Fortenberry and Goalen 2007, 7). The second phase of redecoration and construction implemented during the final years leading up to the destruction of the city remained unfinished. It could be that the Fourth Style redecoration was in response to the large earthquake of AD 62 and other reconstruction events were in response to the later seismic activity leading up to the eruption of AD 79 (Robinson et al. 2008, 25). However, the comparative neglect of the House of the Surgeon is striking as it is the only property in the insula that was not rebuilt in a better and bigger form (Jones and Robinson 2007, 393).

Archaeological excavations revealed that the basic layout of the property plots within Insula VI.I were largely maintained through time, as were the spatial distinctions between residential and commercial/industrial areas. At the time of the eruption in AD 79 this insula was a mix of land usage with both commercial and residential properties. However, it was a spatially structured society (Robinson 1996, 370). The archaeology from Insula VI.I demonstrates new insights into the commonly-held belief that Pompeii was a decaying city, due to the ravages of earthquakes and after shocks, but rather was a city returned to some form of 'normality' before its final destruction in AD 79 (Jones and Robinson 2007, 404). Hence, the results from the excavations reveal a more intensive use of space and a rise in the socio-economic complexity present in Pompeii as the population increased from the 3rd century BC to the eruption (Jones and Robinson 2007, 404).

Romans, Rubbish, and Refuse

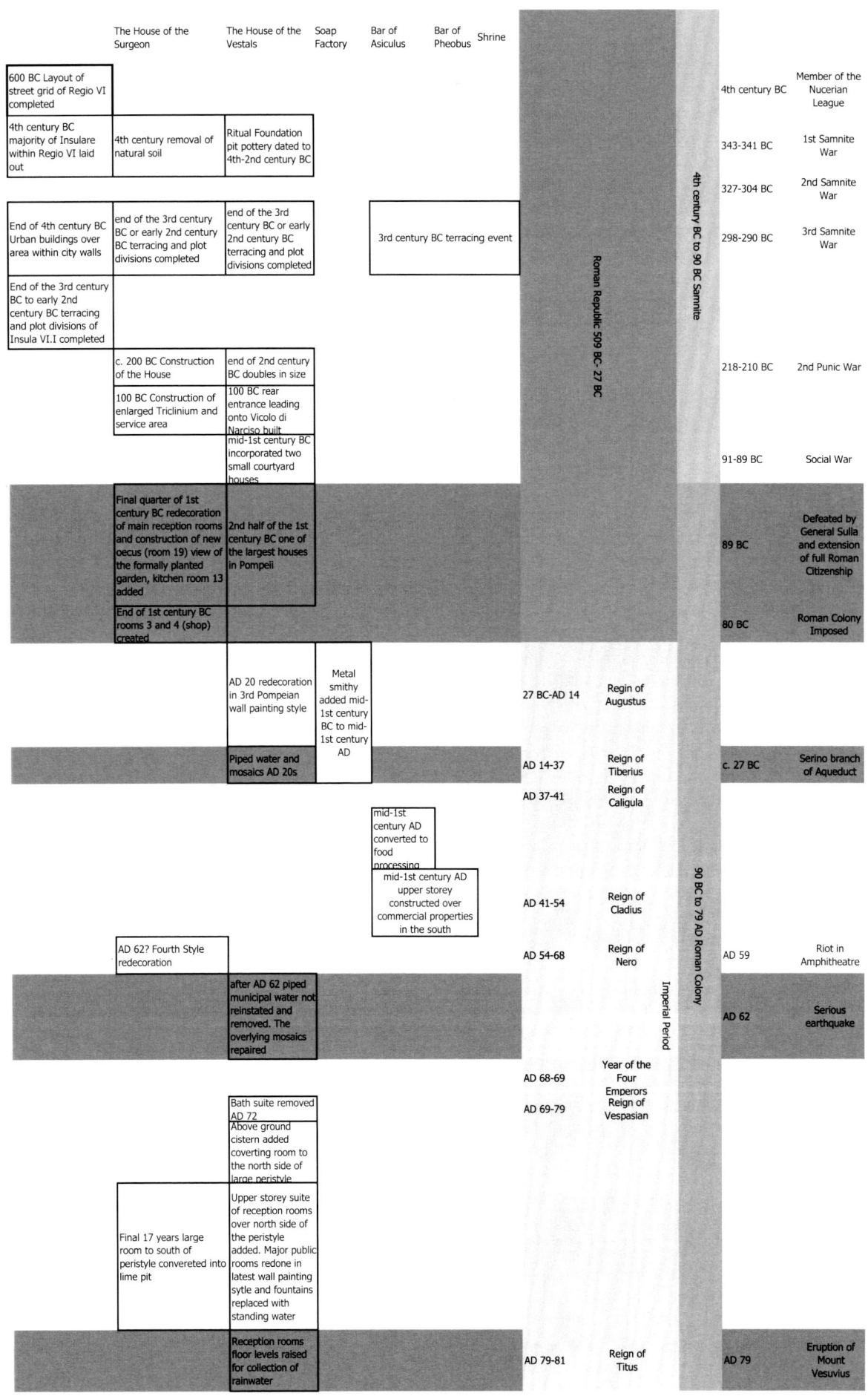

Figure 11: Timeline for Regione VI, insula I

Chapter 3

Methodology

The principal aim of the Anglo-American Project in Pompeii's (AAPP) excavations within Insula VI.I is to establish the social and economic history of the Insula (Jones and Robinson 2004, 107, 128). Therefore, large scale research projects at Pompeii, including the AAPP, British School at Rome excavations directed by Fulford and Wallace-Hadrill (1999), and Ciarallo and De Carolis (2001) with the *Soprintendenza Archeologica di Napoli e Pompei* have employed modern environmental recovery techniques such as flotation and sieving for the recovery of macroscopic and microscopic environmental data (Ciaraldi and Richardson 2000, 75; Robinson 2002, 94; Veal 2009, 2).

3.1 Research methods

Ciaraldi (2001, 40) notes that one of the major recurring themes in Pompeian archaeobotany is the 'lack of thoroughness in the methodological approach and in the analysis'. A detailed environmental sampling and methodological strategy was implemented from the beginning of the AAPP. All archaeobotanical material for examination for this research was collected from Insula VI.I during the past twelve years of fieldwork.

3.2 Recovery methods

Machine flotation was chosen as the primary recovery technique for the ecofacts as large quantities of matrix can be processed fairly easily in the field and easy to use for training field school students. A wide range of preserved botanical material can also be recovered, which allows for the quantitative study of the macrobotanical remains (Pearsall 2000, 14-15). Fish bones and the carbonized and mineralized plant remains were recovered through this process (Ciaraldi 2001, 76; Ciaraldi and Richardson 2000, 75; French 1971 cited in Richardson, Thompson and Genovese 1997, 89-90). Ciaraldi (2001, 143) noted that plant remains were unevenly distributed through the different archaeological horizons and there was a noticeable paucity of archaeobotanical material from earlier defined deposits at the House of the Vestals (VI.I.vii), aside from a few unique contexts.

Near-total retrieval of environmental remains for the entire Insula was employed during the excavation. All stratified deposits were screened through 5 mm mesh screens (Ciaraldi and Richardson 2000, 75). In addition, 20 litre sediment samples (where possible) were taken from every securely stratigraphically sealed unit excavated and other deposits of interest were processed with a modified Ankara-style flotation tank (Ciaraldi 2001, 76). This was originally designed by the British Institute of Archaeology at Ankara in the 1970s to process the light sediments from the Near East and Mediterranean region (Renfrew, Monk and Murphy 1976, 21). Flotation entailed sediment being poured gently into the large tank of the flotation machine. A two horsepower pump was used to force water through the underlying perforated metal pipes inside the tank and allowed for the agitation of the matrix and mechanical circulation was provided by stirring the sediment gently by hand. The water was allowed to overflow into a collection barrel through the spout, which had a 0.5 mm mesh flot bag attached to capture the light fraction. Due to water shortages in southern Italy during the summer months of July and August, a water recycling program was established in which a series of settling tanks were used to minimize water usage. The water in the tank was changed completely after two to three samples were processed, ensuring that the samples were not seriously contaminated by previous samples.

Heavy fractions were collected with a 1 mm mesh inside the flotation machine. Heavy fractions (or residues) collected from the 1 mm mesh were rinsed and allowed to dry. The volume was then measured and approximately 25% was sorted in the field using geological sieves of 2 mm and 1 mm to facilitate sorting. During the 1995 to 2004 field seasons, the sieved heavy fractions <2 mm were labelled and shipped to the University of Bradford for future sorting under a low magnification binocular microscope (Ciaraldi 2007, 57). During the 2005 to 2007 field seasons, all heavy fractions were sorted in the field. If a sample was particularly rich in ecofacts, it was shipped to the University of Bradford for more detailed sorting. Field school students were assisted in this step. All biological material, except faunal remains, were separated, labelled and sent to the University of Bradford for further analysis by specialists.

The light fraction was labelled and air-dried on a clothes line in the shade. Due to the small particle sizes and the low density of charred remains in these samples, all light fractions were sent to University College London for sorting from the 2006 post-excavation season. Light fractions collected in previous field seasons were sent to the University of Bradford to be sorted. In the laboratory, weights and volumes of all fraction samples were recorded. The samples were screened through 2 mm and 1 mm geological sieves for ease of sorting. Using a low-powered binocular microscope, 4.8x to 56x magnification, all biological material was sorted, measured, counted, and entered into an excel database.

3.3 Sampling

Due to a lack of information on the preservation, deposits of interest and density of ecofacts within this section of Pompeii a complete blanket sampling strategy was employed by the AAPP in which all contexts from earlier phases, up to and including the Roman horizon, were collected and examined (Bon et al. 1997, 153; Richardson, Thompson and Genovese 1997, 88). The limitations to this strategy include the time needed to process all samples excavated and its expense based upon limited resources. However, establishing a standard sample size throughout a research project reduces potential sampling bias (Adams and Gasser 1980, 295). This contrasts with previous environmental research at Pompeii which was largely concentrated on 1st century AD deposits, e.g. gardens, vineyards and open spaces sealed by the falling ash and lapilli of the eruption (Ciarallo 1993, 2000; Jashemski 1970, 1974, 1992; Meyer 1980). Sampling was largely small-scale and from discrete areas of the city (Richardson, Thompson and Genovese 1997, 88-89).

During the first few seasons, 1995 to 1998, when excavation was not as extensive and the collection of environmental samples was correspondingly limited, residues with a visibly high concentration of ecofacts and/or charcoal were sorted completely to avoid the creation of a backlog. In other residues, 50% was initially sorted to determine if the quantity of ecofacts warranted the sorting of the other half. However, this sampling policy was revised due to taphonomic factors, incomplete stratigraphic sequence information in spots of the Insula and the lack of features due to the presence of mosaic floors (Ciaraldi 2001, 77, 140).

Only a few charred seeds per litre were observed in the initial analysis of the samples by Ciaraldi (2001, 73). Of note were the low seed densities from the majority of samples which were nearly entirely from secondary fill deposits. Similarly, the recent charcoal analysis from Insula VI.I revealed that the recovered charcoal came mostly from secondary fill (Veal 2009, 59; Veal and Thompson 2008, 5). Specifically, from the House of the Vestals (VI.I.vii) the majority of deposits contained secondary fill, as they contained a variety of pottery types such as domestic wares, transport amphorae, and other artefacts such as glass, lamps, faunal remains and building debris used to level floor surfaces (De Sena and Ikäheimo 2003, 305).

Ciaraldi (2001, 73) regarded these areas as too disturbed to yield meaningful archaeobotanical information. Increasing the volume of the samples would not statistically improve the recovery of palaeobotanical material as seen from the results from this study (Figure 12). Therefore, the revised ecofact sampling policy employed for the remaining field seasons, 1999 to 2006, in which the number of excavation units substantially increased, included sorting the entire light fractions and sampling and sorting 25% of the heavy residues recovered from flotation in cases where ecofacts and artefacts were low upon visual inspection. This latter pattern typically occurred when heavy fractions were composed primarily of building material/rubble and/or plaster and lapilli.

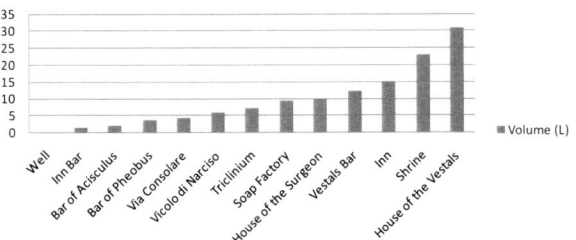

FIGURE 12: TOTAL VOLUME OF RESIDUE (L) EXAMINED PER PROPERTY FROM INSULA VI.I

In exceptional circumstances 50% to 100% of samples were examined in the field from discrete features and deposits. These were completely sorted to provide immediate feedback to excavators in the field. From 2005 onwards, to permit processing of the large volumes of residues accumulated over the course of the excavation, 25% of the total of the heavy residue was sorted in the field. Geological sieves of 2 mm and 1 mm aperture sieve sizes were used to facilitate sorting. Ten percent of the heavy residues with low ecofact density were sorted in the field, e.g. disturbed building material/rubble deposits normally consisting of construction fill, plaster and/or lapilli.

Heavy fractions must be sorted as inorganic materials with specific gravity of ≥ 2.5 generally sinks, e.g. large and dense fruit stones and mineralized archaeobotanical material (Green 1979, 279). Heavier or waterlogged plant remains, despite their relatively low specific gravity, will also sometimes sink into the heavy fraction. In addition, due to the poor preservation of the archaeobotanical remains recovered to date, such as those with surface cracks or large pores, some specimens will become waterlogged and sink into the heavy fraction (Renfrew, Monk and Murphy 1976, 16-17). Thus, the heavy fraction must be examined.

3.4 Distribution of sample number by type

All heavy and light fraction samples stored at the University of Bradford with proper contextual information were sorted including samples from 2007 and earlier (Table 2 and Figure 13). Based upon the sampling strategy and constraints imposed by the archaeology the number of contexts examined was influenced by excavation conditions and the extent within each property that investigations below the AD 79 layer could be made. The number of contexts broadly corresponds with the size of property which in turn is related to how many trenches could be opened. The different property 'types' were assigned based upon archaeological evidence.

Nearly half of the contexts examined from Insula VI.I were from commercial properties which dominated Insula VI.I during the 1st century AD (Figure 14). The next major category, domestic properties, was composed of two large elite houses, the House of the Surgeon (VI.I.x) and House of the Vestals (VI.I.vii). Very few contexts were obtained from ritual deposits and the two roads.

Methodology

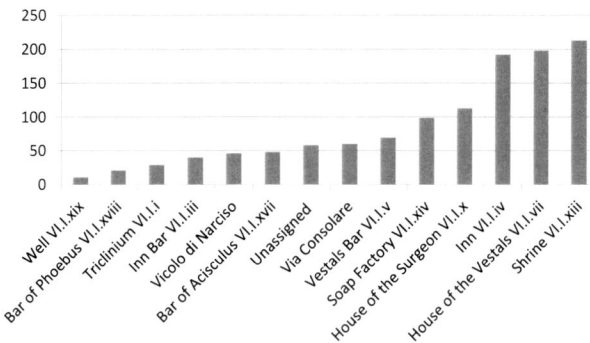

FIGURE 13: NUMBER OF SAMPLES EXAMINED FROM DIFFERENT PROPERTIES WITHIN INSULA VI.I

Insula VI.I	Number of Contexts Examined	Type
Well	11	Ritual
Bar of Pheobus	21	Commercial
Triclinium	29	Commercial
Inn Bar	40	Commercial
Vicolo di Narciso	46	Road
Bar of Acisculus	48	Commercial
Unassigned	58	Unassigned
Via Consolare	60	Road
Vestals Bar	69	Commercial
Soap Factory	99	Commercial
House of the Surgeon	113	Domestic
Inn	192	Commercial
House of the Vestals	198	Domestic
Shrine	213	Ritual
Total	1197	

TABLE 2: NUMBER OF SAMPLES EXAMINED FROM PROPERTIES WITHIN INSULA VI.I

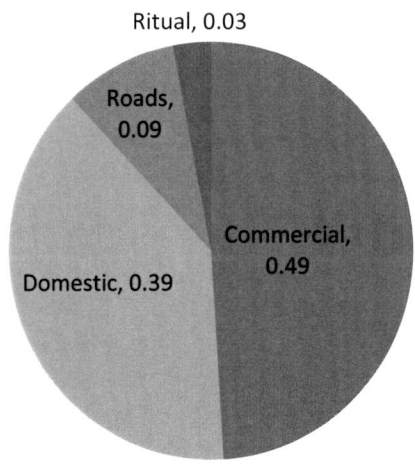

FIGURE 14: NUMBER OF SAMPLES EXAMINED PER PROPERTY TYPE

3.5 Identification

Identifications were made using a low power M6C-R binocular microscope, with a magnification range of 4.8x to 56x to observe the morphological features of the different botanical specimens (Results chap. 4). Identifications were made using the UCL Institute of Archaeology Mediterranean and Near Eastern reference collections and seed atlases such as Cappers, Bekker and Jans's *Digital Seed Atlas of the Netherlands* (2006), and Martin and Barkley's *Seed Identification Manual* (2000). Specimens were photographed using a digital camera. Some samples were further examined and photographed using SEM microscopy. Taxonomy used follows Zohary and Hopf (2001) for crops and fruits. Tentative species level identifications are noted by 'cf ' (*confer*) (Cappers 2006, 50).

3.6 Quantification

Ciaraldi (2001, 165) was looking at specific deposits within Insula VI.I, such as the drain, toilet and foundation deposit from the House of the Vestals (VI.I.vii), which had exceptional preservation and were therefore extremely rich in ecofacts. Given the paucity of archaeobotanical remains from the remaining contexts of Insula VI.I, Ciaraldi's original quantification strategy was modified slightly to account for this limitation. In Ciaraldi's (2001, 83) quantification plan only the anatomically diagnostic parts were counted, e.g. only the beaks of grapes and the apex of olive stones, walnuts and hazelnuts were counted as diagnostic features. In the present study grape pips are counted as one regardless of whether a beak is present because whether preserved through carbonisation or mineralisation, the beaks were usually broken off. A recent study by Margaritis and Jones (2006, 785) noted that some of the grape pips had what appeared to be modern breaks, which were believed to have been caused during the excavation process or the flotation procedures.

3.7 Fruit

Due to the paucity of archaeobotanical fruit remains all recovered endocarp (stone) fragments or seeds were counted as one.

FIGURE 15: MINERALISED *MALUS DOMESTICA* (APPLE) SEED FROM INSULA VI.I

Grape

As with olives the identification of cultivated versus wild grapes has proven problematic and has been a topic of heated debate among archaeobotanists for the last half century since Stummer (1911) initially pointed out that those pips from Vitis sylvestris were small, short and broad, whereas pips of cultivated grapes had longer stalks and were narrower in relation to their length (Renfrew 1973, 127-129; Runnels and Hansen 1986; Smith and Jones 1990, 317). Traditionally, the established distinction between archaeological specimens of wild (*Vitis vinifera* ssp. *sylvestris*) and domesticated (*Vitis vinifera* ssp. *vinifera*) grapes has been based on the overall dimensions and other morphological characteristics of the grape pips. However, due to the fact that there is considerable overlap between the cultivated and wild types 'pip shape cannot be regarded as a safe diagnostic trait for distinguishing between wild and cultivated Vitis remains in archaeological excavations' (Zohary and Hops 2001, 153).

A number of studies on both cultivated and wild charred grape remains has demonstrated that the moisture content in the plant and fruit and the oxygen level present during the time of charring could potentially influence the final dimensions and morphology of the charred grape seeds and ultimately affect their preservation and survival in the archaeobotanical record (Margaritis and Jones 2006, 791-792). Pragmatically, the stalks from grape pips often break off due to their slender size (Figure 16, Figure 17). Figure 18, a SEM of a complete carbonised grape pip, has a fracture along the beak and body of the pip. In this study all grape pips, based upon the time period and the known cultivation of grapes in the region, will be assumed to be cultivated. Grape pips were counted as one if they were relatively complete, i.e., if an oval shaped chazal scar or 'shield' on the dorsal side and two halves of the symmetrical pip with the two narrow furrows on the ventral side with a central 'bridge' were roughly intact (Renfrew 1973, 127).

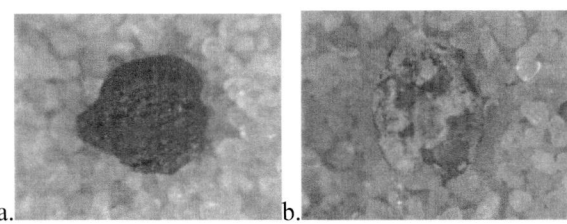

FIGURE 16: A. CARBONISED *VITIS VINIFERA* (GRAPE) PIPS AND B. MINERALISED *VITIS VINIFERA* (GRAPE) PIPS (BOTH DORSAL VIEW) FROM INSULA VI.I

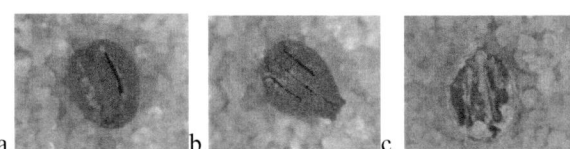

FIGURE 17: A. AND B. CARBONISED *VITIS VINIFERA* (GRAPE) PIPS AND C. MINERALISED *VITIS VINIFERA* (GRAPE) PIP (ALL VENTRAL VIEWS) FROM INSULA VI.I

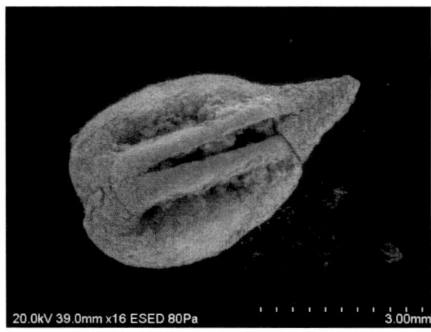

FIGURE 18: SEM OF CARBONISED *VITIS VINIFERA* (GRAPE) PIP FROM INSULA VI.I

The pips were highly eroded probably due to the porous nature of the matrix from the AD 79 layer, especially the large quantities of lapilli. A separate category for preserved grape pit half fragments was constructed, which were divided by two, and fragments of grape pips, which were divided by four, for both carbonised and mineralised specimens. Another separate category was created for charred grape pips embedded in mesocarp, which were counted as one whole grape (Figure 19b). Grape petioles or the stems attaching the fruit to the vine were given a separate category (Figure 20).

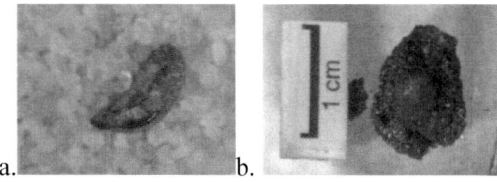

FIGURE 19: A. CARBONISED HALF FRAGMENT OF *VITIS VINIFERA* (GRAPE) PIP AND B. CARBONISED GRAPE PIP EMBEDDED IN MESOCARP FROM INSULA VI.I (SCALE IN CENTIMETRES)

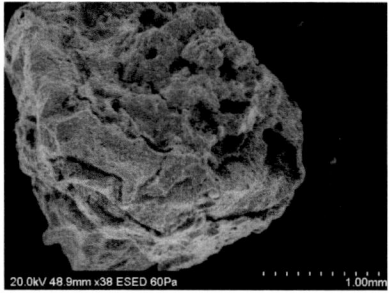

FIGURE 20: SEM OF CARBONISED *VITIS VINIFERA* (GRAPE) PETIOLE FROM INSULA VI.I

Olives

Olives possess a high variability in the size of the endocarp, and it impossible to distinguish between the wild forms and those of cultivated varieties (Kislev 1996; Liphschitz et al. 1991, 441; Zohary and Hopf 1988 cited in Galili et al. 1997, 1141). However, based upon the narrow time period and the ancient literary sources regarding olive cultivation in this study it will be assumed that all olives recovered have been cultivated.

Methodology

Warnock (2007, 80), looking at olive fracture pattern, observed that whole stones had fracture lines on a quarter section of the stone. The majority of other olive stones were fragmented from these two initial stone fragments. It was decided that to avoid an over-representation of carbonised olive fragments due to their high preservation and recovery rates, a correction factor was needed to avoid skewing the results. Thus, the weight of three whole charred archaeological olive stones (Figure 21) from the Insula was recorded and divided by three to obtain an Average Weight. To produce an estimate of the percentage of whole olive stones by weight, the weight of charred olive stone fragments from other deposits were divided by the Average Weight for an olive stone. Whole or complete olive stones were counted as one and olive stones split along suture lines were counted as half.

FIGURE 21: WHOLE CARBONISED *OLEA EUROPEA* (OLIVE) (SCALE IN CENTIMETRES) FROM INSULA VI.I (SCALE IN CENTIMETRES)

Fig

In terms of carbonised fruits, Ciaraldi (2001, 82) only counted fragments of mesocarp or endocarp > 1 cm and no estimates were made of their total volume. In contrast, in this study the volume of fig mesocarp (Figure 22, Figure 23) and fig achenes (Figure 24, Figure 25) was measured, if present in large quantities, to avoid an overrepresentation of this taxon due to the abundance of mineralised fig achenes. Figs can posses several hundred seeds per fruit (Dennell 1976, 231; Robinson et al. 2006, 218). A known volume of only mineralised fig achenes was counted and used to estimate other large volumes of this taxon.

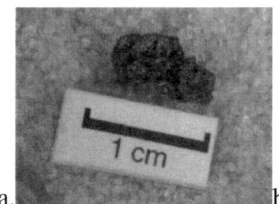

FIGURE 22: *FICUS CARICA* (FIG) CARBONISED MESOCARP FROM INSULA VI.I

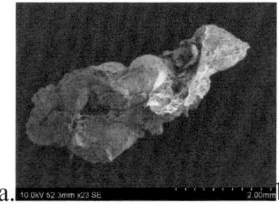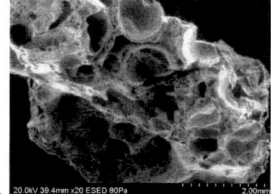

FIGURE 23: A. AND B: SEMs OF CARBONISED *FICUS CARICA* (FIG) MESOCARP FROM INSULA VI.I

In part, due to their extremely small size, it is impossible to distinguish between the pips of wild and cultivated figs (Zohary and Hops 2001, 164). Mineralised fig pips retain their characteristic pyriform shape (Figure 39 and Figure 40) (Jashemski et al. 2002, 109).

FIGURE 24 A. AND B: MINERALISED *FICUS CARICA* (FIG) FROM INSULA VI.I (SCALE IN CENTIMETRES)

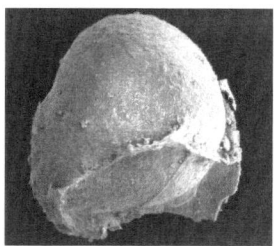

FIGURE 25: SEM OF ONE MINERALISED *FICUS CARICA* (FIG) FROM INSULA VI.I

Nutshells

All nutshell fragments were recorded as fragments and counted whether diagnostic features were present or not. The category of 'nuts (generic)' was created as many carbonised fragments of nutshells lack diagnostic features.

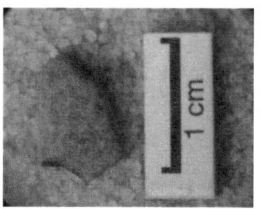

FIGURE 26: A. CARBONISED *PINUS PINEA* (PINE NUT) SHELL FRAGMENT A. INSIDE VIEW B. OUTER SHELL VIEW FROM INSULA VI.I

3.8 Cereals

Recovered cereals are almost exclusively carbonised (van der Veen 2008, 26). Ciaraldi (2001, 82) counted whole grains of cereals as one and she grouped cereal fragments together to obtain an estimate of the number of entire grains. All chaff components, including glumes, and rachis fragments, were counted as one specimen, while spikelet forks were counted as a half (Figure 27). In this study a similar quantification scheme is followed except that all fragments of *Triticum* are counted as one due to the small numbers recovered. This included the category *Triticum* sp. (Figure 28) which could not be identified down to species.

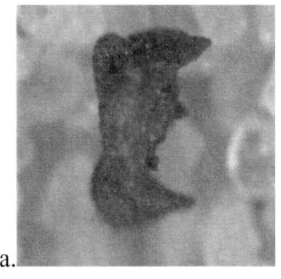 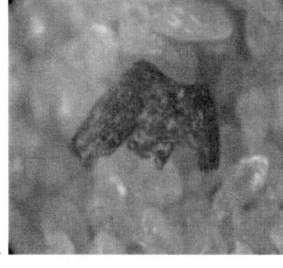

FIGURE 27: A. SPIKELET FORK OF *T. DICOCCUM* (EMMER) B. SPIKELET FORK OF *T. DICOCCUM* (EMMER) FROM INSULA VI.I

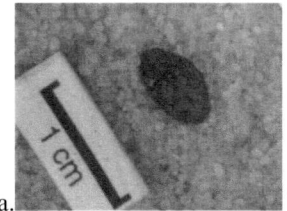 

FIGURE 28: A. *HORDEUM VULGARE* (BARLEY) (DORSAL VIEW) AND B. *HORDEUM VULGARE* (BARLEY) (VENTRAL VIEW) FROM INSULA VI.I

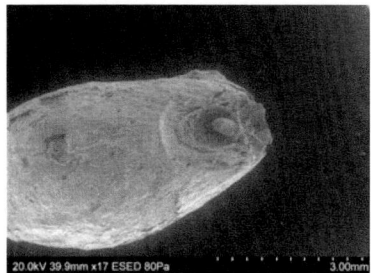

FIGURE 29: SEM OF CARBONISED *HORDEUM VULGARE* (BARLEY) (DORSAL VIEW) FROM INSULA VI.I

FIGURE 30: BADLY DAMAGED CARBONISED *TRITICUM* SP. FROM INSULA VI.I

Using controlled experimental conditions Boardman and Jones (1990, 1-4) found that some cereals are more prone to carbonization or destruction than other. The results revealed that bread wheat (*Triticum aestivum/durum*) and barley (*Hordeum vulgare*) grains were the first specimens to become carbonised and subsequently destroyed amongst einkorn (*Triticum monococcum*), emmer (*Triticum dicoccum*), spelt (*Triticum spelta*), bread wheat and 6-row barley.

Bread wheat suffered the most distortion even under relatively low temperatures, probably because bread wheat has lax glumes and barley grains agglomerated together at temperatures above 350° C. This may explain why so few cereal grains of these two species were recovered at Pompeii. Interestingly, cereal grains appear to withstand the greatest range of preservation conditions. Cereal grains survive over chaff, which is usually burnt away. There appears to be a corresponding relationship between the high temperatures of the fire and the decreasing state of preservation for cereal grains; glumes were observed to provide some protection (Boardman and Jones 1990, 8). The degree of distortion of cereal grains provides insight into the complicated reactions which occur during the charring process. An increasingly oxidizing environment and higher temperatures (i.e quick burning) can mean greater distortion and swelling whereas the lack of distortion can signal low heat conditions (Boardman and Jones 1990, 8; Braadbaart 2008; Braadbaart and van Bergen 2005; Braadbaart et al. 2004).

3.9 Millet

All recovered millets were counted as one in this study. *Panicum* was identified based upon its plump, round caryopsis and small embryo indentation. Setaria is more ovate in shape with a narrow and longer embryo indentation (Figure 31).

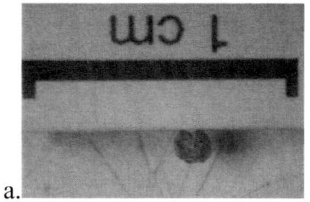 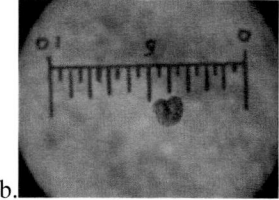

FIGURE 31: A. CARBONISED *PANICUM* B. CARBONISED SETARIA BOTH FROM INSULA VI.I (SCALE IN MILLIMETRES)

3.10 Pulses

Small seeds, whether whole or incomplete were also counted as one. The genus *Vicia* is difficult to identify as pulses contain a large amount of moisture and are often badly distorted when they come into contact with intense heat (Figure 32) and their tendency to split into cotyledons (or halves) (Figure 33) (Renfrew et al. 1976). Therefore, whole *Vicia* specimens were counted as one and individual cotyledons were counted as half. In addition, fractions of specimens of *Vicia* were also counted, including 'greater than two thirds' and 'less than one third' to obtain a more accurate estimate.

FIGURE 32: CARBONISED *VICIA* SP. FROM INSULA VI.I (SCALE IN CENTIMETRES)

 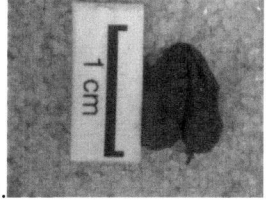

FIGURE 33: A. CARBONISED *VICIA SATIVA* AND B. *VICIA FABA* FROM INSULA VI.I

3.11 Limitations of the data

Taphonomy

The archaeological record at Pompeii was formed by a complex series of events that influenced its preservation over the past two thousand years (Bon 1997, 12). To a certain extent, it was influenced by the fairly recent direction of its rediscovery and excavation. The archaeobotanical assemblage from the AD 79 destruction layer, if left undisturbed, would have been excellent examples of *de facto* refuse, extremely rare in the archaeobotanical record (Miksicek 1987, 219). Therefore, theoretically these layers would not have been exposed to the usual taphonomic processes (Ciaraldi 2001, 138). Hence, it would have contained a well preserved, fairly complete cross-section of Roman food through carbonization (Meyer 1980, 435).

Ascher was the first scholar to mention the erroneous concept of the Pompeii premise – Pompeii as a city captured perfectly at one moment in time - which was being used implicitly in the archaeological literature of the 1960s. He stated: 'What the archaeologist disturbs is not the remains of a once living community, stopped as it were, at a point in time' (Ascher 1961, 324). Binford (1981) otherwise attributed the Pompeii premise to Schiffer, despites Schiffer's insistence that it was a 'counterproductive' and an 'insidious' concept. This was largely based upon Schiffer's category of Pompeii-like assemblages of *de facto* refuse which he defined as consisting 'of artifacts from the systemic inventory, often still useable, left behind on occupation surfaces when people abandon activity areas, structures, and settlements' (Schiffer 1985, 18).

The substantive disagreement between Binford and Schiffer lies in the fact that Schiffer (1985, 20) accused the new archaeology of either ignoring or underestimating the effects of formation processes. He argued that these assumptions overlooked processes that may have affected a site and thus scholars were presupposing a Pompeii premise in their archaeological analyses. As a counter, Binford claimed that for behavioural archaeologists, like Schiffer, 'Pompeii is the most desirable condition of the archaeological record' (Binford 1981, 205). Schiffer's (1977, 250) disillusionment with the complexities of the archaeological record in not uncovering a 'series of little Pompeiis only in need of 'dusting off' in order to yield 'ethnographies', or complete pictures of the past' caused Schiffer to attack 'the new archaeology' claiming that new archaeologists believed complete cultural systems were present and visible on archaeological sites. Pragmatically, Binford states that by 'holding up the Pompeii premise as an ideal insures that the imperfect world of 'dirt archaeology' will almost always be a frustrating one' (Binford 1981, 205). Despite their disagreements, both Schiffer and Binford agree on the point that most archaeological sites are not 'little Pompeiis' (Schiffer 1985, 38).

However, despite Binford's use of Pompeii as a prototype of the behaviour snapshot, it has come to be recognized that '…even Pompeii does not conform to the 'Pompeii Premise'' (Allison 1991, 49). The destruction horizon of AD 79 at Pompeii ideally would have provided unique information about a small Italian community during the third quarter of the 1st century AD (Ling 2005, 97). In reality, the archaeology of the AD 62-79 destruction layer was in 'chaos' due to earth tremors, as well as the eruption and pyroclastic activity (Cooley and Cooley 2004, 1). In the contexts below the AD 79 destruction horizon, similar taphonomic factors would have affected the Pompeian archaeological record as at other archaeological sites, including cultural, natural and post-depositional factors (Bon 1997, 7).

Modern taphonomic factors

The soil matrix around the Bay of Naples is largely composed of volcanic lapilli, created during the repeated eruptions of Mount Vesuvius. The thickness of volcanic deposits covers the soil to a depth of two to five metres in some areas (Ling 2005, 15). It presents a challenge to flotation because of the low specific gravity of the spectrum of size particles of lapilli that float on top of the water's surface. Samples containing a large quantity of lapilli can clog the flotation machine and can accumulate in the light fraction, ultimately resulting in bulky residue samples. The sorting process was impeded by the presence of lapilli to the extent that Ciaraldi (2001, 77) noted that it hampered the collection of plant remains and resulted in the re-evaluation of the original sampling strategy for the AAPP. Indeed, Pliny the Elder's failed rescue mission to Herculaneum during the eruption in AD 79 was due in part to the massive quantities of lapilli floating on the water's surface on the Bay of Naples, preventing the launching of his ship. The porous nature of the lapilli from the AD 79 layer also effects the preservation of the macrobotanical remains, which are easily damaged by the lapilli particles during the recovery process of excavation and flotation.

Like the majority of archaeological excavations contamination of soil samples by modern seeds is a problem. Indeed, most aerobic soils contain modern seeds to a considerable depth (Renfrew, Monk and Murphy 1976, 14; Keepax 1977, 226). Due to the porous nature of the tephra (fragmentary remains from previous volcanic eruption) it can be assumed that modern seeds would quickly travel through the tephra material, and intrusive disturbances from various sources including looting, and contaminant lapilli from the 79 AD eruption, which raises the possibility of a downward movement of archaeobotanical material into earlier archaeological layers (Fuller and Weber 2005, 95).

The extremely poor preservation conditions at Pompeii is evidenced by the most recent attempt at pollen analysis by Veal (2009, 96), in collaboration with *Direttrice of the Laboratorio delle Ricerche Applicate at the Soprintendenza di Pompeii, Dott.ssa* Annamaria Ciarallo, and researchers Prof. Tullio Pescatore and Prof. ssa Rosaria Senatore with the *Università degli Studi di Sannio* in Benevento, Campania, samples taken both inside and just outside the city of Pompeii failed to yield any pollen. However, unlike the pollen samples taken by Dimbleby and Grüger (2002, 211) in which the protective 79 AD layer of lapilli was still present and allowed for the preservation of the ancient pollen, this pollen study reinforces the notion that any intact organic material which is not either mineralised or carbonised is likely to be modern (Veal 2009, 96). Therefore, as is the case in the majority of archaeobotanical analyses, only botanical remains which are either carbonised or mineralised are regarded as archaeologically relevant and other seeds as modern and intrusive (Fuller and Weber 2005, 95; Keepax 1977; Minnis 1981; Pearsall 2000).

What archaeologists have termed the second destruction of Pompeii is currently taking place due to human created modern taphonomic factors during the latter half of the 20th century. Pompeii has particularly suffered; large areas are exposed to the elements and it has been subjected to serious earthquakes. It has even been subject to Allied bomb damage during WWII when in August and September 1943 over 150 bombs hit the city when it was thought that Germans soldiers were hiding there. Other problems include modern vegetation, including the planting of ornamental trees for aesthetic reasons and the ever-present threat of weeds, as well as the stress caused by over two million tourists visiting each year, atmospheric pollution from one of the most heavily populated areas in the world, the numerous excavations themselves, vandalism, theft and the expensive and on-going demand for repair and maintenance of the site have all taken their toll (Bon 1997, 8). In 1996 Pompeii was placed upon the World Monuments Fund's list of 100 most endangered archaeological sites (Foss 2007, 36). In 2008 the Italian government declared a year-long state of emergency for the site due to prolonged lack of investment, bureaucratic corruption and scandals, littering and theft (BBC news July 4, 2008 http://newsvote.bbc.co.uk/mpapps/pagetools/print/news.bbc.co.uk/2/hi/europe/7490735.stm). On November 6, 2010 the House of the Gladiator (VI.V.iii) collapsed (The Guardian, November 12, 2010 http://www.pasthorizons.com/index.php/archives/11/2010/neglected-ruins-of-pompeii-declared-a-disgrace-to-italy). Hence, although renowned for its exceptional preservation, Pompeii has ironically become an archaeological site with exceptional taphonomic issues.

3.12 Historical factors

Seismic activity beginning in AD 62 likely initiated the piece-meal abandonment of the city. Allison's (1991, 54) study of artefact assemblages from a number of Pompeian residences shows a recurring pattern of hoarding and multiple-activity areas. Allison also revealed new insights into room function. The process of abandonment was more complex than previously thought. Allison argues that not all destruction/repair/alteration events and changes in the composition of the population should be assumed to date to the 5th of February AD 62 earthquake recorded by the historian Tacitus and the philosopher Seneca or the hypothetical one of AD 70 proposed by Zanker (1998, 4) (Allison 1991, 54-56; Allison 2004, 25). Hence, evidence of restoration and disturbance may not be related to this possible recent disaster. Based upon Allison's work on artefact assemblages in Pompeian houses it is evident that the archaeological processes of deposition are as complicated as at any other archaeological site (Laurence 1995, 63; Laurence 2007, xi). Houses 11 and 12 from Regione I, insula IX illustrate the need to examine each house in Pompeii individually in order to appreciate the individual variation and the relationship between the domestic architecture and the household artefact assemblages.

Hence, Pompeii was not an 'untouched' city 'frozen in time' upon its rediscovery in the 18th century (Cooley 2003, 13). Widespread post-eruption intrusion, including holes in walls and ceilings, disturbance of wall decoration, graffiti and lamps from later periods indicates that robbing, looting and treasure hunting occurred after its destruction (Bon 1997, 8-10; Jashemski 1979a, 612-613). Datable objects belonging to 2nd, 3rd, or 4th centuries AD and even into the early Christian and Medieval periods confirms disturbance continued after the eruption (Ling 2005, 13, 156-157).

With its parallel development through the early stages of the discipline of archaeology Pompeii's 'excavation and scholarship prove a nightmare of omissions and disasters' (Wallace-Hadrill 1994, 65). Pompeii, along with the surrounding Vesuvian sites, were victims of two centuries of archaeological experiment (Wallace-Hadrill 1990, 153) and became

> '*de facto* laboratory for excavation and recording methods, publication standards, the conservation and analysis of the entire range of artefacts and ecofacts, the study of human-environment relations, and cultural resource management. Together, the buried cities of Vesuvius form the longest continuously excavated archaeological zone in the world- an honor they can never lose, and a burden they can never shift' (Foss 2007, 28-29).

Ironically, the influence of Pompeii has not permeated into present-day academic archaeology due to the fact that little analytical work has been attempted. One of the reasons for this lack of analytical research might be the long-standing view that Pompeii is a microcosm of Roman life and therefore not the domain of archaeological investigation (Laurence 1995, 63).

The earliest antiquarian hunts for treasures at Pompeii were supported under the patronage of Charles Bourbon, founder of a new royal dynasty in Naples. The prestige and wealth from the recovered antiquities were used to solidify his position and international reputation. The clearance of the city was subject to the political turmoil of the 19th century with the expulsions of the Bourbons in 1806, their replacement by the French under Napoleon Bonaparte in 1808 and the return in 1815 and subsequent exile again of the Bourbons under Garibaldi in1859. Giuseppe Fiorelli directed excavations at Pompeii from 1860 to 1875 and was the first to usher in a systematic approach to the clearance and documentation of Pompeii (Cooley 2003, 13-14; Francissen 1987, 116). Despite these improved recording measures the early excavators at Pompeii had no interest in any of the environmental remains recovered during the initial clearance of the volcanic debris. Even during the late 19th and early 20th century, environmental remains barely warranted a brief note in the excavation journal (Cooley 2003, 99).

Largely everything of monetary value uncovered during the primary clearance of Pompeii was removed by the first excavators. Their quest for statuary and attractive artefacts, destined to furnish the private residences of the elites of Western Europe, was driven by personal financial gain. The majority of artefact assemblages from Pompeii are now acknowledged to be incomplete. Further, based upon expectations drawn from Classical sources of where items should be found before the actual excavations had commenced, early archaeologists complicated the artefact assemblage by removing and re-positioning artefacts (Allison 1991, 50, 52; Bon 1997, 10).

3.13 Current research

Excavators for over a hundred years have focused mainly on the AD 79 eruption level and standing structures to reveal the last layer of the town with little interest in pre-Roman history. Starting in the 1930s selective excavations beneath the AD 79 layer were made on several houses and in public areas (Cooley 2003, 113; Ling 2005, 27). However, it has only been under the relatively recent direction in 1995 of Prof. P. Guzzo, former *Soprintendente di Pompei* that major stratigraphic excavations below the AD 79 destruction layer have revealed an emerging chronological sequence for the urbanization of and the Roman influence on, the town (Robinson 2002, 93-94).

New research at Pompeii is exploring below the AD 62-79 layer with evidence from every phase contributing equally to the history of the site (Bon 1997, 12). New archaeological research at Pompeii '… will require sophisticated methodologies, careful consideration of post-depositional processes and imaginative new archaeologically based models of socio-economic development' (Dyson 1997, 155). This relatively recent interest in the urban development of Pompeii over its entire occupation pragmatically addresses one of the pressing issues facing Pompeii, the conservation of what has already been subjected to modern taphonomic processes. Therefore, the unexcavated areas are being left untouched whilst new archaeological research is allowed to carry on (Cooley 2003, 14-15).

3.14 The field school

A number of difficulties present themselves in data collection whilst running a field school. Among them is the training of inexperienced students in excavation and flotation. The various methods for the collection of macrobotanical remains will affect the quantity and types of ecofacts that are recovered (Pearsall 2000, 11). Practically, dirt clings to charred macrobotanical remains which can easily be missed by inexperienced excavators whilst trowelling and destroyed in the process. In addition, to a greater extent than is typically acknowledged, the results of analysis depend on the flotation sample size, how the sample is measured and processed and how well the plant material within the sample is able to withstand the rigors of flotation (Wright 2005, 19). Carbon is chemically inert and important to protect from mechanical damage at this stage. With the teaching of numerous field school students, the quality of each flotation sample was difficult to maintain due to the varying levels of abilities of the students.

3.15 Contexts

Secondly, the possibility that the archaeobotanical remains recovered from construction fill or rubble found in certain contexts were brought in from different locations other than the rooms within the Insula cannot be dismissed. However, circumstantial evidence such as secondary deposits containing matching plaster from nearby internal walls and based upon the fragile nature of the charcoal assemblage within these deposits lends credence to the theory that these construction fill deposits originated from within the house and were created and deposited during the numerous construction and redecoration events occurring during the occupation of this Insula rather than having been transported over a distance to be re-deposited within Insula VI.I (Veal 2009, 33). Veal (2009, 33) cites the particularly interesting charcoal deposit consisting entirely of young, one to three years of ring growth, cuttings from *Vitis vinifera* (grapevine). The excellent preservation and the fragile nature of the charcoal assemblage led to the conclusion that it could not have been re-deposited from another location.

The re-deposition of carbonised archaeobotanical remains likely took place through standard routes including wind, rain, animal trampling and human activities. The re-deposition and incorporation of these charred archaeobotanical remains as a background scatter become what Fuller and Weber (2005, 93) term 'general habitational fill samples' or similarly 'Type C' samples by Hubbard and Clapham (1992). Such deposits are abundant in this study, as they are on most archaeological sites, and it is difficult to account for all the post-depositional processes that acted upon these types of samples. Despite the fact that these samples do not appear to come from discrete contexts they can be quantitatively useful. These types of samples tend to show recurrent patterns in their composition, which Fuller and Weber (2005, 93, 103) suggest relate to recurrent formation processes.

The majority of contexts from Insula VI.I contained both carbonised and mineralised archaeobotanical remains. These assemblages typically included mineralised fig achenes, carbonised olive endocarp fragments, and carbonised indeterminate vasicular tissue. This generalised mix of different archaeobotanical remains (both mineralised and carbonised) would appear to support the hypothesis: the majority of contexts examined in this study came from re-deposited or secondary fill deposits due to their lack of specific depositional events (e.g evidence of ritual, crop-processing waste or storage). Therefore, these assemblages are most likely composed of everyday routine food activities which have become incorporated into the archaeological matrix and then were transferred and re-deposited becoming secondary deposits.

3.16 Preservation

Ciaraldi (2001, 171) found the intra-site comparison of plant assemblages from Pompeii difficult due to numerous differences in the various archaeological contexts, preservation conditions and the taphonomy of the excavated deposits. A direct comparison between the charred and mineralised assemblages was not possible due to their obvious differences and therefore only a few general hypotheses were generated by Ciaraldi (2001, 171) based upon their analysis. To attempt to overcome this problem in the present study counts are employed.

Carbonisation

Numerous theories have been proposed to explain the process of the carbonisation of the archaeobotanical remains at Pompeii, particularly of those from the AD 79 destruction layer. Volcanic eruptions are well-known for their role in the formation of the geological record. The intense heat generated from volcanic emissions can char organic material in the absence of oxygen created by a pyrolytic mechanism (Hatcher 2002, 217). It has long been held that the archaeobotanical remains recovered from the AD 79 destruction layer at Pompeii were carbonised by the heat and by-products of the volcanic eruption, including the lapilli, ash and gases. This appears to be the case for the carbonised archaeobotanical material from other sites such as Herculaneum and the *villa rustica* at Boscoreale, specifically the AD 79 destruction layer (Meyer 1980, 402; Robinson 2002, 93). It is also possible that some of the carbonised material from the AD 79 layer could have originated as domestic refuse from this time period.

In other areas around Mount Vesuvius, where the heat was less intense, partially carbonised material is present (Jashemski 2002b, 82). Indeed, Robinson (2002, 93) argues that at Pompeii the pyroclastic debris was not sufficiently hot enough to carbonise organic material. Evidence for undamaged pollen grains appears to support this theory that the carbonisation of organic remains was not due to excessive heat. To resolve this issue, Dr. Breger analysed carbonised figs, olives and hay from three different sites destroyed by Mount Vesuvius and found that only the hay samples from the villa at Torre Annunziata had high enough carbon levels to indicate burning (Meyer 1980, 403).

Another proposed route of carbonisation of botanical material, not necessarily from the AD 79 layer, includes burnt offerings. These intentionally burnt ritual food items represent an interesting archaeobotanical case study in the ancient world as edible items such as fruits and bread were often deliberately burned and preserved through this method. Robinson (2002, 93) argues that burnt offerings present a taphonomic problem as 're-worked charred material from burnt offerings was likely to have been a major component of other charred assemblages from the sites [around Pompeii]'.

In addition, archaeobotanical material recovered from the earlier stratigraphic layers within Insula VI.I presumably became incorporated into the archaeological record through the standard route of carbonisation found at other sites, normally involving deliberate or accidental exposure to fire. Subsequently, these carbonised archaeobotanical remains were exposed to similar taphonomic processes found on other archaeological sites in the Mediterranean region and incorporated into the archaeological record.

Mineralisation

Limited research has been done on the factors influencing the process of mineralisation of macrobotanical remains on archaeological sites. A few studies have focused on mineralisation from archaeological contexts, including the common process of calcium carbonate ($CaCO_3$) replacement in arthropods, such as woodlice, millipedes and centipedes (Girling 1979, 309; McCobb et al. 2004) and a few studies on seeds (Ciaraldi 2001, 2007; Green 1979; Kenward and Hall 2000; Marshall et al. 2008).

At Pompeii, the mineralisation of organic remains happens through the process of phosphatic mineralisation. Mineralisation is a naturally occurring complex process in which phosphate and mineral deposits replace minerals from decaying organic matter. Phosphatisation, the releasing of phosphate ions, from a variety of sources including faunal remains, food residues and excrement, is thought to be caused by microbial decay and the movement of the dissolved ions facilitated by fluctuating water levels within the archaeological deposit (McCobb et al. 2001; Marshall et al. 2008, 860; Robinson et al. 2006, 214). Within an archaeological setting, possible sources of phosphate and calcium include human faecal material from specific archaeological deposits such as middens, cess and latrine pits, mammal bones, fish bones and scales, and plant material (Carruthers 2000 cited in Ciaraldi 2001, 66). The phosphate ions react with the calcium ions from calcium carbonate, which can come from different sources within the deposit including limestone or mortar fragments, human faecal material, shells of marine molluscs, fish scales and bones, faunal remains, or the leaching of natural substrate in contact with the deposit (Robinson et al. 2006, 214). The result is a cast of the organic material called a pseudomorph, which is reproduced in various degrees of perfection (Green 1979, 282, 284).

The process of phosphate mineralisation requires a liquid medium to be able to fully engulf an item so that the ions can diffuse and become concentrated and result in the precipitation of calcium phosphate (Marshall et al. 2008, 860). Due to their small size, seeds which possess an intact seed coat and fly puparia, with tough larval skin, are particularly well suited to preservation through this process (Robinson et al. 2006, 215). Additional factors such as the original percentage of calcium minerals in, and thickness of the seed's testa, the presence and amount of cuticles, lignin and tannins within the cellular structure and the decomposition of organic material all impact upon the overall mineralisation process (Carruthers 2000; Ciaraldi 2007, 49-50; Girling 1979; McCobb et al. 2003; Miksicek 1987). Seeds which have become mineralised typically consist of light inorganic material with a poorly defined crystalline apatite structure. The main constituent is calcium phosphate, $Ca_4(PO_4)_6(OH)_2$, visible using X-ray diffraction, which is normally light brown in colour (Green 1979, 279, 281). There appears to be a bias towards seeds which are swallowed intact, such as fig and grape achenes (Cool 2006, 119; Robinson et al. 2006, 215). Hence, seeds with intact thin cell walls such as fig and apple seeds which also passed through the body are much more likely to be preserved through mineralisation.

Ciaraldi (2001, 150-151, 2007, 49-50) explains that the unique taphonomy created by the copious ash and lapilli from the numerous volcanic eruptions and the alkaline volcanic soils surrounding Mount Vesuvius resulted in an archaeological matrix rich in calcium phosphate minerals. Pedological analysis shows that similar soil compositions are present today as in the past, loamy sand to sandy loam, with a high percentage of extractable nutrients and moisture retention capabilities with abundant ash and pumice present (Ciaraldi 2001, 66; Ciaraldi 2007, 49; Foss 1988, 133; Foss et al. 2002, 77-78). Hence, one theory regarding mineralisation is that calcium in the subsoil is dissolved into solution by a reaction of percolating ground water on available sources of phosphate (Green 1979, 281).

Ciaraldi's (2001, 66) analysis revealed mixed preservation, both carbonisation and mineralisation, in the archaeobotanical assemblage from different sections within the city of Pompeii. The mineralisation process was likely facilitated by a combination of mineral and volcanic ash, as well as partial water logging and leaching of unique contexts of drains and cesspits. Exposure to large quantities of calcium phosphates produces higher quantities of mineralised material. This can be seen in specific deposits from the House of the Vestals (VI.I.vii) as identified and analysed by Ciaraldi (2001, 162-163), including the drains which span over a century, and the toilet feature. From Insula VI.I, a characteristic yellowish-green stain was observed on amphorae in a pit from the House of the Surgeon (VI.I.x) and on the drains, the toilet and cesspits from the House of the Vestals (VI.I.vii), along with mineralised archaeobotanical remains (Robinson et al. 2008, 3).

According to Green (1979, 281), a number of factors influence the process of mineralisation including soil type, type of organic refuse present and sufficient circulation of groundwater. Kenward and Hall (2000, 519; 2008, 586) suggest that urban occupation deposits, with their variation in sedimentology and chemistry on a minute centimetre scale, present a more complex set of problems, in regards to the process of mineralisation, than rural sites. Within urban occupation deposits organic material is often preserved via anoxic water logging. It appears that the process of mineralisation is enhanced by the partial water logging of botanical materials under anaerobic conditions (Briggs and Wilby 1996 cited in Ciaraldi 2007, 49-50); although no firm scientific preservation model for the mineralisation of organic deposits has been established to date (Kenward and Hall 2008, 588). Complete water logging otherwise impedes the process of mineralisation.

Ciaraldi (2001, 67; 2007, 49-50) argued that some samples from the House of the Vestals (VI.I.vii) underwent water logging during stages of mineralisation as a differential density of the mineralised archaeobotanical material was visible through the depth of a deposit. She hypothesized that a liquid medium facilitated this downward movement of heavy preserved botanical material. All seeds were counted but no mention was made of either the weight or size of the seeds in this analysis. Ciaraldi (2007, 49-50) also observed aggregations of calcium phosphate minerals having vague shapes of seeds within waterlogged samples from the *Villa Vesuvio*. Again this suggests a possible correlation between mineralisation and water logged botanical material. High temperatures are also thought to facilitate the deposition of minerals on organic material and thus the high temperatures during the summer months in Campania may have accelerated this process of preservation within the upper soil horizon (Ciaraldi 2001, 67; Ciaraldi 2007, 50) (see Results, Chapter 4, for further discussion of mineralisation).

No middens or refuse pits have been discovered to date within the city walls of Pompeii (Veal 2009, 68-69). There is evidence of ancient cities having rubbish collection systems. In Mainz and Cologne it is believed that a rubbish designation space was allocated in front of the city walls on the Rhine side from where it could be swept away during periods of flooding (Cremaschi et al. 1994, 148-149). It is likely that Rome had a rudimentary public refuse collection service (Wilson 2002, 480). Hence, it is probable that Pompeii may have possessed a similar rubbish collection system. At Pompeii, like most Roman towns, it is believed that much household rubbish left the house through the window at night. It would seem that Romans became desensitized to the dirt and smell. Stone floors and mosaic pavements were probably swept clean but earthen or straw-covered floors containing domestic refuse may have stayed in the house and were simply covered over with fresh earth or straw (Liebschuetz 2000, 54, 61).

Pompeii likely possessed many cesspits and a scattered network of sewers. Rainwater and waste water were led out of the town by way of the paved streets. The steep slope in the direction of the Bay of Naples leading to the sea and the River Sarno facilitated the disposal of sewage from the city. The geological subsoil consisted of porous volcanic material which could absorb rainwater right away. A cesspit is a kind of soakaway: liquid material seeped into the ground and solid material remained in the pit. It is possible that the content of these pits were placed on fields outside the city or gardens within the city based upon evidence of small fragments of bones and pottery in some of the gardens (Jansen 2000, 37-39; Jashemski 2002b, 27). This would also present another explanation for the low quantities of environmental remains.

Chapter 4

Results

4.1 Contexts examined from Insula VI.I

The total number of contexts, from the properties within Insula VI.I, examined in this study was 1,294. The number of contexts obtained from each property was determined roughly by the size and archaeology of the property and the accessibility of putting in excavation trenches safely.

4.2 Contexts by time period

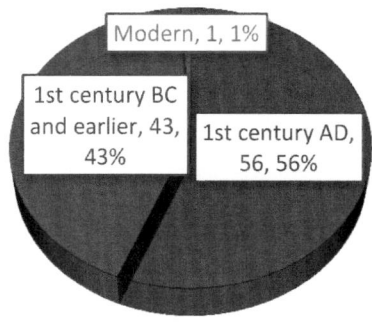

FIGURE 34: TOTAL NUMBER OF CONTEXTS EXAMINED GROUPED BY TIME PERIOD

The majority (56%) of contexts examined in this study were from the 1st century AD.

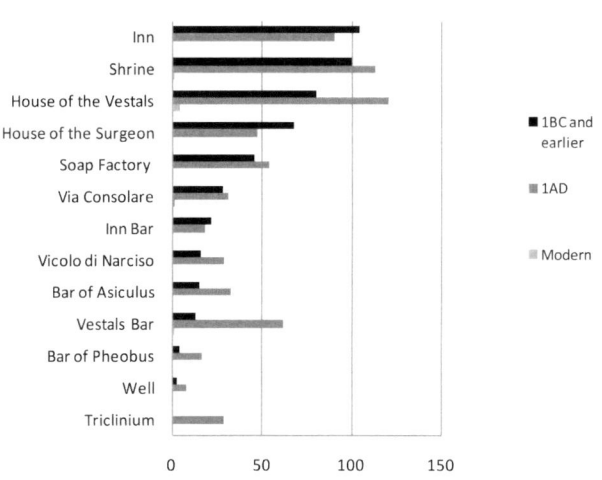

FIGURE 35: NUMBER OF CONTEXTS EXAMINED BY TIME PERIOD SHOWN BY PROPERTY FROM INSULA VI.I

Few modern contexts were included in this study as they were largely identified during the excavation process and during flotation and excluded. Few contexts from Insula VI.I could be securely dated to earlier than the 1st century BC. The Triclinium area from the Inn is the only property from Insula VI.I in which no contexts from earlier than the 1st century AD were retrieved.

4.3 Total archaeobotanical remains

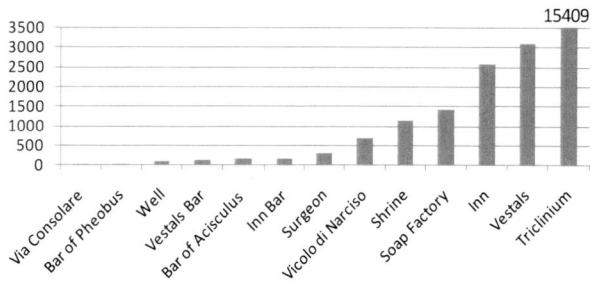

FIGURE 36: TOTAL COUNT OF ARCHAEOBOTANICAL REMAINS (CARBONISED AND MINERALISED REMAINS) PER PROPERTY FROM INSULA VI.I

Five times as many archaeobotanical remains were recovered from the Triclinium [n=15,402] than from the House of the Vestals [n=2264.2], which had the second highest number of recovered botanical remains from Insula VI.I.

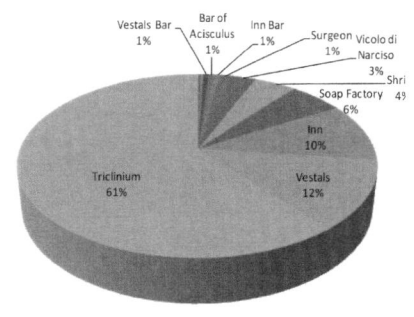

FIGURE 37: PERCENTAGE OF TOTAL COUNT OF ARCHAEOBOTANICAL REMAINS RECOVERED FROM INSULA VI.I BY PROPERTY

Over half (61%) of the archaeobotanical remains recovered from Insula VI.I came from the Triclinium area.

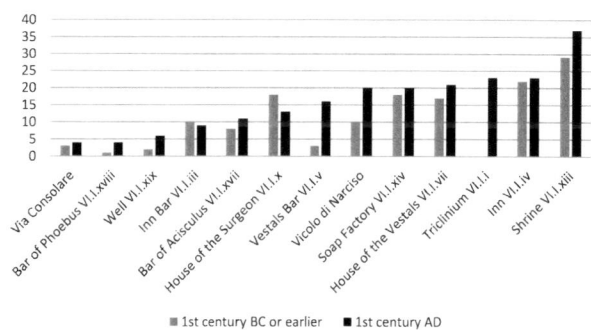

FIGURE 38: TOTAL NUMBER OF TAXA (CARBONISED AND MINERALISED) PER PROPERTY FROM INSULA VI.I

There appears to be a trend towards a slight increase in number of taxa through time, from the 1st century BC to the 1st century AD, aside from the House of the Surgeon and Inn Bar (although not significant - 10 to 9 taxa).

Despite the fact that the Shrine only has 4% of the total number of archaeobotanical remains from the Insula it has the highest number of taxa for both time periods. This suggests that something different is occurring within this property.

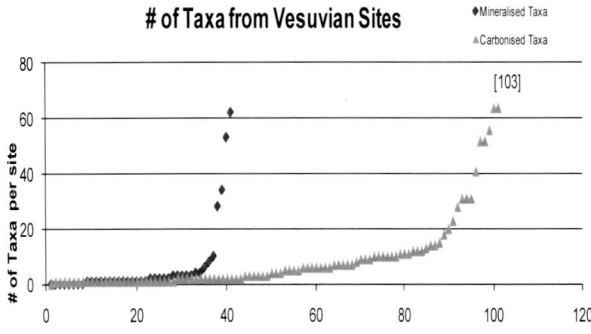

FIGURE 39: NUMBER OF MINERALISED AND CARBONISED TAXA FROM PUBLISHED SOURCES ON VESUVIAN SITES INCLUDING DATA FROM CIARALDI (2001) PHD THESIS; MATTERNE AND DERREUMAUX 2008; ROBINSON 2002

Figure 39 illustrates the fact that a greater number of recorded Vesuvian sites possessed carbonised taxa as opposed to mineralised taxa. This may be due to the fact that specific preservation conditions for mineralisation are required. It is also possible that mineralised remains were not recognised in the past during the early stages of clearance/excavation of Pompeii. Whereas carbonised remains, particularly large deposits of recognisable fruits, nuts and bread, would have been easily identified and in some cases fortuitously retained and stored e.g. amphorae and *dolia* containing charred pulses, fruits and the famous round bread divided into sections found in the oven.

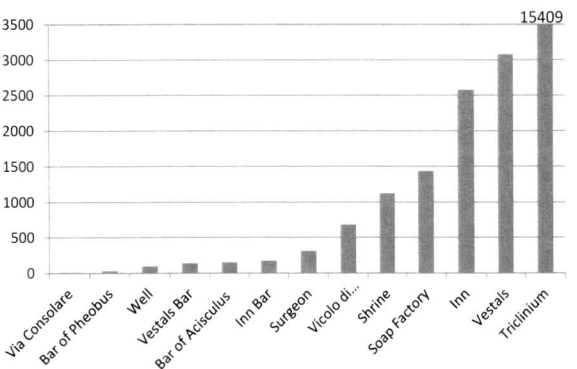

FIGURE 40: TOTAL NUMBER OF ARCHAEOBOTANICAL (CARBONISED AND MINERALISED) REMAINS PER CONTEXT EXAMINED BY PROPERTY FROM INSULA VI.I

The highest numbers of archaeobotanical remains per context were recovered from Vicolo di Narciso (n=453.2)

and the Triclinium (n=146.8 from the 1st century BC and n=112.3 from the 1st century AD). Both properties have unique deposits with significant mineralised components in the botanical assemblages. Other archaeobotanical remains from Insula VI.I were recovered from secondary fill, which produced fewer (<25) archaeobotanical specimens per context.

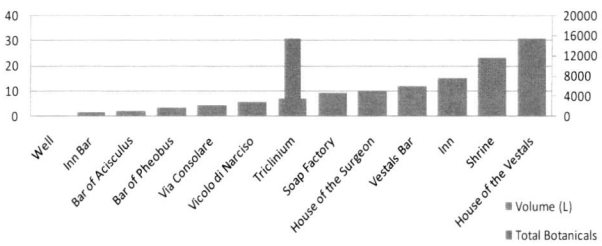

FIGURE 41: TOTAL VOLUME OF RESIDUE EXAMINED PLOTTED ALONGSIDE TOTAL COUNT OF ARCHAEOBOTANICAL (CARBONISED AND MINERALISED) REMAINS PER PROPERTY FROM INSULA VI.I

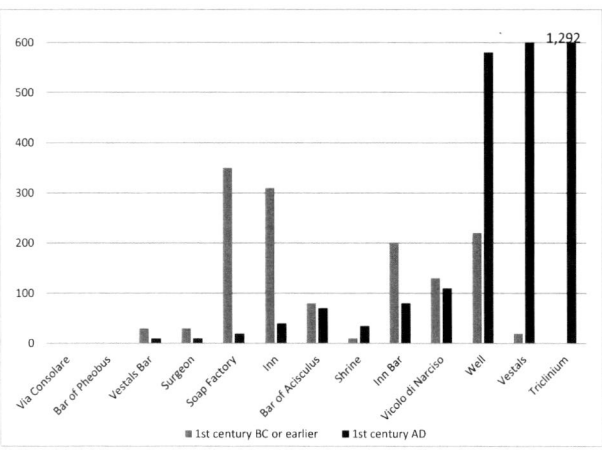

FIGURE 42: ARCHAEOBOTANICAL REMAINS (CARBONISED AND MINERALISED) RECOVERED PER L OF RESIDUE EXAMINED PER PROPERTY FROM INSULA VI.I

The Vestals Bar [n=46.33], House of the Surgeon [n=46.48], Soap Factory [n=362.84], Inn [n=325.63], Bar of Acisculus [n=80.29], Inn Bar [n=205.61], Vicolo di Narciso [n=146.81] have higher numbers of archaeobotanical remains per L for the 1st century BC or earlier. These values drop substantially for the Soap Factory [n=29.53], Inn [n=41.32], and Inn Bar [n=76.5] in the 1st century AD. The Well [n=573.33], House of Vestals [n=1292.81] and Triclinium [n=2173.36] are dominated by archaeobotanical remains per L from the 1st century AD. Low levels of archaeobotanical remains per L were obtained from the Via Consolare [n=3.64], Bar of Pheobus [n=10.5], Vestals Bar [n=10.71], Bar of Acisculus [n=61.4] and the House of the Surgeon [n=13.11]. Archaeobotanical remains per L were low from the Shrine [n=25.09 from 1st century BC and n=61.55 from 1st century AD] for both time periods due to the large volume (over 23 L) of residue examined from this property. Due to the exceptional preservation conditions created by the toilet feature in the Triclinium, in which a large number of mineralised fig seeds were recovered, there is a huge spike in number of seeds (mineralised and carbonised) per L of residue examined.

4.4 Preservation

A diachronic comparison of plant assemblages from the properties within Insula VI.I is problematic due to differences in the archaeological contexts including preservational conditions and taphonomic history. Therefore, a direct comparison between the charred and mineralised assemblages is not possible due to obvious differences in their preservation.

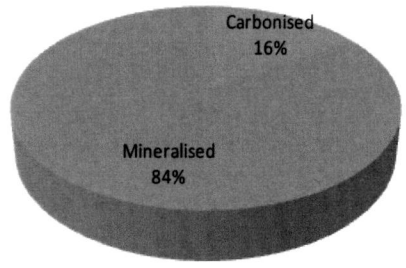

FIGURE 43: PERCENTAGE OF ARCHAEOBOTANICAL REMAINS RECOVERED FROM INSULA VI.I BY PRESERVATION

The majority of archaeobotanical remains were mineralised (84%) (Figure 43). The total count of mineralised archaeobotanical remains from Insula VI.I was 20,414. The total count of carbonised archaeobotanical remains identified from insula VI.I was 4,104. The total combined count of identified archaeobotanical remains (mineralised and carbonised) was 24, 519.

4.5 Carbonised archaeobotanical remains from Insula VI.I

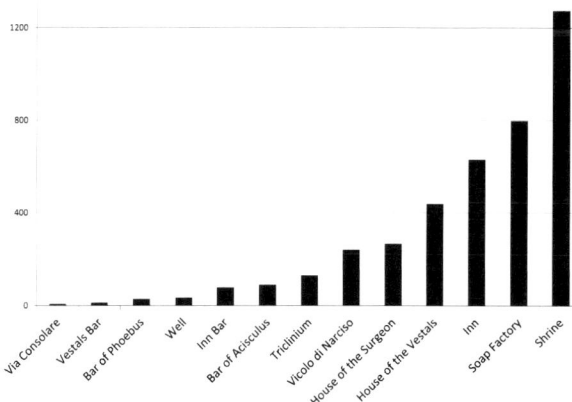

FIGURE 44: TOTAL COUNT OF CHARRED BOTANICAL REMAINS PER PROPERTY FROM INSULA VI.I

The Shrine produced the highest count of carbonised archaeobotanical remains for contexts within Insula VI.I: 1273 charred remains or 31% of the charred assemblage (Figure 44, Figure 45, Figure 46). For a brief period during the 1st century AD this feature was likely a roadside shine

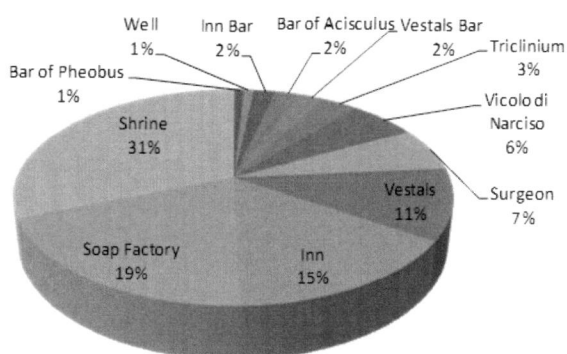

FIGURE 45: TOTAL COUNT OF CARBONISED ARCHAEOBOTANICAL REMAINS PER PROPERTY FROM INSULA VI.I

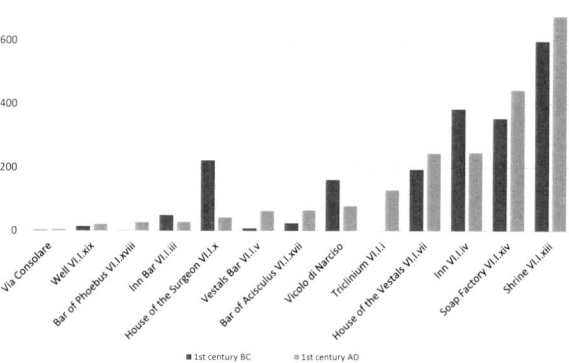

FIGURE 46: TOTAL COUNT OF CHARRED BOTANICAL REMAINS PER PROPERTY FROM INSULA VI.I BY TIME PERIOD

used for ritual purposes by passers-by entering or leaving the town along the Via Consolare through the Herculaneum Gate. The absence of further development of this feature was probably a factor in the favourable preservation of carbonised botanical remains here. Unlike other properties within Insula VI.I, where archaeobotanical remains became incorporated into the archaeological matrix as secondary fill as a result of construction repair, no construction of the Shrine is evident during the 1st century AD. Thus, it provides a unique example of ritual offerings from Insula VI.I.

Low numbers of carbonised archaeobotanical remains were recovered from the Well [n=13.8 from 1st century BC and n=21 from 1st century AD], located at the southern tip of Insula VI.I and on the main street, Via Consolare. The Well is thought to have been briefly used as a ritual site for burnt offerings in the 1st century AD (Jones and Robinson 2004). Both properties, the Well and Via Consolare would have been exposed to the same taphonomic processes, including heavy pedestrian and cart traffic in the past, and tourist traffic since the 18th century (when Insula VI.I was cleared). The bar areas, Vestals Bar, Bar of Pheobus, Inn Bar and Bar of Acisculus also produced low numbers of charred botanical remains [n=91.8, n=28.2, n=78, n=89.9 respectively]. This is surprising because areas where food was made and/or sold would be expected to contain evidence of food preparation, consumption and/ or storage. However, it should be noted that all four bar

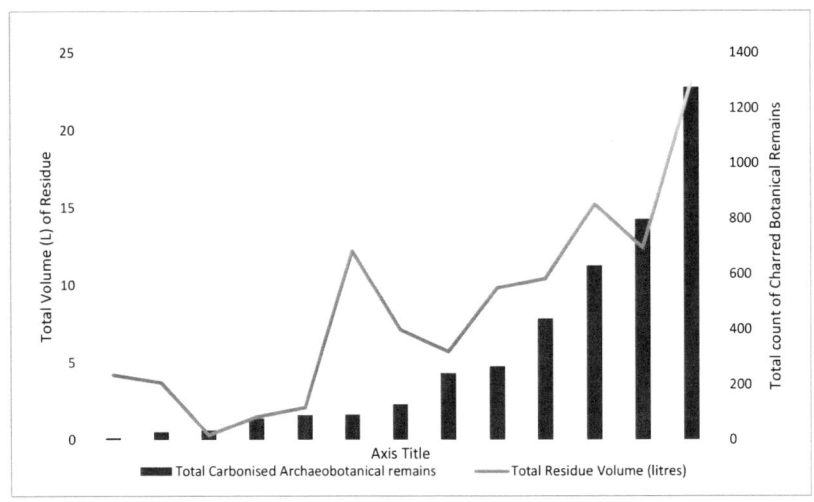

FIGURE 47: TOTAL COUNT OF CARBONISED BOTANICAL REMAINS AND TOTAL VOLUME OF RESIDUE EXAMINED PER PROPERTY FROM INSULA VI.I

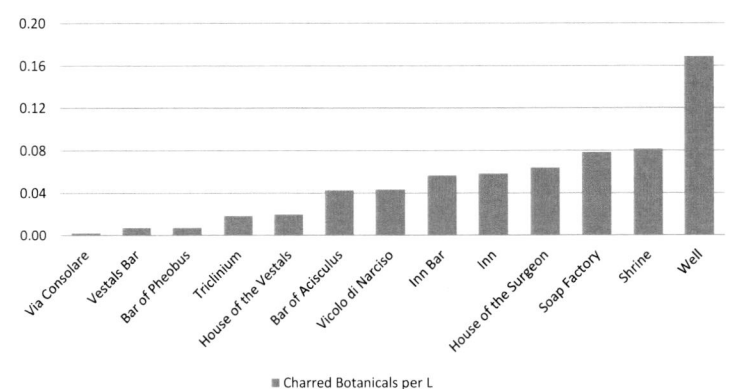

FIGURE 48: CHARRED BOTANICAL REMAINS PER L OF RESIDUE EXAMINED PER PROPERTY FROM INSULA VI.I

properties were small (in terms of square footage) in area and therefore customers probably carried away purchased food or drink.

The positive correlation between total residue volume examined per property and the number of charred botanical remains recovered may be due to their having been recovered from secondary deposits (Figure 47). Such secondary deposits probably represent a range of activities for each property. Vestals Bar is the exception having a higher ratio of carbonised botanical remains recovered by the volume of residue examined [n=7.5]. One possible explanation for this pattern is that Vestals Bar, which is situated within the House of the Vestals, occupies a greater area and therefore produced a larger sample of residue [Vol=12.2L].

In general, a very low recovery rate of carbonised botanical remains per litre of residue examined was obtained from all the properties from Insula VI.I, <0.8 charred botanical remains per L (Figure 48). The well is unique among these properties, having a small physical size and produced a correspondingly low volume of sample (Vol=0.212L).

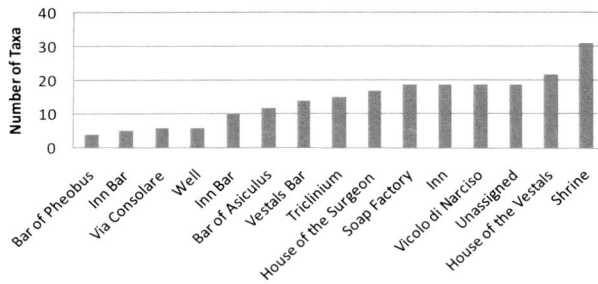

FIGURE 49: NUMBER OF CARBONISED TAXA PER PROPERTY FROM INSULA VI.I

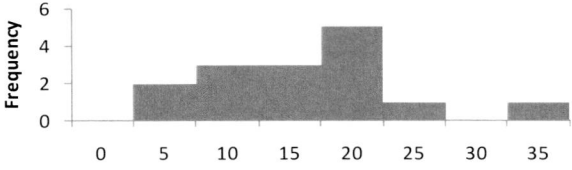

FIGURE 50: HISTOGRAM OF CARBONISED TAXA FROM INSULA VI.I

The Shrine produced the highest number of carbonised taxa [n=31] (Figure 49). It is unique among the sampled properties of Insula VI.I in terms of the range of carbonised taxa. This is probably because the plant remains represent

Figure 51: Number of carbonised taxa from Vesuvian sites (additional data from Ciaraldi (2001) PhD thesis)

purposely burnt ritual offerings. In contrast, low numbers of carbonised taxa were recovered from the Bar of Pheobus [n=4], Inn Bar [n=5], Via Consolare [n=6], and Well [n=6] (Figure 49). In part these patterns may be due to the small areas that they provide for sampling.

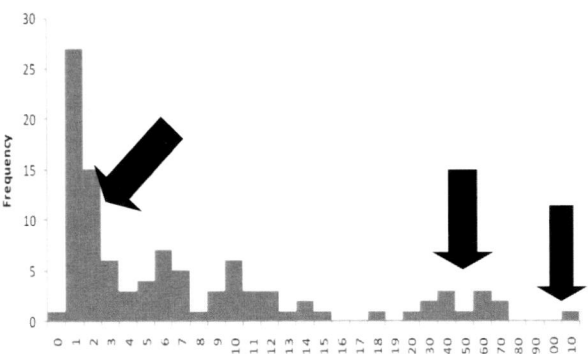

FIGURE 52: HISTOGRAM OF CARBONISED TAXA FROM VESUVIAN SITES (ADDITIONAL DATA FROM CIARALDI (2001) PHD THESIS)

Three distinct distribution patterns are visible in Figure 52. The first arrow to the left represents the majority of known Vesuvian sites in which one to two carbonised taxa were recovered. It is likely that this peak represents results from secondary fill in which few carbonised taxa were found. The central arrow indicates the frequency of unique deposits, including: the House of the Vestals (VI.I.vii) drain and toilet feature; *Porta Nocera, Necropolis* burnt funerary offerings; *dolium* contents from the *Villa Vesuvio* at *Scafati*; and burnt fodder from the House of the Chaste Lovers. The third arrow indicates the carbonised hay from *Oplontis* [n=103 carbonised taxa], a rare archaeobotanical find due its unique preservation, which is a significant outliner on this graph.

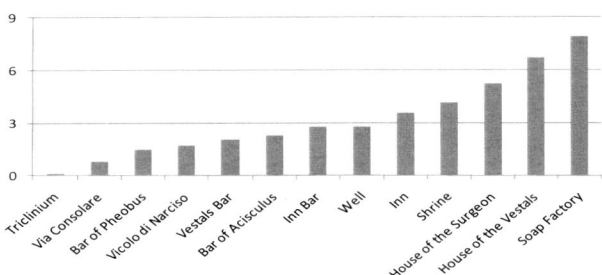

FIGURE 53: CARBONISED ARCHAEOBOTANICAL REMAINS PER CONTEXT EXAMINED PER PROPERTY FROM INSULA VI.I

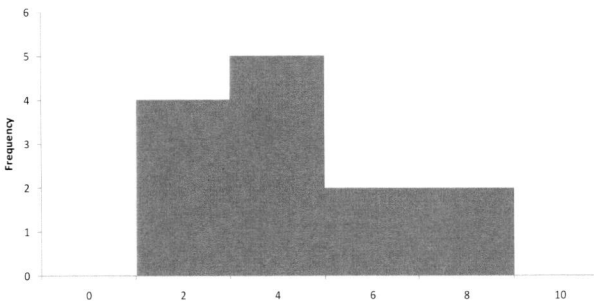

FIGURE 54: HISTOGRAM OF RECOVERED CARBONISED BOTANICAL REMAINS PER CONTEXT FROM INSULA VI.I

Carbonised botanical remains per context from the properties within Insula VI.I were low (Figure 53, Figure 54).

4.6 Mineralised archaeobotanical remains from Insula VI.I

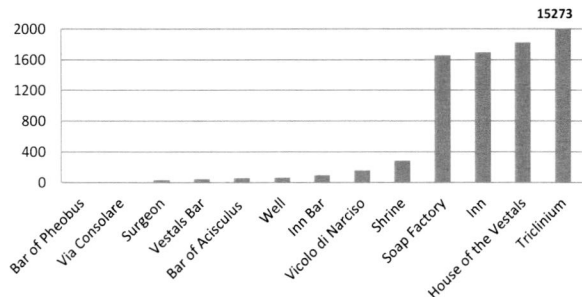

FIGURE 55: TOTAL COUNT OF MINERALISED ARCHAEOBOTANICAL REMAINS PER PROPERTY FROM INSULA VI.I

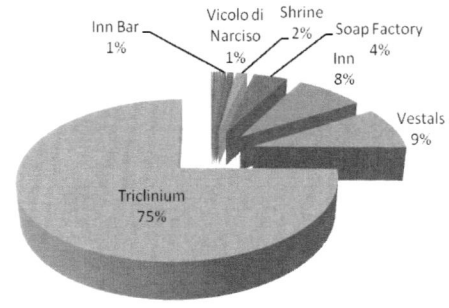

FIGURE 56: TOTAL COUNT OF MINERALISED ARCHAEOBOTANICAL REMAINS BY PROPERTY FROM INSULA VI.I

The majority (75%) of mineralised archaeobotanical remains came from the Triclinium area, which possessed a toilet feature.

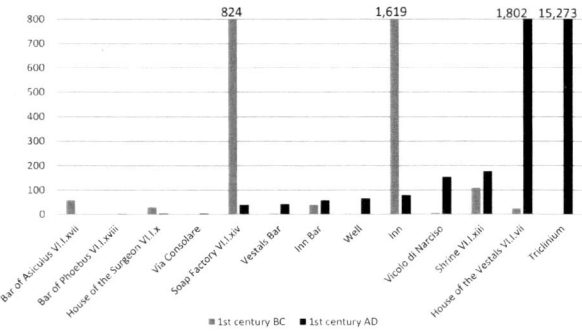

FIGURE 57: TOTAL COUNT OF MINERALISED BOTANICAL REMAINS RECOVERED PER PROPERTY FROM INSULA VI.I BY TIME PERIOD

There is a slight increase over time in the number of mineralised archaeobotanical remains in the Vestals Bar [n=3.8 1st century BC, n=41.4 1st century AD], Inn Bar [n=38.9 1st century BC, n=58.2 1st century AD], Well [n=0 1st century BC, n=65 1st century AD], Vicolo di Narciso [n=5 1st century BC, n=154.6 1st century AD], Shrine [n=108.8 1st century BC, n=177.2 1st century AD], and the House of the Vestals [n=23.8 1st century BC, n=1802.4]. The House of the Surgeon [n=29.56 1st century BC, n=4 1st century AD], the Inn [n=1619.33 1st century BC, n=79.06 1st century AD] and the Bar of Acisculus [n=58.61 1st century BC, n=1.06 1st century AD] were the only properties to show a decrease in mineralised remains in the 1st century AD. The Inn showed a high number of mineralised botanical remains [n=1619] from the 1st century BC or earlier time period. The 1st century AD House of the Vestals [n=1802.39] and the Triclinium [n=15273] were the only properties within Insula VI.I that produced substantial numbers of recovered mineralised plant remains.

Acisculus [n=27.88] and Vicolo di Narciso [n=27.95]. The number of mineralised plant remains recovered from the Well [n=309.52] can probably be attributed to the small volume of sample collected (Vol=0.212 L sediment).

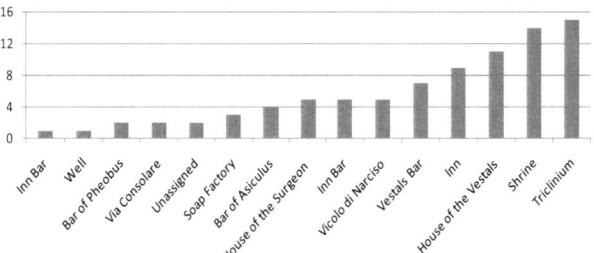

FIGURE 60: NUMBER OF MINERALISED TAXA PER PROPERTY FROM INSULA VI.I

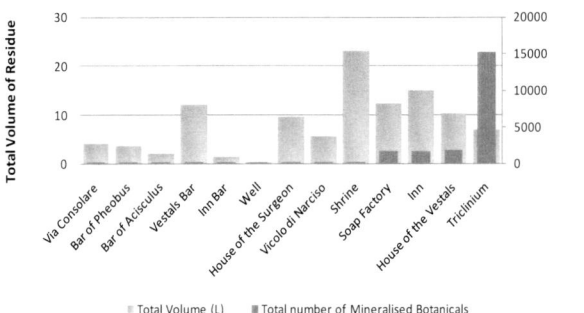

FIGURE 58: TOTAL COUNT OF MINERALISED BOTANICAL REMAINS RECOVERED AND TOTAL VOLUME OF RESIDUE EXAMINED PER PROPERTY FROM INSULA VI.I

All samples from Insula VI.I properties yielded lower numbers of recovered mineralised plant remains when compared with the volume of residue examined; the Triclinium being the exception. The large number [n=15, 095] of mineralised fig achenes recovered from the toilet feature skews the results from the Triclinium. The Shrine is also distinct, having few mineralised botanical remains compared with the large volume of sediment examined. In this case the ratio of mineralised remains to sediment [n=12.35 mineralised remains per L] is distinct from that of the carbonised remains [n=0.008 carbonised remains per L] (Figure 58). The differences between the charred and mineralised assemblages can possibly be explained by taphonomic processes, given that mineralisation requires specific conditions (explained in chapter 3), which did not occur in the majority of properties within Insula VI.I.

The Triclinium yielded the largest total count of mineralised remains [n=15273] and the highest number of mineralised taxa [n=13] (Figure 59, Figure 60). The advantageous preservational conditions of this toilet feature probably account for the high number of mineralised specimens. The Soap Factory had a high total count of mineralised remains [n=867.77] but representing only a few taxa [n=3] which suggests that a narrow range of plants were brought to this commercial property. In contrast, the Shrine possessed a low count of mineralised remains [n=286.05] but a high number of mineralised taxa [n=14], a pattern that suggests that a greater diversity of plants were brought to this site but conditions did not favour mineralisation (Figure 60).

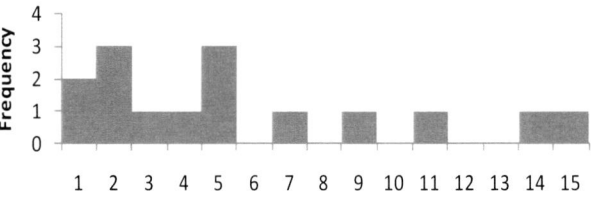

FIGURE 61: HISTOGRAM OF MINERALISED TAXA FROM INSULA VI.I

The results from Insula VI.I with regards to the number of mineralised taxa are not unusual for the Vesuvian area as the majority of contexts had one mineralised botanical recorded per context. The highest number of mineralised taxa came from the late 1st century BC toilet [n=56] and drain [n=64] features from Insula VI.I as noted by Ciaraldi (2001, 2007).

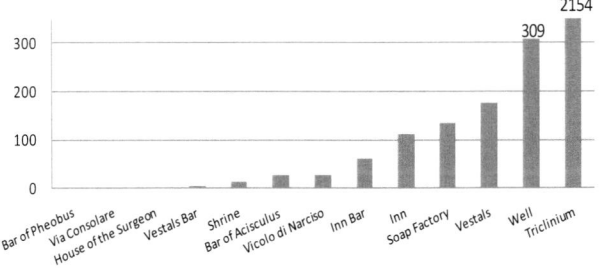

FIGURE 59: MINERALISED BOTANICAL REMAINS RECOVERED PER L OF RESIDUE EXAMINED PER PROPERTY FROM INSULA VI.I

Relatively low quantities of mineralised remains per L of residue examined were found in the Bar of Pheobus [n=0.83], Via Consolare [n=0.98], House of the Surgeon [n=3.45], Vestals Bar [n=3.69], Shrine [n=12.35], Bar of

The Triclinium, located within the Inn, has exceptional preservation due to the presence of the toilet feature, resulting in 324.9 mineralised botanical remains per context examined. The other Insula VI.I properties produced significantly lower ratios of mineralised archaeobotanical remains per context examined [Average mineralised remains per context n=4.13]. Likewise, low ratios of mineralised plants were recovered from the Bar areas [Bar of Pheobus n=0.24; Bar of Acisculus n=0.25; Inn Bar n=1.2; Vestals Bar n=2.05], Via Consolare [n=0.05] and the House of the Surgeon [n=0.76] probably due to the absence of conditions suitable for mineralisation to occur.

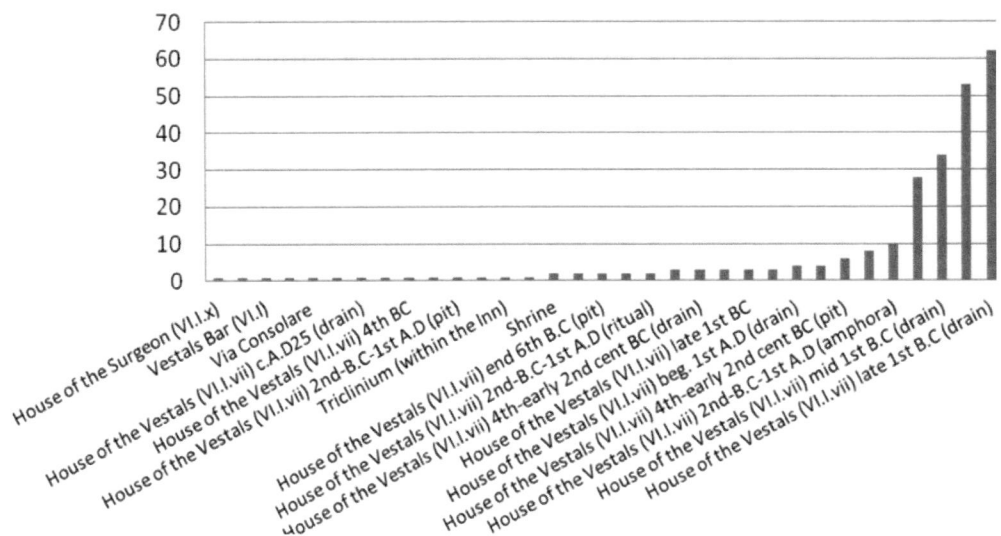

FIGURE 62: NUMBER OF MINERALISED TAXA FROM VESUVIAN SITES (ADDITIONAL DATA FROM CIARALDI (2001) PHD THESIS)

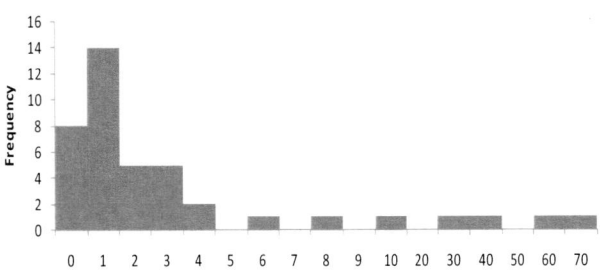

FIGURE 63: HISTOGRAM OF MINERALISED ARCHAEOBOTANICAL REMAINS FROM PUBLISHED SOURCES ON VESUVIAN SITES (ADDITIONAL DATA FROM CIARALDI (2001) PHD THESIS)

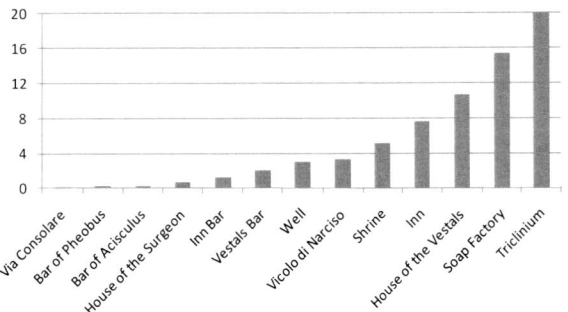

FIGURE 64: MINERALISED BOTANICAL REMAINS PER CONTEXT PER PROPERTY FROM INSULA VI.I

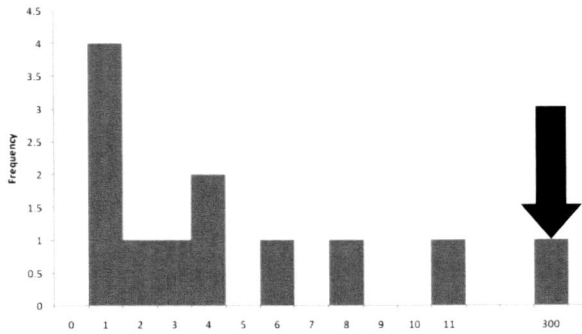

FIGURE 65: HISTOGRAM OF RECOVERED MINERALISED BOTANICAL REMAINS PER CONTEXT FROM INSULA VI.I

The arrow points to the outlier of the Triclinium results, in comparison with the number of mineralised botanical per context from the result of Insula VI.I, shown in the other bars.

4.7 Residue volume

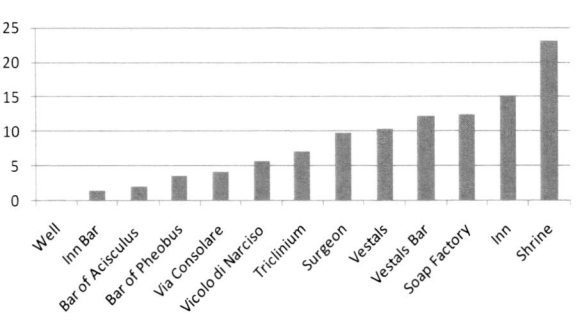

FIGURE 66: TOTAL VOLUME (L) OF RESIDUE EXAMINED PER PROPERTY FROM INSULA VI.I

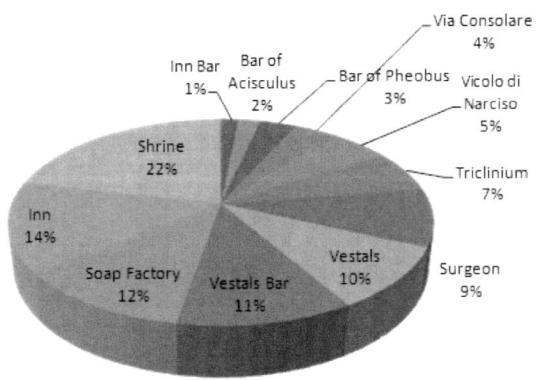

FIGURE 67: TOTAL RESIDUE VOLUME (L) PER PROPERTY FROM INSULA VI.I

The Shrine produced the largest volume of residue from contexts within Insula VI.I [n=23.16L]. In part this was due to the format of the feature and the excavation processes used. The Shrine was subject to a larger open excavation

RESULTS

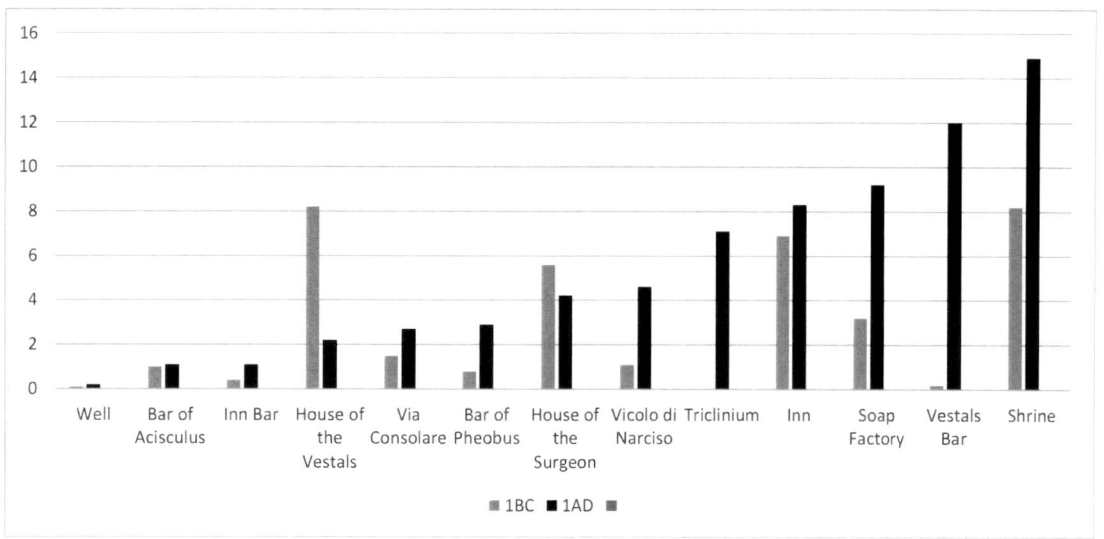

FIGURE 68: BREAKDOWN OF RESIDUE VOLUME (L) BY TIME PERIOD

area with little mosaic or floors to impede excavation below the AD 79 layer. Other properties within Insula VI.I occupied larger areas, such as the Inn [n=15.16L], Soap Factory [n=12.45L], Vestals Bar [n=12.24L] and House of the Vestals [n=10.37]. However, they also had smaller rooms and due to the presence of features e.g. unstable structural masonry walls, mosaics and intact floors, it was often impossible to excavate beyond the AD 79 layer and collect environmental samples. The domestic properties, *the* House of the Vestals (VI.I.7) and the House of the Surgeon (VI.I.10) were the only properties that had a higher volume of residue from the 1st century BC or earlier than from the 1st century AD. The plots for these two domestic houses are thought to have been the earliest identifiable structures within this Insula and therefore offer the greatest possibility of providing the earliest evidence from Insula VI.I (see Chapter 2).

4.8 Unknowns/indeterminate archaeobotanical material from Insula VI.I

Carbonised unknowns/indeterminate archaeobotanical material from Insula VI.I

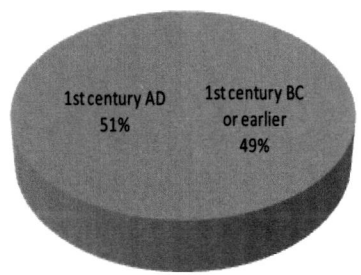

FIGURE 69: PERCENTAGE OF TOTAL CARBONISED ARCHAEOBOTANICAL UNKNOWNS/INDETERMINATES FROM INSULA VI.I BY TIME PERIOD

There was nearly equal percentage of carbonised unknowns/indeterminates from both time periods from Insula VI.I.

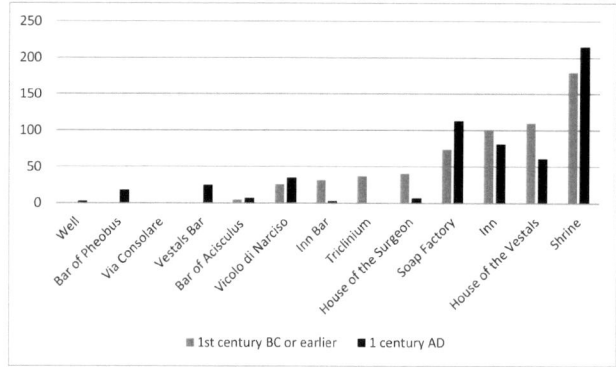

FIGURE 70: COUNT OF CARBONISED INDETERMINATES/UNKNOWNS BY PROPERTY AND TIME PERIOD FROM INSULA VI.I

The Shrine also produced the highest counts of carbonised indeterminates/unknowns [n=395]. This could be due to the fact the ritual offerings were deliberately burnt and therefore difficult to identify due to fact that they were carbonised and highly fragmented.

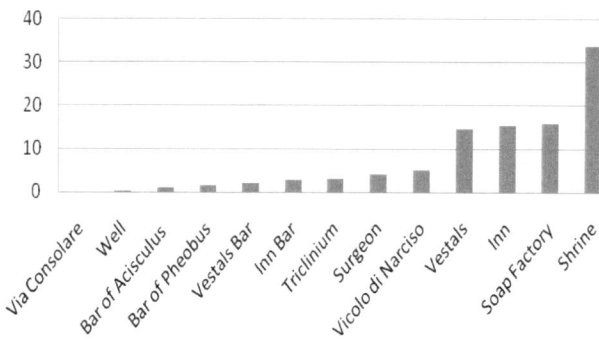

FIGURE 71: PERCENT OF CARBONISED ARCHAEOBOTANICAL REMAINS PER PROPERTY FROM INSULA VI.I THAT WERE UNKNOWN/INDETERMINATE

The Shrine produced the highest percentage [n=33.6%] of carbonised unknowns/indeterminates. The Soap Factory produced the second highest percentage of carbonised unknowns [n=15.91%], probably due to the extremely high temperatures to which the plants were exposed, evidenced by iron working debris including iron filings and slag.

Mineralised unknowns/indeterminate archaeobotanical material from Insula VI.I

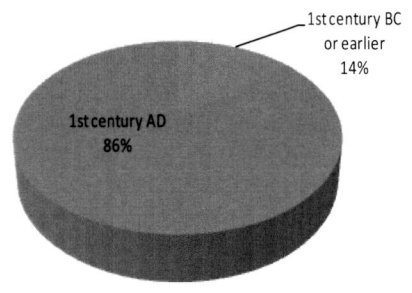

FIGURE 72: PERCENT OF TOTAL COUNT OF MINERALISED UNKNOWNS/INDETERMINATES FROM INSULA VI.I

A significantly higher proportion of mineralised unknowns/indeterminates occurred in samples from the 1st century AD [n=86%] than from the 1st century BC or earlier time period [n=14%].

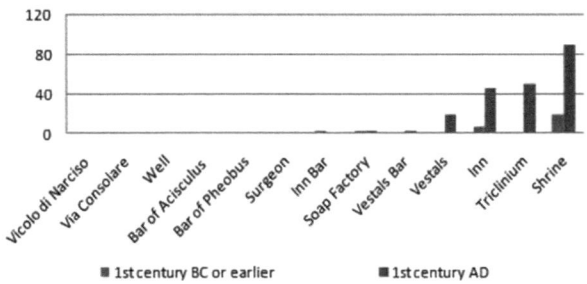

FIGURE 73: TOTAL COUNT OF MINERALISED INDETERMINATE BY PROPERTY BY TIME PERIOD FROM INSULA VI.I

Few mineralised unknowns/indeterminates were recovered from the 1st century BC or earlier period [n=33]. There is a substantial jump in mineralised unknowns/indeterminates in the 1st century AD from the House of the Vestals [n=1 1st century BC, n=19 1st century AD], the Inn [n=7 1st century BC, n=46 1st century AD] and the Shrine [n=19 1st century BC, n=90 1st century AD].

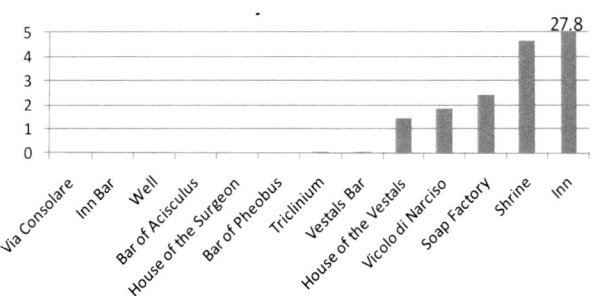

FIGURE 74: PERCENTAGE OF UNKNOWN MINERALISED REMAINS PER PROPERTY FROM INSULA VI.I

The Inn produced the highest percentage of unknown mineralised remains [n=27.8%], possibly due to their being damaged in construction and alterations to this structure. The Triclinium, which had the largest count of mineralised archaeobotanical remains from Insula VI.I [n=15,273], had few unknowns [n=49 or 0.32% of the total mineralised assemblage] which testify to the excellent preservation conditions created by the toilet feature.

From every property within Insula VI.I, aside from the Inn, the percentage of mineralised unknown/indeterminates is lower than the percentage of carbonised unknown/indeterminates (Figure 74). This may in part be due to the fact that mineralised remains from specific taxa are easily recognisable e.g. grapes. Also, a large portion of the mineralised remains are from specific deposits rather than secondary fill. Hence, taphonomic differences between primary and secondary deposits probably explain the difference between the carbonised and mineralised unknowns/indeterminates.

Veal (2009, 153) put forward a possible hypothesis that increased archaeological matrix depth is related to an increased charcoal fragmentation (based upon her recent findings from different areas within the city of Pompeii, including within Insula VI.I). A variety of factors are known to weaken charcoal structure and influence fragmentation including the flow of water through earlier contexts resulting in increased interstitial mineral deposition, increased compaction from soil weight above and higher rates of bioturbation (observed at Porta Stabia, Pompeii). Although there is no direct archaeobotanical equivalent of fragmentation in charcoal the percentage of unknowns/indeterminates from this study provided a standard for rough comparison. The same spatial distribution pattern was not observed in the archaeobotanical unknown/indeterminate assemblage from Insula VI.I. Carbonised unknowns/indeterminates remain fairly steady through time; in contrast, counts of mineralised unknowns/indeterminates increased in the 1st century AD (Figure 75). Possible explanations for these patterns are that the more fragile specimens from earlier deposits were lost, and the fact that specific deposits where mineralised remains are abundant, such as the toilet feature, date to the late 1st century BC to 1st century AD, and thus are later in date.

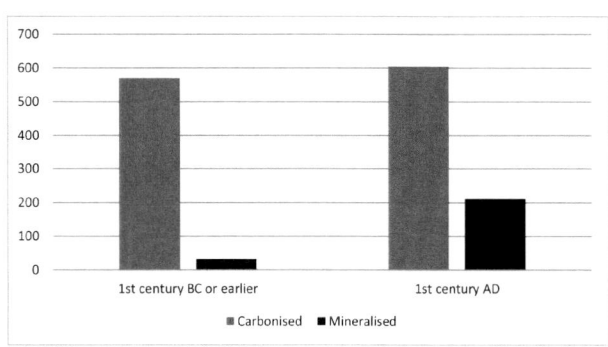

FIGURE 75: TOTAL COUNT OF CARBONISED AND MINERALISED UNKNOWNS FROM INSULA VI.I BY TIME PERIOD

Results

Properties within Insula VI.I	Vicolo di Narciso		Via Consolare		Tricli-nium	Well		Bar of Acisculus		Bar of Pheobus		Shrine		Soap Factory		House of the Vestals		House of the Surgeon		Inn		Vestals Bar		Inn Bar	
Time Period	1BC	1AD	1BC	1AD	1AD	1BC	1AD	1BC	1AD	1BC	1AD	1BC	1AD	1BC	1AD	1BC	1AD	1BC	1AD	1BC	1AD	1BC	1AD	1BC	1AD
Number of Contexts	16	32	28	31	34	4	9	17	37	4	18	115	127	50	104	87	133	76	50	106	97	13	65	22	19
Total Residue Volume (litres)	1.1	4.6	1.5	2.7	7.1	0.1	0.2	1	1.1	0.8	2.9	8.2	14.9	3.2	9.2	8.2	2.2	5.6	4.2	6.9	8.3	0.2	12	0.4	1.1
Total Number of Archaeobotanical Remains	165	515	1.6	9.7	15409.1	13.8	86	82.5	68.4	1	30.2	206	920	1178	482	232	2851	259	54.6	2236	343	11.4	129	89.1	86.4
Number of Archaeobotanical Remains per Context	147	112	0.1	0.3	453.2	3.5	9.6	4.9	1.8	0.3	1.7	1.8	7.2	23.3	2.6	2.7	21.4	3.4	1.1	21.1	3.5	0.9	2	4	4.5
Number of Taxa	10	20	3	4	23	2	6	8	11	1	4	29	37	18	20	17	21	18	13	22	23	3	16	10	9
Total Carbonised Archaeobotanical Remains	161	79.7	1.6	5.6	129	13.8	21	23.9	66	1	27.2	597	676	353	443	194	245	222	43.6	383	247	7.6	84.2	50.2	27.8
Total Mineralised Archaeobotanical Remains	5	155	0	4.1	15273	0	65	58.6	1.1	0	3	109	177	825	38.9	23.8	1802	29.6	4	1619	79.1	3.8	41.4	38.9	58.2
Total Carbonised Indeterminate	26	35	0	0	37	0	3	5	7	0	18	180	215	74	113	110	61	41	7	101	81	1	25	32	3
% Carbonised Indeterminate	16.1	43.9	0	0	28.7	0	14.3	20.9	10.6	0	66.2	30.1	31.8	20.9	25.5	56.9	24.9	18.4	16.1	26.4	32.8	13.2	29.7	63.8	10.8
Total Mineralised Indeterminate	0	0	0	0	49	0	0	0	0	0	0	19	90	3	3	1	19	0	0	7	46	0	3	3	1
% Mineralised Indeterminate	0	0	0	0	0.3	0	0	0	0	0	0	17.5	50.8	0.4	0.4	4.2	1.1	0	0	0.4	58.2	0	7.2	7.7	1.7

TABLE 3: SUMMARY OF RESULTS FROM INSULA VI.I BY PROPERTY BY TIME PERIOD

4.9 Recovery method flot/residue/*in situ*/dry sieved

House of the Surgeon (VI.I.x) (*Casa della Chiurugo*)

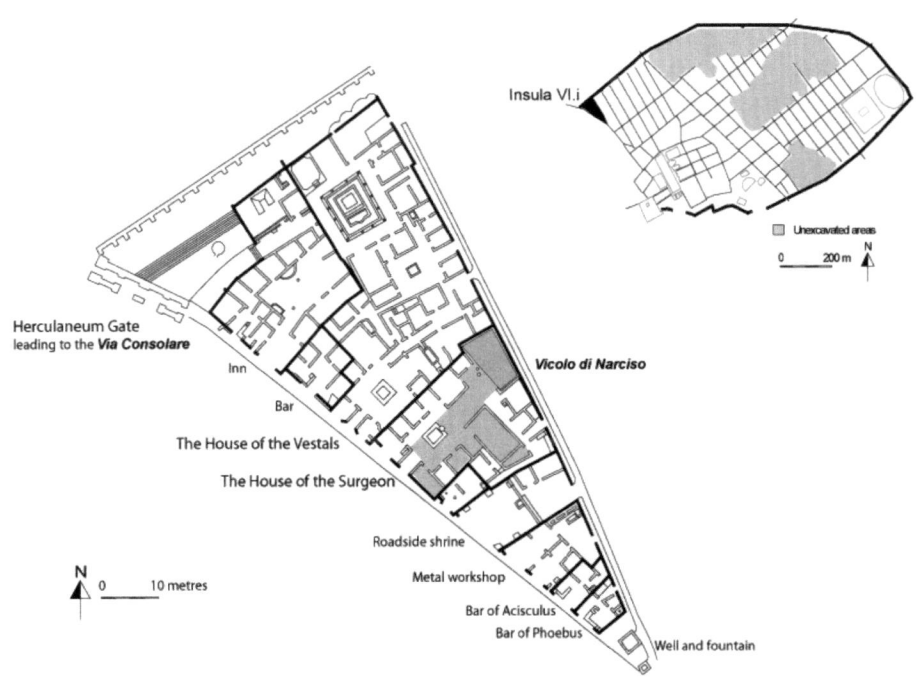

FIGURE 76: EXCAVATION AREAS FOR THE HOUSE OF THE SURGEON (VI.I.x) (AAPP05 RESOURCE HANDBOOK P. 27)

House of the Surgeon (VI.I.x) 1st century BC

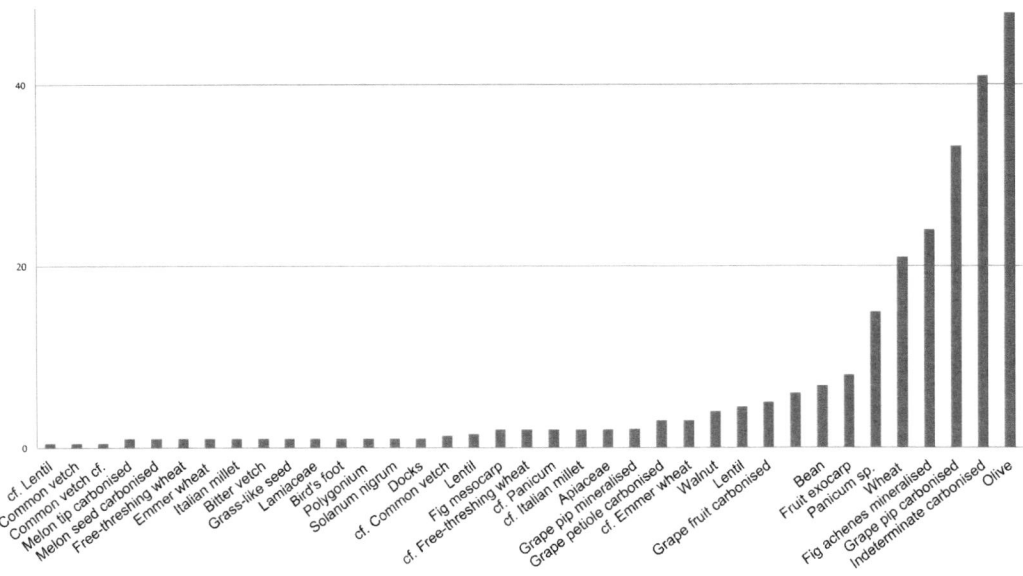

FIGURE 77: 1ST CENTURY BC ARCHAEOBOTANICAL REMAINS FROM THE HOUSE OF THE SURGEON (VI.I.x) (REMAINS CARBONISED UNLESS STATED OTHERWISE)

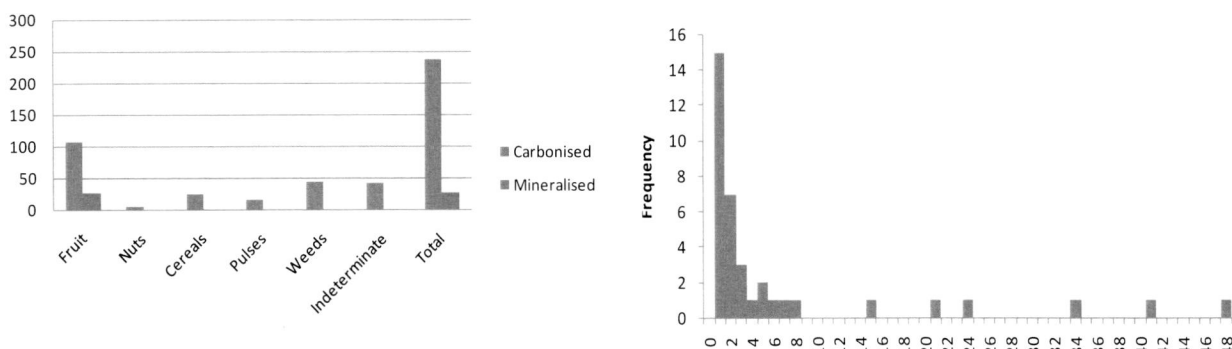

FIGURE 78: COUNT OF ARCHAEOBOTANICAL REMAINS FROM THE HOUSE OF THE SURGEON (VI.I.x) BY CATEGORY FROM THE 1ST CENTURY BC

FIGURE 79: HISTOGRAM OF COUNT OF ARCHAEOBOTANICAL REMAINS FROM THE 1ST CENTURY BC FROM THE HOUSE OF SURGEON (VI.I.x)

House of the surgeon (VI.I.x) 1st century AD

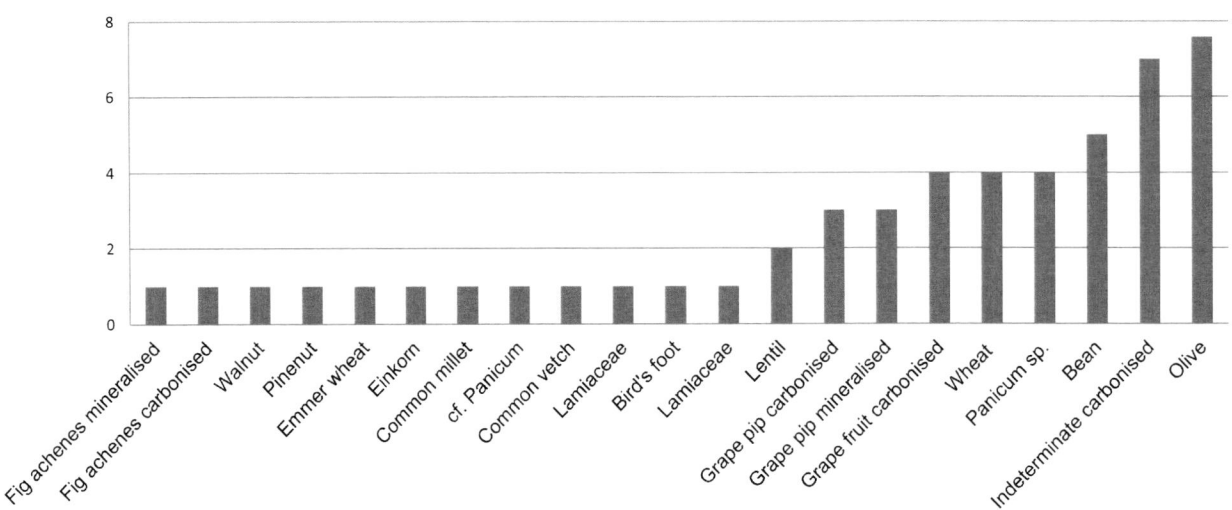

FIGURE 80: COUNT OF ARCHAEOBOTANICAL REMAINS FROM THE HOUSE OF THE SURGEON (VI.I.x) FROM THE 1ST CENTURY AD (REMAINS CARBONISED UNLESS STATED OTHERWISE)

Results

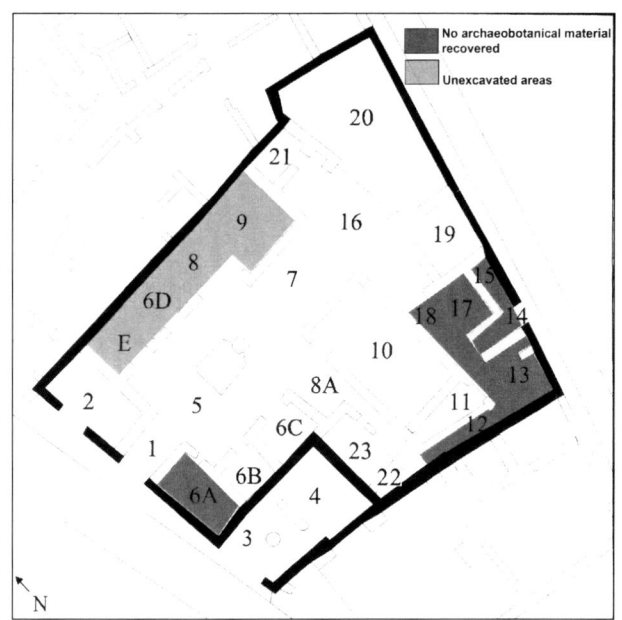

There is a high ubiquity of olive endocarp fragments within the House of the Surgeon (VI.I.x).

Room 22 by far had the largest number of archaeobotanical remains from this property. Room 22 is thought to have been a kitchen area with a pit feature, likely a cesspit which would account for the large mineralised archaeobotanical assemblage recovered from this room.

All archaeobotanical material recovered from the House of the Vestals, excluding the latrine and ritual deposits, from the 1995 to 2003 field seasons was analysed by the author.

FIGURE 81: HOUSE OF THE SURGEON (VI.I.x) AREAS WHERE NO ARCHAEOBOTANICAL REMAINS WERE RECOVERED AND AREAS NOT EXCAVATED (AAPP05 RESOURCE BOOK P. 30)

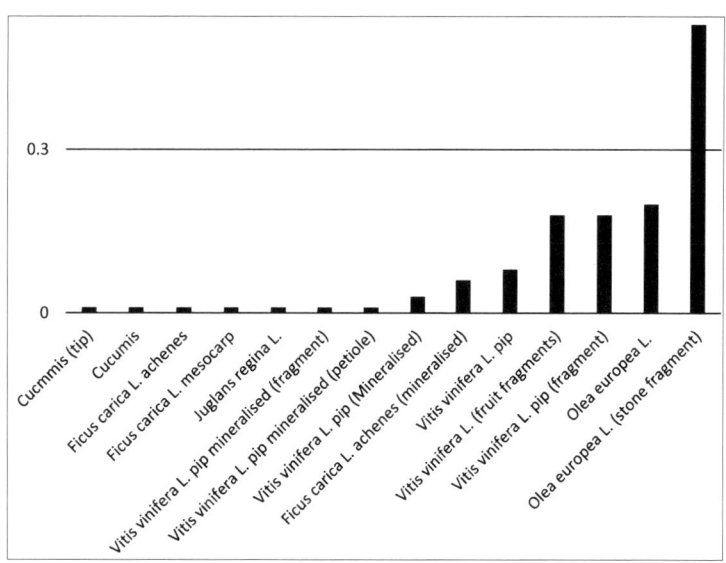

FIGURE 82: UBIQUITY OF ARCHAEOBOTANICAL REMAINS FROM HOUSE OF THE SURGEON (VI.I.x) FROM FLOTATION

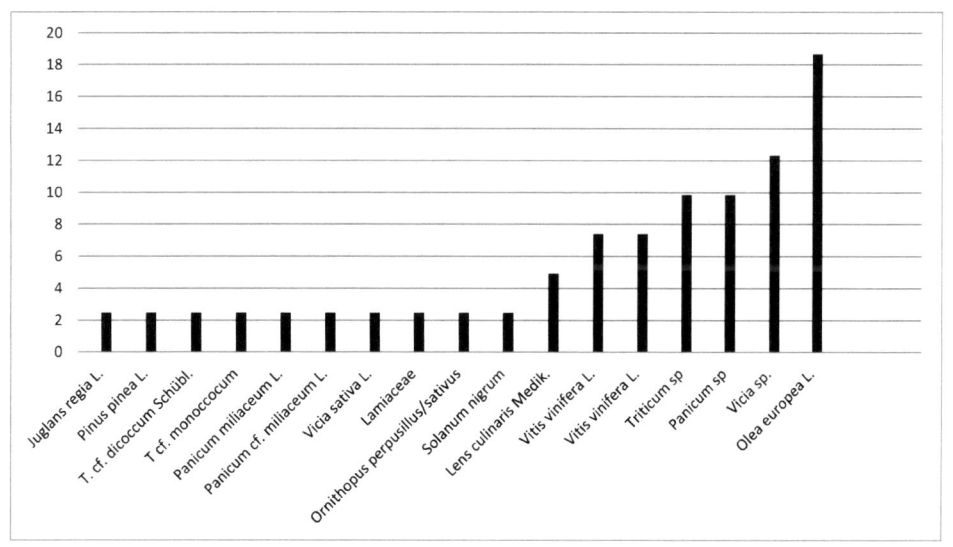

FIGURE 83: SHOWS THE DISTRIBUTION OF THE TOTAL COUNT OF ARCHAEOBOTANICAL (CARBONISED AND MINERALISED) REMAINS PER ROOM.

House of the Vestals (VI.I.vii) (*Casa delle vestali*)

FIGURE 84: EXCAVATION AREAS FOR THE HOUSE OF THE VESTALS (VI.I.VII) (AAPP05 RESOURCE HANDBOOK P. 27)

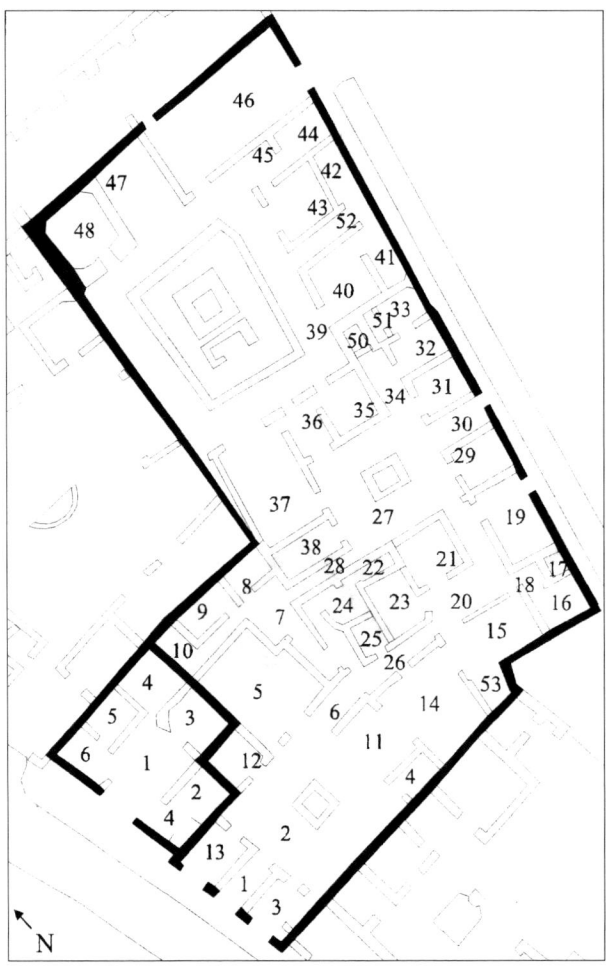

FIGURE 85: HOUSE OF THE VESTALS (VI.I.VII) ROOM NUMBERS
(AAPP05 RESOURCES HANDBOOK P. 32)

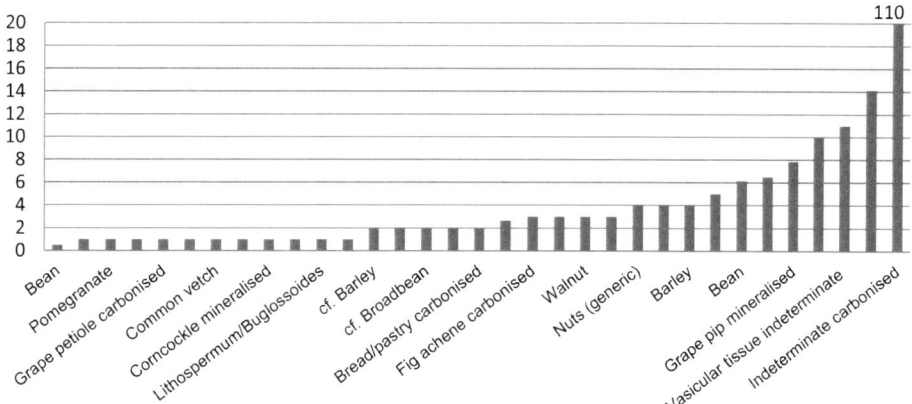

FIGURE 86: 1ST CENTURY BC ARCHAEOBOTANICAL REMAINS FROM THE HOUSE OF THE VESTALS (VI.I.VII) (REMAINS CARBONISED UNLESS STATED OTHERWISE)

House of the Vestals (VI.I.vii) 1st century BC

Carbonised indeterminates dominate [n=50%] the assemblage in the 1st century BC.

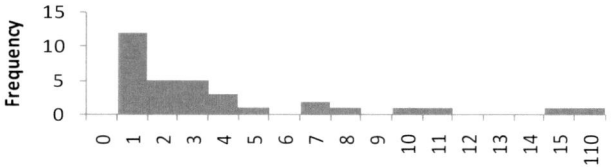

FIGURE 87: 1ST CENTURY BC HISTOGRAM OF ARCHAEOBOTANICAL REMAINS FROM THE HOUSE OF VESTALS (VI.I.VII)

House of the Vestals (VI.I.vii) 1st century AD

Mineralised fig achenes were the most numerous archaeobotanical remain recovered from the House of the Vestals from the 1st century AD [n=1703].

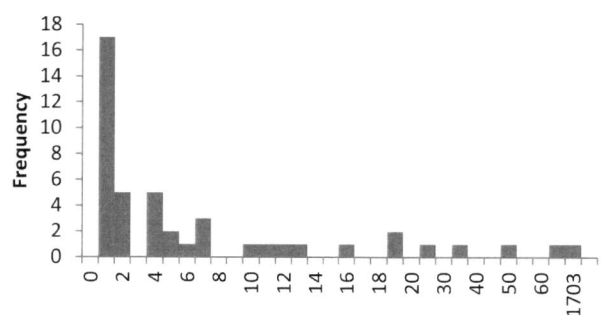

FIGURE 89: 1ST CENTURY AD HISTOGRAM OF ARCHAEOBOTANICAL REMAINS FROM THE HOUSE OF THE VESTALS (VI.I.VII).

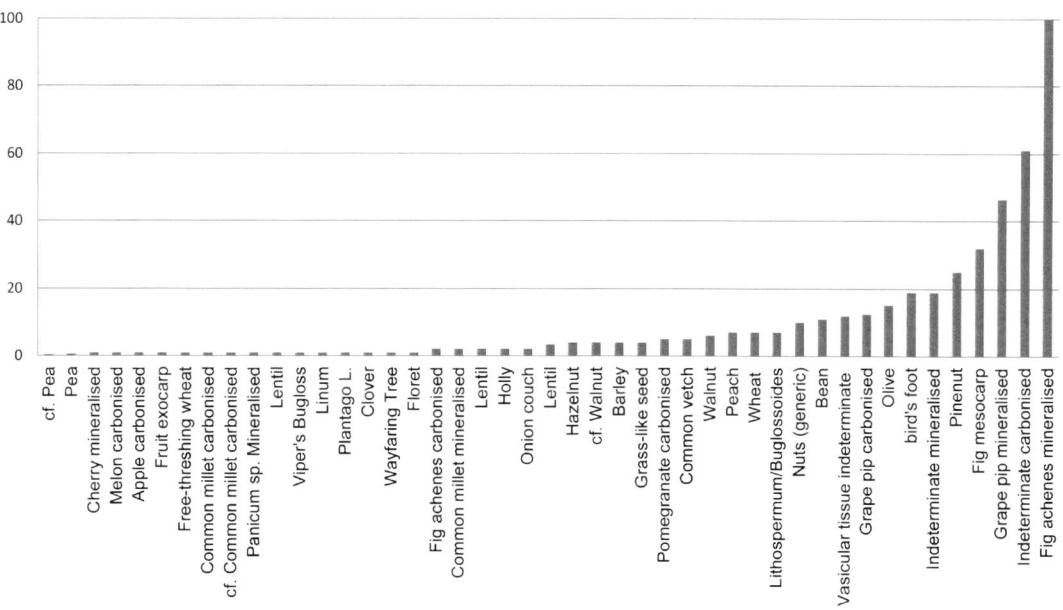

FIGURE 88: TOTAL COUNT OF ARCHAEOBOTANICAL REMAINS (MINERALISED AND CARBONISED) RECOVERED DATING FROM THE 1ST CENTURY AD FROM THE HOUSE OF THE VESTALS (VI.I.VII)

Shrine (VI.I.xiii)

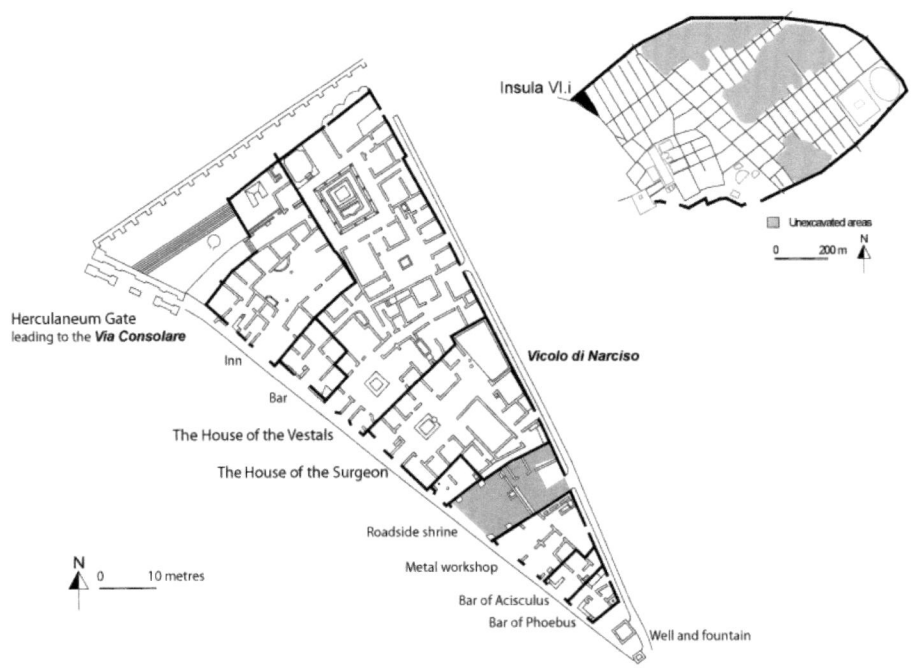

FIGURE 90: EXCAVATION AREAS FROM THE SHRINE (VI.I.XIII) (AAPP05 RESOURCE HANDBOOK P. 27)

Shrine (VI.I.xiii) 1st century BC

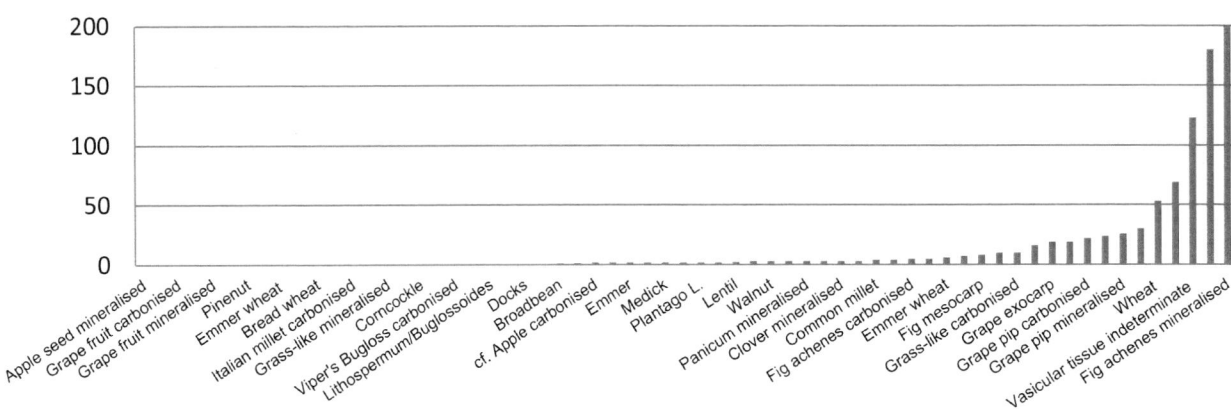

FIGURE 91: ARCHAEOBOTANICAL REMAINS RECOVERED FROM THE SHRINE (VI.I.XIII) FROM THE 1ST CENTURY BC OR EARLIER (REMAINS CARBONISED UNLESS STATED OTHERWISE)

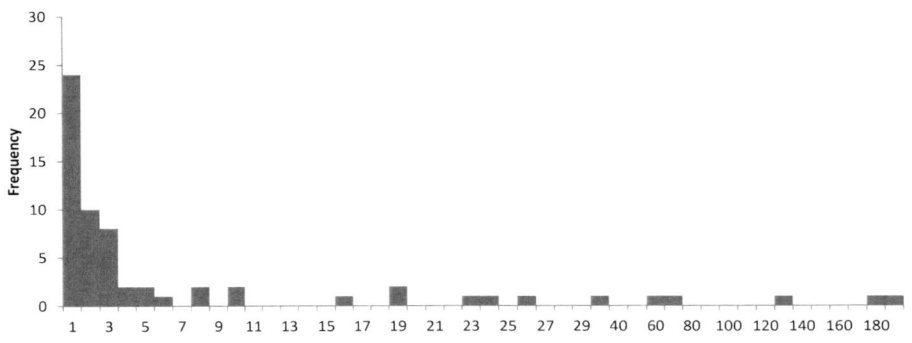

FIGURE 92: HISTOGRAM OF ARCHAEOBOTANICAL REMAINS FROM THE SHRINE (VI.I.XIII) FROM THE 1ST CENTURY BC OR EARLIER

Shrine (VI.I.xiii) 1st century AD

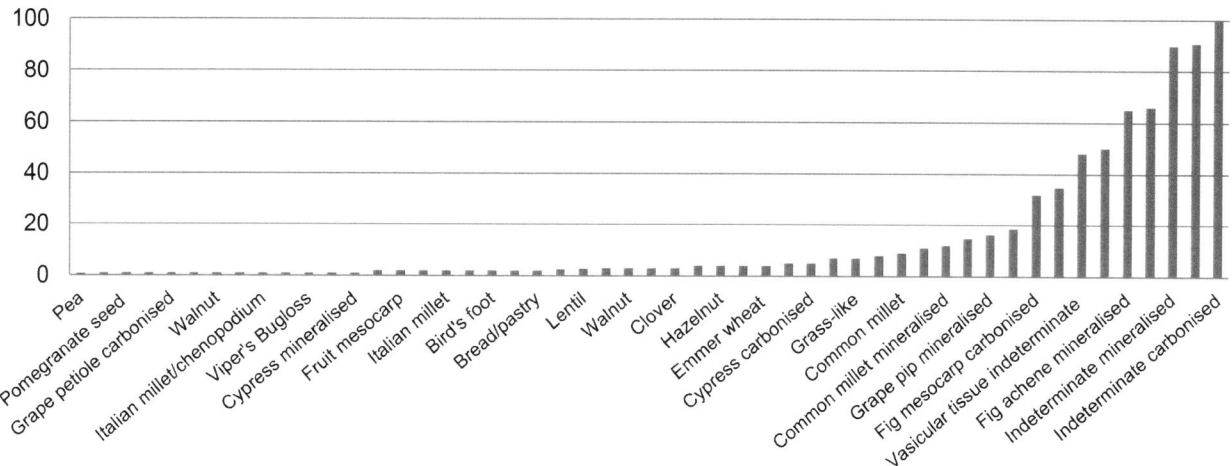

Figure 93: 1st century AD archaeobotanical remains from the Shrine (VI.I.xiii) (remains carbonised unless stated otherwise)

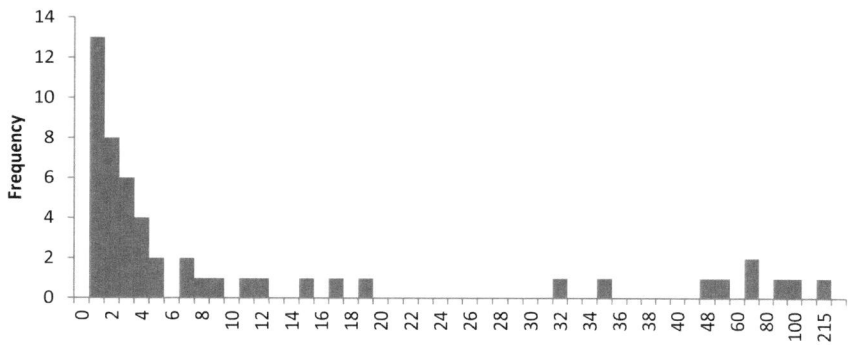

Figure 94: Histogram of archaeobotanical remains recovered from the Shrine (VI.I.xiii) from the 1st century AD.

Soap factory (VI.I.xiv)

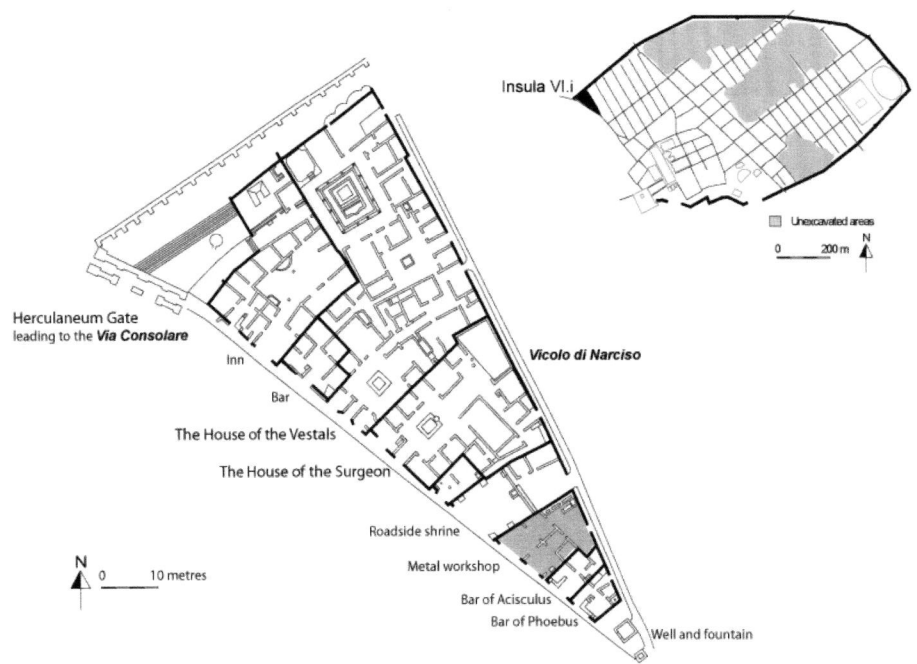

FIGURE 95: EXCAVATION AREAS FROM THE SOAP FACTORY (VI.I.XIV) (AAPP05 RESOURCE HANDBOOK P. 27)

Soap factory 1st century BC

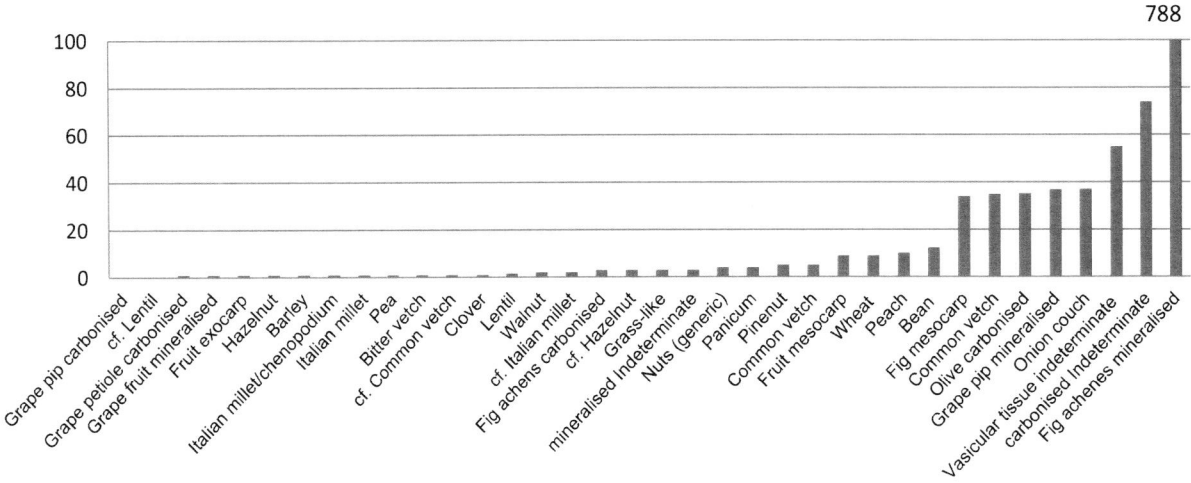

FIGURE 96: ARCHAEOBOTANICAL REMAINS FROM 1ST CENTURY BC FROM THE SOAP FACTORY (VI.I.XIV) (REMAINS CARBONISED UNLESS STATED OTHERWISE)

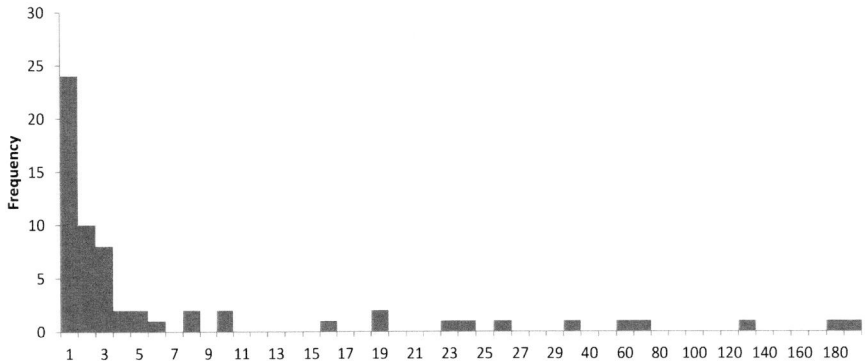

FIGURE 97: HISTOGRAM OF 1ST CENTURY BC ARCHAEOBOTANICAL REMAINS FROM THE SOAP FACTORY (VI.I.XIV)

Soap factory 1st century AD

Carbonised indeterminates [n=113] represent the outliner from the 1st century AD from this property.

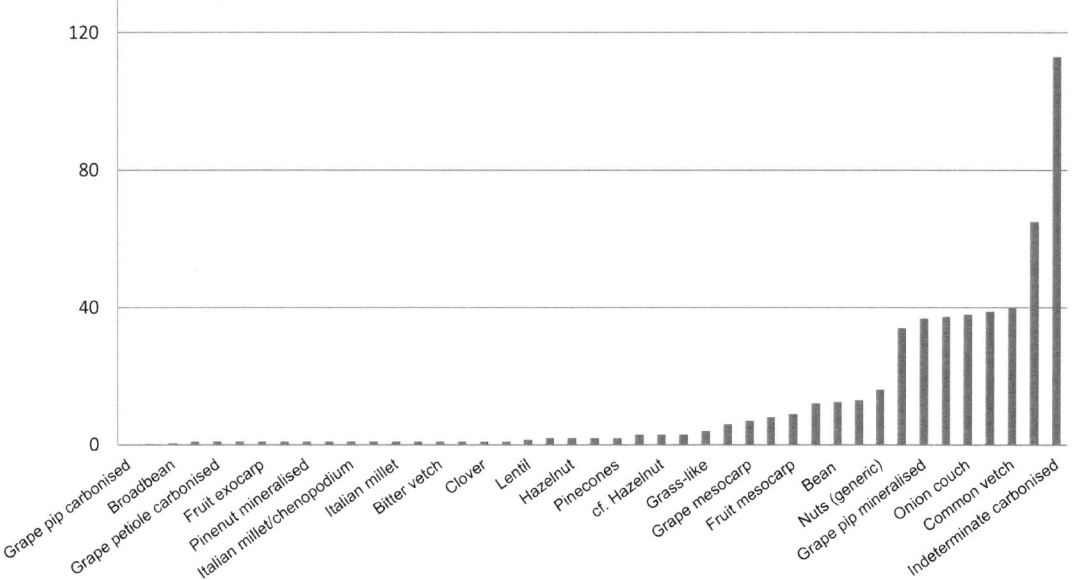

FIGURE 98: ARCHAEOBOTANICAL REMAINS FROM 1ST CENTURY AD FROM THE SOAP FACTORY (VI.I.XIV) (REMAINS CARBONISED UNLESS STATED OTHERWISE)

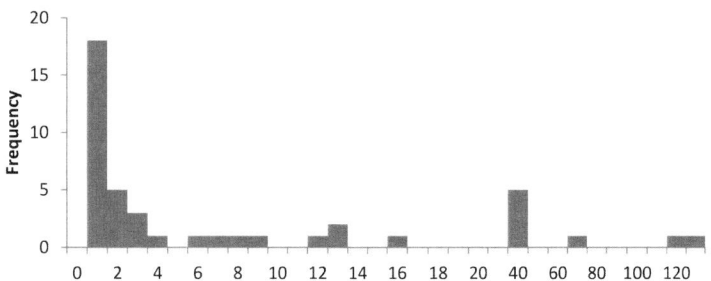

FIGURE 99: HISTOGRAM OF COUNT OF ARCHAEOBOTANICAL REMAINS FROM THE SOAP FACTORY (VI.I.XIV) FROM THE 1ST CENTURY AD

Triclinium

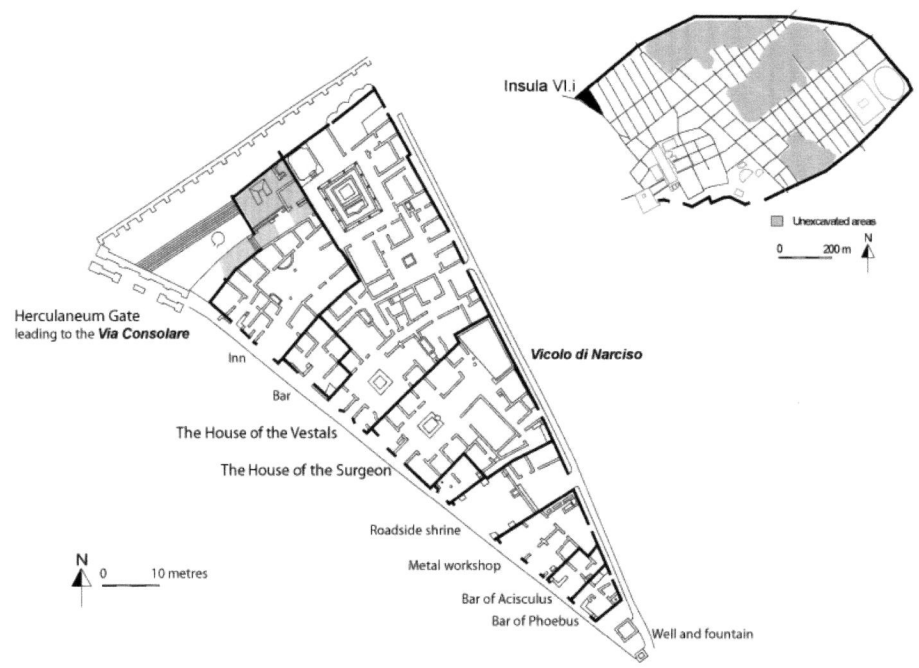

FIGURE 100: EXCAVATION AREAS FROM THE TRICLINIUM LOCATED WITHIN THE INN (AAPP05 RESOURCE HANDBOOK P. 27)

Triclinium 1st century AD

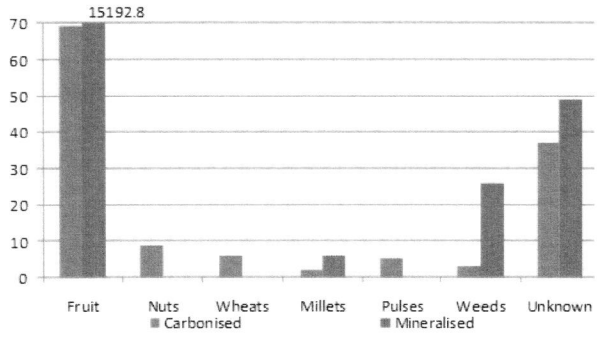

FIGURE 101: ARCHAEOBOTANICAL REMAINS RECOVERED FROM THE TRICLINIUM BY PRESERVATION AND GENERAL CATEGORY FROM INSULA VI.I

None of the Triclinium area contexts examined here can be dated to earlier than 1st century AD. Mineralised fig achenes dominate this assemblage totalling 15095 achenes out of a total of 15273 mineralised remains [n=98%]. A large proportion of the fig achenes were from the Ramp area.

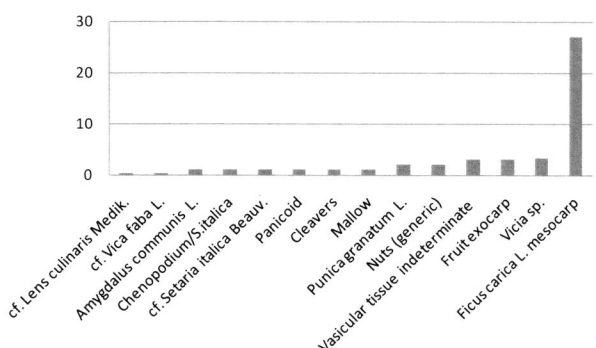

FIGURE 102: CARBONISED ARCHAEOBOTANICAL REMAINS FROM THE TRICLINIUM FROM INSULA VI.I (ARCHAEOBOTANICAL REMAINS CARBONISED UNLESS STATED OTHERWISE)

Via Consolare

FIGURE 103: EXCAVATION AREAS FROM VIA CONSOLARE (AAPP05 RESOURCE HANDBOOK P.27)

Via Consolare 1st century BC

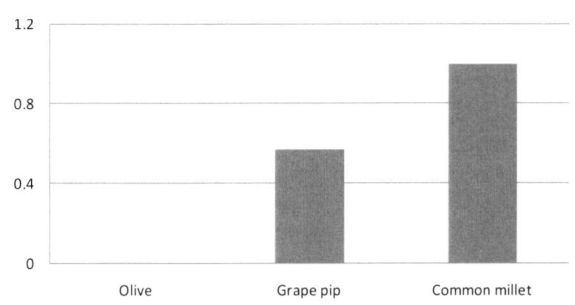

FIGURE 104: ARCHAEOBOTANICAL REMAINS FROM 1ST CENTURY BC FROM VIA CONSOLARE (REMAINS CARBONISED UNLESS STATED OTHERWISE)

From the 1st century BC fragments of charred grape pips, olive endocarp and one carbonised common millet was recovered.

Via Consolare 1st century AD

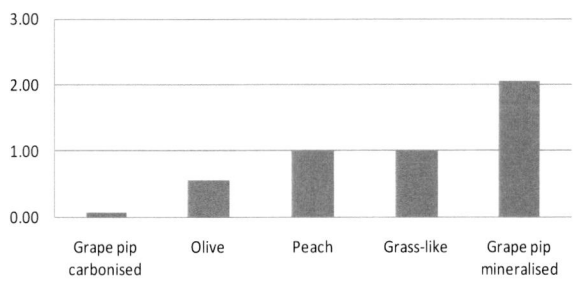

FIGURE 105: ARCHAEOBOTANICAL REMAINS FROM 1ST CENTURY AD FROM VIA CONSOLARE (REMAINS CARBONISED UNLESS STATED OTHERWISE)

The lowest number [n=11.26] of archaeobotanical remains from Insula VI.I came from the Via Consolare. From the 1st century AD a few charred olive endocarp fragments, carbonised and mineralised grape pips and one carbonised weedy panicoid seed were found. A few modern archaeobotanical remains were recovered including one cherry stone (*Cerasus avium*), two apricot pits (*Prunus armenicara*) and two fragments of peach pits (*Prunus persica*). Although all these taxa are known to have been present in Pompeii in ancient times, based upon previous archaeobotanical analysis (Meyer 1980, 1988), none were carbonised or mineralised and were from very late or questionably modern contexts. Therefore, they were deemed modern and excluded from this study. It is likely these items were eaten and discarded in recent times by careless tourists or custodi, illustrating issues of contamination and littering present at the site.

Vicolo di Narciso

FIGURE 106: EXCAVATION AREAS FROM VICOLO DI NARCISO (AAPP05 RESOURCE HANDBOOK P. 27)

Vicolo di Narciso 1st century BC

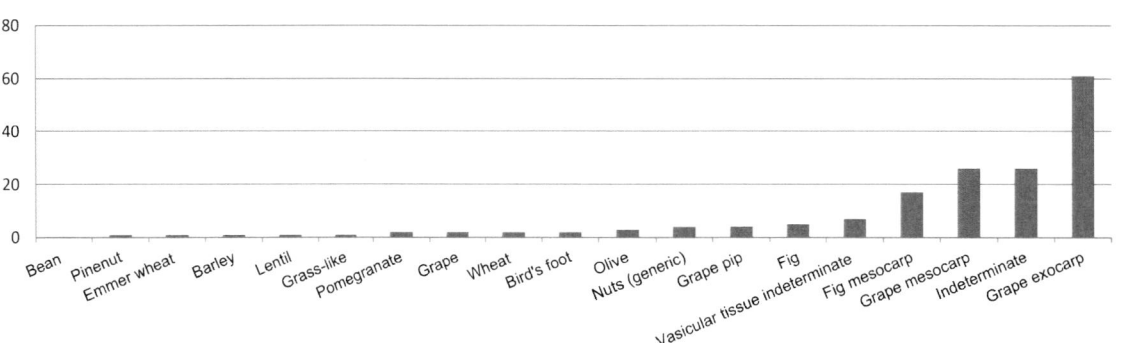

FIGURE 107: 1ST CENTURY BC ARCHAEOBOTANICAL REMAINS FROM VICOLO DI NARCISO (REMAINS CARBONISED UNLESS STATED OTHERWISE)

Vicolo di Narciso 1st century AD

Fragments of carbonised grape exocarp dominated this assemblage.

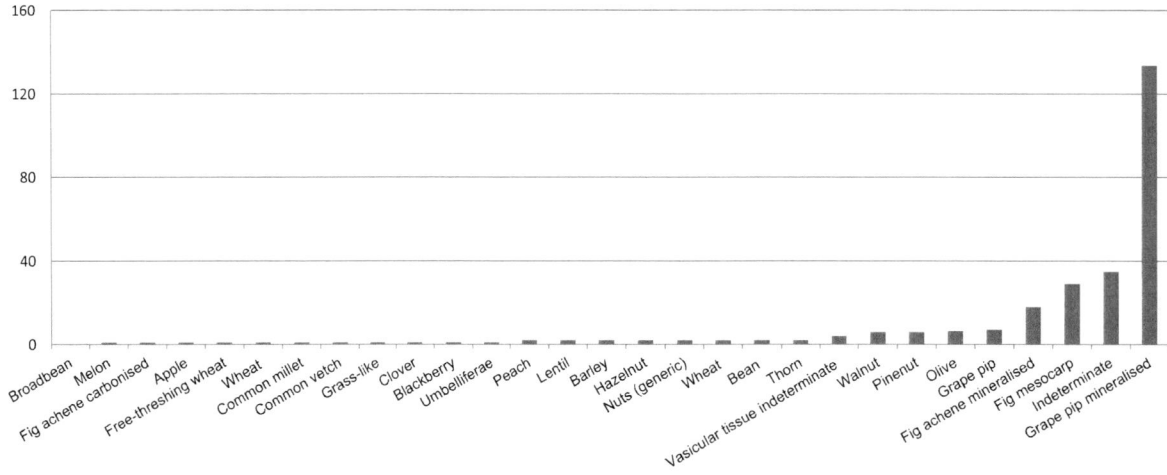

FIGURE 108: 1ST CENTURY AD COUNT OF ARCHAEOBOTANICAL REMAINS FROM VICOLO DI NARCISO (REMAINS CARBONISED UNLESS STATED OTHERWISE)

Comparison of archaeobotanical assemblage from roads

The Vicolo di Narciso and the Via Consolare plant assemblages contrast notably with the latter having a paucity of archaeobotanical remains [n=11.26] and the former being fairly rich [n=680.31]. The Vicolo di Narciso was particularly rich in mineralised plant remains [n=159.57], most likely due to the drain emptying onto this back street from the House of the Vestals. It was probably more socially acceptable to allow drainage onto the backstreet rather than the front street with the grand entrance of the House of the Vestals and the House of the Surgeon. Today this backstreet is roped off to tourists because the walls remain unsecured. Via Consolare (as mentioned previously) was one of the first streets of Pompeii to be cleared of volcanic debris during the 18th century. During former times, when the city was inhabited, and in modern times, since the rediscovery of the city 200 years ago this road served as a busy thorough-fare used to exit the city leading towards Herculaneum. Today it is used by tourists to visit the famed *Villa dei Misteri* (Villa of the Mysteries). Predictably, the majority of the archaeobotanical remains have been lost due to this early clearance and trampling.

Well

FIGURE 109: EXCAVATION AREAS FROM THE WELL (AAPP05 RESOURCE HANDBOOK P. 27)

Well 1st century BC

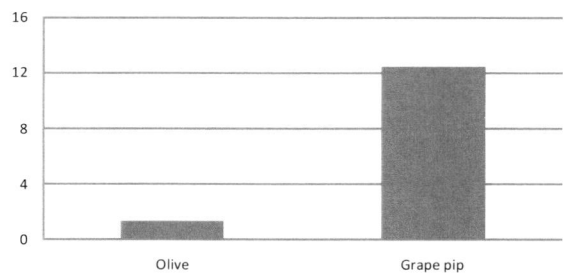

FIGURE 110: 1ST CENTURY BC ARCHAEOBOTANICAL REMAINS FROM THE WELL FROM INSULA VI.I (REMAINS CARBONISED UNLESS STATED OTHERWISE)

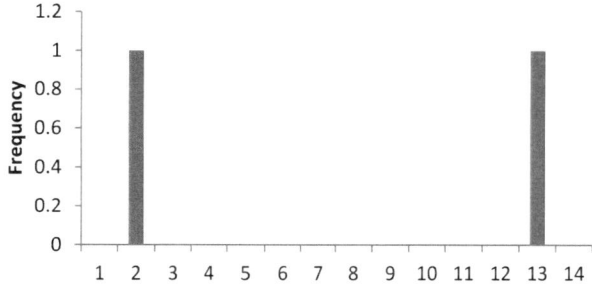

FIGURE 111: HISTOGRAM OF COUNT OF ARCHAEOBOTANICAL REMAINS FROM THE WELL FROM THE 1ST CENTURY BC

From the 1st century BC a few carbonised fragments of olives and grape pips were recovered.

Well 1st century AD

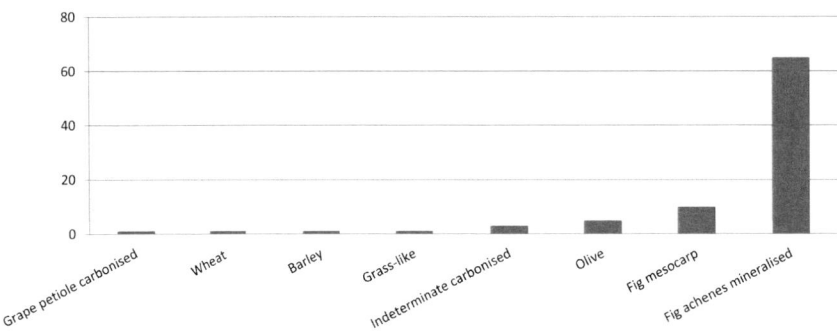

FIGURE 112: 1ST CENTURY AD ARCHAEOBOTANICAL REMAINS FROM THE WELL FROM INSULA VI.I
(REMAINS CARBONISED UNLESS STATED OTHERWISE)

Mineralised fig achenes dominate the 1st century AD assemblage from the Well. A few fragments of carbonised fig mesocarp, one grain of wheat and barley and one weedy panicoid seed were also present. It may have been the presence of water at the Well which facilitated the mineralisation of the botanical remains. No archaeobotanical remains were recovered from the North side of the Well from the 1st century BC.

4.10 Unique deposits from insula VI.I

All archaeobotanical remains from the latrine and ritual deposits from the House of the Vestals were analysed exclusively by Ciaraldi (2001).

Ritual deposit

The ritual deposit in Room 11 from the House of the Vestals (VI.I.vii) was interpreted as a foundation offering. The archaeobotanical assemblage from this deposit was analysed by Ciaraldi (2001). It was found to be similar to other domestic ritual deposits recovered from various locations throughout Pompeii. Ciaraldi (2001, 171) argued that in comparison with the ritual deposit from the House of Hercules' Wedding (VII.IX.xlvii), the House of the Vestals (VI.I.vii) ritual assemblage revealed a greater diversity in terms of fruits and spices. Ciaraldi and Richardson (2000, 81) suggest that this may indicate a greater availability of food resources and wealth of the inhabitants to purchase such perishable status items.

Latrine

At the rear of the House of the Vestals (VI.I.vii) a pair of rooms flank the far end of the front entrance passageway. The larger of these two rooms was interpreted to have been used as a service area (Area 50/Room 19) as it had a low cooking surface. It appears to have been in use until the early 1st century BC (Ciaraldi 2001, 150; Jones and Robinson 2007, 402). The archaeobotanical assemblage recovered was analysed by Ciaraldi (2001) and was probably from a latrine deposit that was contemporary with, and adjacent to the kitchen. From this semicircular feature, 70 plant taxa were identified by Ciaraldi (2001, 150). Although Ciaraldi (2001, 165) observed no change in the stratigraphy of the latrine deposit she claims that the recovered plant assemblages from the six samples showed a clear gradation in seed density towards the lower levels, possibly the result of gravitational forces.

Toilets in the kitchen area were a common feature of Pompeian households (Hobson 2009; Jansen 2001, 38). They were strategically placed in or near kitchens allowing for the disposal of burnt kitchen waste. The presence of carbonised material including nuts, weeds and cereals testifies to this fact. Columella mentions the usefulness of ashes as a 'disinfectant', 'cleaning agent' and 'odor suppressant'. Deposition of ashes from the cooking area into the latrine may have contributed to the mineralisation process as potash, which contains potassium or potassium carbonates in water soluable form, which along with the high percentage of organic matter in the latrine would have facilitated mineralisation (Ciaraldi 2001, 67; Green 1979, 279). This could have contributed to the number of mineralised taxa recovered from this deposit. The mineralised latrine assemblage from the House of the Vestals (VI.I.vii) showed similar characteristics with the mineralised deposits from the House of Hercules' Wedding (VII.IX.xlvii) in which cereals were virtually absent and fruits and weeds were abundant. This is typical of the preservation found among mineralised contexts (Ciaraldi 2001, 149-152).

A variety of mineralised weeds were also recovered by Ciaraldi (2001, 151) from her analysis of the latrine deposit. Cereal crop weeds such as the large and poisonous arable flax weed, *Agrostemma githago* L. (corn cockle) were recovered from the latrine deposit (Figure 113). The seeds of corn cockle are normally handpicked from a cereal crop before processing and consumption. They may also have had medicinal purposes in antiquity as Dioscorides claims they were employed as a purgative and treatment for scorpions bites (Mat.Med. III. 115 cited in Ciaraldi 2001, 160).

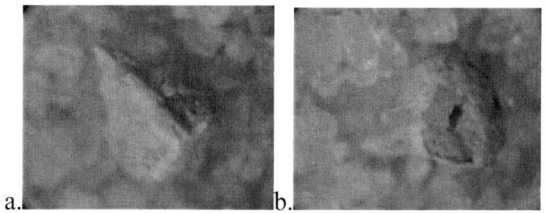

FIGURE 113: MINERALISED *AGROSTEMMA GITHAGO* L. (CORN COCKLE) FROM INSULA VI.I (NOT FROM THE LATRINE DEPOSIT)

Among the arable weed seeds recovered and identified by Ciaraldi (2001, 159) were Pimpernel (*Anagallis* sp.), both scarlet pimpernel and blue pimpernel, in both charred and un-charred forms. These were probably brought in on clothes or shoes from the gardens or else brought into the house with the cereals. In contrast, the recovered ruderal, Queen Anne's lace or wild carrot (*Daucus carota* L.), commonly grows in orchards, vineyards and fodder crops and rarely with cereal crops (Hanf 1983, Pignatti 1982). It may have been an accidental inclusion or otherwise brought in intentionally for medicinal purpose. Riddle et al. (1994) cites its use as a contraceptive in ancient times. Seeds from the wayfaring tree (*Viburnum lantana* L.) and buckthorn (*Rhamnus catharticus* L.) were also recovered from the latrine deposit. Both species may have been used both as ornamentals and for their medicinal properties (Ciaraldi 2001, 159). Specifically, the bark of buckthorn was used as a purgative and may have been used to extract a yellow colour for dying fabrics. The presence of buckthorn seeds may otherwise have resulted from the disposal of the entrails of birds or other animals after they were cleaned for subsequent cooking and consumption (Ciaraldi 2001, 156).

A range of spices and herbs were also recovered and identified from the latrine deposit by Ciaraldi (2001, 160) such as the seeds and leaves of Origanum sp. and the seeds of the families *Umbellifeae* and *Labiatae*. As some of the herbs recovered were fragmentary, Ciaraldi (2001, 150) proposed that they may have been cut up and used for flavouring foods.

Ciaraldi (2001, 153) found that the most common mineralised fruit was fig achenes. These achenes are often over-represented due to the large number of seeds produced by each fruit (Ciaraldi 1990; Cool 2006, 119; Dennell 1976; Green 1979; Robinson et al. 2006). Ciaraldi (2001, 153) found fig achenes and grape pips in nearly every deposit from the House of the Vestals (VI.I.vii), which she argued constituted an archaeobotanical 'background noise' to each context examined. Apples, pears or quinces (Malus/Pyrus/Cydonia), sorb (*Sorbus domestica*) and Prunus sp. were also commonly recovered mineralised fruits.

Drain

Other sections of the drain were excavated in other areas of the House of the Vestals (VI.I.vii). However, no other areas of the drain yielded mineralised or charred archaeobotanical remains. Therefore, it is likely that no concentrated organic matter, such as fecal material, was moving through these parts of the drain. Sea urchin fragments, some bones and pottery sherds were recovered in fairly large numbers from other parts of the drain (Ciaraldi 2001, 170). Similar to the archaeobotanical assemblage from the toilet, Ciaraldi (2001, 173) found that there was a lack of cereals except for one mineralised grain of millet. Large concentrations of fruits, weeds and spices were found. Ciaraldi (2001, 173) found a number of carbonised indeterminate remains were recovered from Vicolo di Narciso, possibly representing kitchen waste.

Ciaraldi (2001, 148) found a sizable number of mineralised archaeobotanical remains [n=159.57] were recovered from Vicolo di Narciso. There was a notable concentration of mineralised lentils from the drain deposit analysed by Ciaraldi (2001, 165-166) from Area 53/Room 20. Archaeologically, pulses are almost exclusively found in a carbonised state (van der Veen 2008, 26). Lentils, as they are often soaked before cooking, are less likely to become charred during cooking. Finding mineralised lentils archaeologically is rare (Figure 115). However, they have been found in excavations of Roman sites in York and London (Hall and Kenward 1990, 289; Giorgi pers comm cited in Robinson et al. 2006, 218). Ciaraldi (2001, 168) noted that the majority of the lentils were recovered from a single sub-sample, which she suggests represent a single disposal event i.e. a few stray lentils were thrown into or fell down the drain.

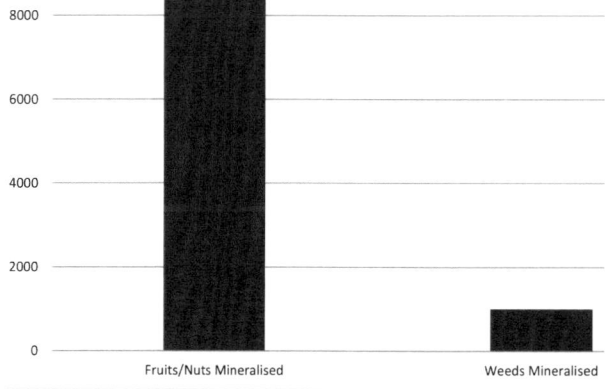

FIGURE 114: TOTAL COUNT OF MINERALISED BOTANICAL REMAINS RECOVERED FROM THE LATE 1ST CENTURY BC TOILET FROM THE HOUSE OF THE VESTALS (VI.I.VII) (DATA FROM CIARALDI 2001)

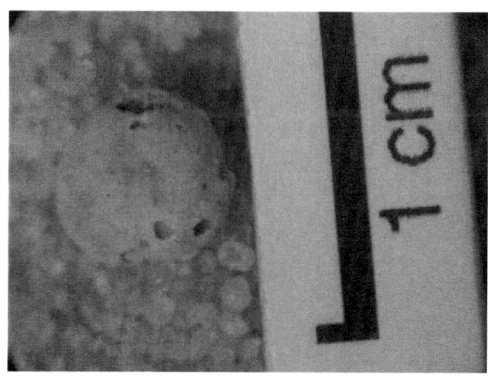

FIGURE 115: MINERALISED LENTIL RECOVERED FROM INSULA VI.I

As with the toilet deposit from Area 50/Room 19, the majority of the charred plant material was most likely from kitchen waste or thrown down the drain with the disposal of cooking ashes. Comparison of the assemblages from the toilet and drain deposits show that they were similar in terms of the range of preserved food plants present (Ciaraldi 2001, 166).

Based upon the results from the drain deposit Ciaraldi (2001, 173) argued that charred crop waste may have been represented by charred seeds from buttercups (*Ranunculus* sp.), the group of *Trifolium/Medicago/Melilotus*, pale flax (*Linum bienne* L.) and canary grass (*Phalaris* sp.). The main drain deposit, like the toilet deposit, contained well-preserved spices including the bay tree (*Laurus nobilis* L.), oregano (*Origanum vulgare* L.), *Mentha* cf. *arvensis*, from *Satureja/Thymus/Mentha*, coriander (*Coriandrum sativum* L.) and fennel (*Foeniculum vulgare* L.). Spices are rarely recovered from archaeological contexts and in this instance their unique preservation can largely be attributed to the rare mineralisation conditions present. The bay tree is native to the Mediterranean basin and is considered to possess medicinal properties. It is common throughout the Vesuvian region where it has been cultivated since Classical times (Jashemski et al. 2002b). Oregano was identified based upon the cell pattern of recovered leaf fragments (Ciaraldi 2001, 169). The diversity of spices recovered also reinforces the theory of an exotic and high-status diet for the inhabitants of the House of the Vestals (VI.I.vii) (Ciaraldi 2001, 169; Ciaraldi and Richardson 2000, 81).

The cesspit in the service area/kitchen of the House of the Vestals (VI.I.vii) emptied onto Vicolo di Narciso, the back street servicing Insula VI.I (Ciaraldi 2001, 69). A large concentration of mineralised plant material was recovered from this deposit. The pattern was similar to those of other mineralised contexts containing portions of mineralised fruit seeds and a few weed species. Plant remains from a 15th century medieval barrel-latrine from Worcester (U.K), although dating from more recent times, comprise a comparable assemblage to the Pompeian latrine in the House of the Vestals (VI.I.vii). The contents of the barrel-latrine, examined by Greig (1981) have a similar archaeobotanical signature to the House of the Vestals cesspit analysed by Ciaraldi (2001, 69), with the preservation of twenty identified edible plant taxa, consisting mainly of fruits, including a few potential exotics (to Britain) such as fig and grape. Also present were a variety of weeds from arable land which suggests that the contents of this cesspit likely included some straw with associated weeds, although few straw macrofossils were discovered (Greig 1981, 273-274).

The majority [n=96%] of the mineralised assemblage came from the 1st century AD with very little mineralised material dating to the 1st century BC. This suggests that the latrine was still in use during the 1st century AD or else that deposits in the drain continued to be flushed out onto Vicolo di Narciso via the drain subsequent to its abandonment.

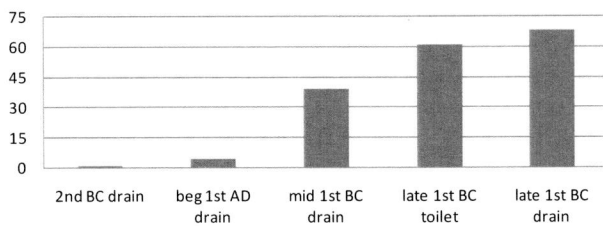

FIGURE 116: TOTAL NUMBER OF MINERALISED AND CARBONISED TAXA FROM THE HOUSE OF THE VESTALS (VI.I.VII) (CIARALDI 2001 DATA)

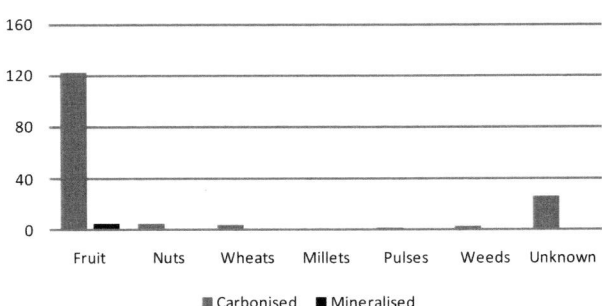

FIGURE 117: ARCHAEOBOTANICAL MATERIAL RECOVERED FROM VICOLO DI NARCISO FROM THE 1ST CENTURY BC

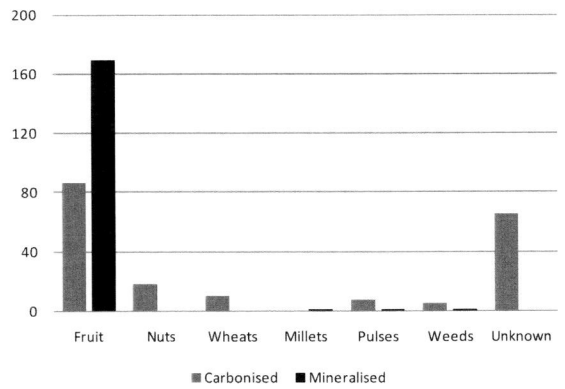

FIGURE 118: ARCHAEOBOTANICAL MATERIAL BY CATEGORY RECOVERED FROM VICOLO DI NARCISO FROM THE 1ST CENTURY AD

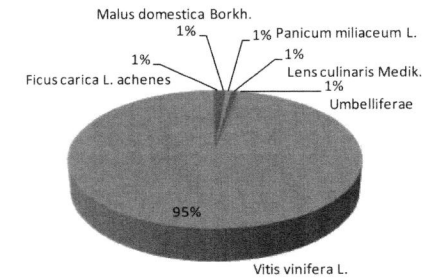

FIGURE 119: MINERALISED REMAINS FROM THE 1ST CENTURY AD FROM VICOLO DI NARCISO

The majority (95%) of the mineralised remains were composed of grape pips (*Vitis vinifera*). The few other mineralised remains from this property included one weed seed, one grain of millet, a fig achene and an apple seed. The carbonised remains from the 1st century AD were dominated by fruit and nuts. Quite a few carbonised indeterminates/unknowns were present. This could be due to their location and associated preservational issues. One interesting deposit of note was AA316, SU 236, from the rear of the Soap Factory. 130.57 mineralised grape pips were recovered. Twenty percent of all the archaeobotanical remains from this time period came from this context within Vicolo di Narciso.

It is believed that the kitchen was in use until the early 1st century BC (Jones and Robinson 2004). This could explain the large number of carbonised archaeobotanical remains recovered from the Vicolo di Narciso during the 1st century BC and the paucity of archaeobotanical remains from the 1st century AD (Figure 119). The bulk of the charred archaeobotanical material consisted of fruit including fig mesocarp, olive, grape (pips, fragments of grape exocarp and mesocarp) and indeterminate vasicular tissue. Weeds from this back street included one weedy panicoid seed and two *Ornithopus perpusillus/sativus* seeds (Figure 120).

Recovered carbonised cereals included one of the few *T. dicoccum* (emmer wheat) spikelet forks from the entire archaeobotanical assemblage from Insula VI.I, dating to the 2nd to 1st century BC. The only mineralised archaeobotanical material recovered were the ubiquitous fig achenes.

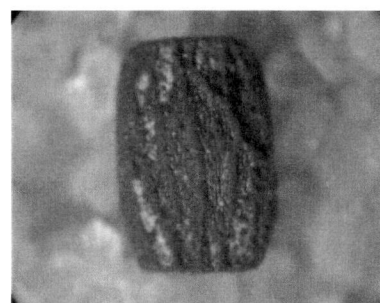

FIGURE 120: CARBONISED *ORNITHOPUS PERPUSILLUS/SATIVUS* SEED FROM INSULA VI.I

4.11 Unique deposits from the House of the Vestals (VI.I.vii)

Mineralised archaeobotaanical remains

The highest number of mineralised taxa came from the late 2nd century BC and the late 1st century BC toilet feature analysed by Ciaraldi (2001). It is likely that this feature

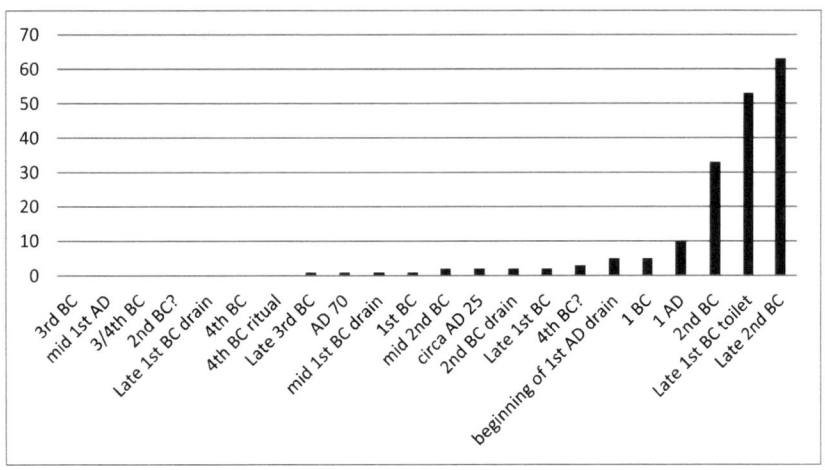

FIGURE 121: COUNT OF MINERALISED TAXA FROM THE HOUSE OF THE VESTALS (VI.I.VII) (ADDITIONAL DATA FROM CIARALDI (2001) PHD THESIS)

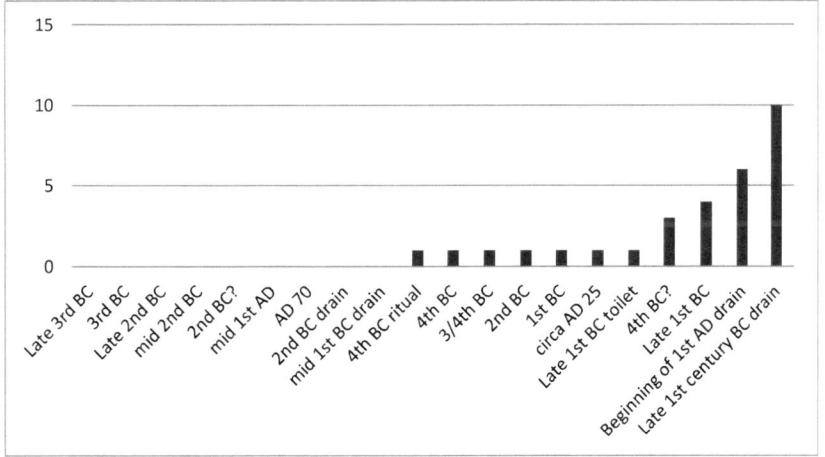

FIGURE 122: COUNT OF MINERALISED REMAINS OF FRUITS AND NUTS FROM THE HOUSE OF THE VESTALS (VI.I.VII) (ADDITIONAL DATA FROM CIARALDI (2001) PHD THESIS)

was in use during 2nd century and continued into the late 1st century BC.

There was a substantial jump in the number of mineralised remains recovered from the late 1st century BC drain [n=9630] analysed by Ciaraldi (2001, 170) compared with other contexts examined [<6]. This may be an artefact of specific preservational conditions that promoted the mineralisation of seeds and increased the number of specimens likely to be recovered. The specific conditions that promote mineralisation, which occurred in a few of the contexts and deposits examined here, undoubtedly substantially increased the number of preserved taxa present. The House of the Vestals (VI.I.vii) clearly exemplifies the effect of cess on promoting differential preservation of distinct taxa via mineralisation of certain types of fruit seeds, particularly fig achenes and a few weedy species. Ciaraldi (2001, 170) found that the drain and toilet feature from the House of the Vestals (VI.I.vii) had a substantially higher number of taxa than the rest of other contexts examined. From secondary fill deposits from the rest of the House of the Vestals (VI.I.vii) 21 taxa were identified from the 1st century AD and 17 from the 1st century BC or earlier were recovered.

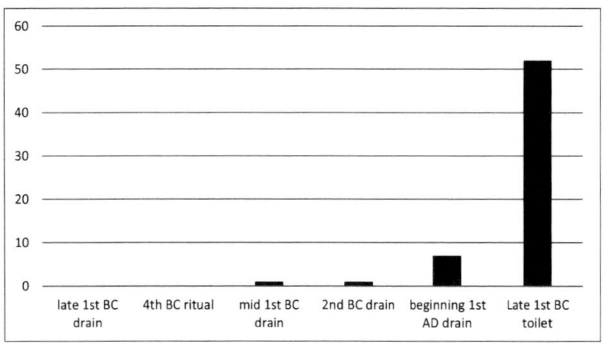

FIGURE 123: MINERALISED TAXA FROM SPECIFIC CONTEXTS WITHIN THE HOUSE OF THE VESTALS (VI.I.VII) (ADDITIONAL DATA FROM CIARALDI (2001) PHD THESIS)

Without sewage the occurrence of mineralisation is extremely low and carbonisation is the dominant mode of preservation (Figure 123). The ratio of carbonised to mineralised archaeobotanical remains for many of the properties illustrates the fact that cess was confined to specific areas or deposits within Insula VI.I.

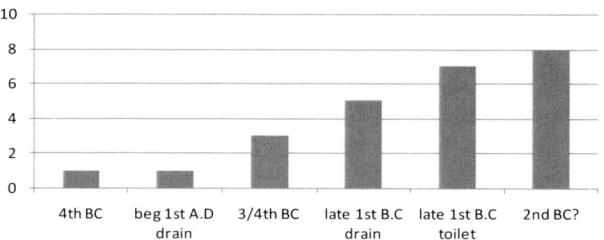

FIGURE 124: MINERALISED WEED SEEDS FROM SPECIFIC DEPOSITS FROM THE HOUSE OF THE VESTALS (VI.I.VII) (ADDITIONAL DATA FROM CIARALDI (2001) PHD THESIS)

Carbonised taxa from the House of the Vestals (VI.I.vii)

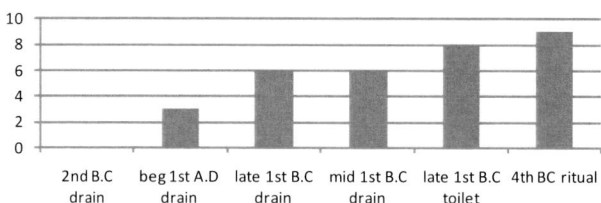

FIGURE 126: CARBONISED TAXA FROM SPECIFIC DEPOSITS WITHIN THE HOUSE OF THE VESTALS (VI.I.VII) (ADDITIONAL DATA FROM CIARALDI (2001) PHD THESIS)

Figure 126 illustrates the fact that the same deposits from the House of the Vestals (VI.I.vii) were not heavily influenced in terms of carbonisation as they were with mineralisation. There is no spike in the number of carbonised taxa and there is an equal number of carbonised taxa from the 1st century BC and AD.

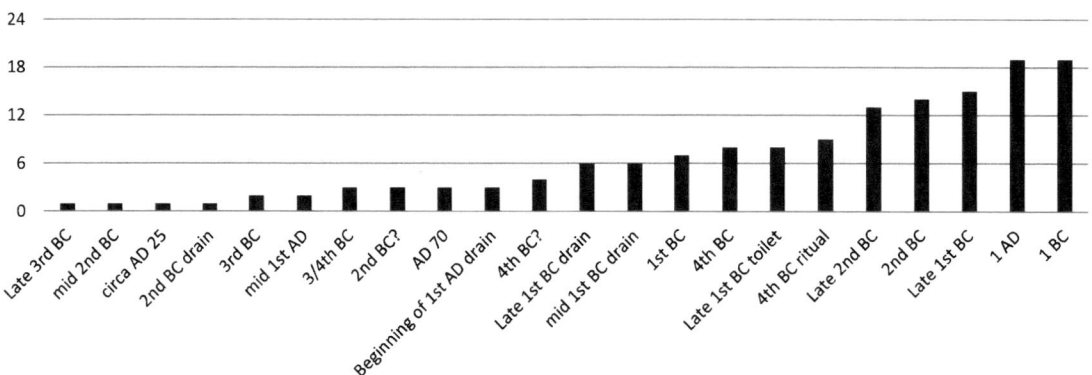

FIGURE 125: CARBONISED TAXA FROM THE HOUSE OF THE VESTALS (VI.I.VII) (ADDITIONAL DATA FROM CIARALDI (2001) PHD THESIS)

TABLE 4: PRESENCE AND ABSENCE TABLE FOR ROADS FROM INSULA VI.I

			Preservation	Via Consolare	Vicolo di Narciso
Cerasus avium	Cherry	half stone	Mineralised	x	-
Cucumis melo L.	Melon	seed	Carbonised	-	x
Ficus carica L.	Fig	achene	Mineralised	-	x
Malus domestica Borkh	Apple	Seed	Mineralised	-	x
Olea europea L.	Olive	endocarp	Carbonised	-	x
Punica granatum L.	Pomegranate	Seed	Carbonised	-	x
Prunus armenicara L.	Apricot	pit	Carbonised	x	-
Prunus persica (L.) Batsch.	Peach	whole	Carbonised	x	x
Vitis vinifera L.	Grape	fruit	Carbonised	-	x
Vitis vinifera L.	Vascular tissue indeterminate	fragment	Carbonised	-	x
Nuts					
Corylus avellana L.	Hazelnut	shell fragment	Carbonised	-	x
Juglans regia L.	Walnut	shell fragment	Carbonised	-	x
Pinus pinea L.	Pinenut	shell fragment	Carbonised	-	x
Nut shell	Nuts (generic)	shell fragment	Carbonised	-	x
Cereals					
T. cf. aestivum/T.dicoccum L.	Free-threshing wheat	caryopsis	Carbonised	-	x
T.dicoccum Schübl.	Emmer wheat	Spikelet fork	Carbonised	-	x
Triticum sp.	Wheat	caryopsis	Carbonised	-	x
cf. Hordeum vulgare L.	Barley	caryopsis	Carbonised	-	x
Panicum miliaceum L.	Common millet	whole	Carbonised	x	x
Panicum miliaceum L.	Common millet	whole	Mineralised	-	-
Pulses					
Lens culinaris Medik.	Lentil	Whole	Carbonised	-	x
cf. Lens culinaris Medik.	Lentil	Whole	Carbonised	-	-
Lens culinaris Medik.	Lentil	whole	Mineralised	-	-
cf. Vicia faba L.	Broadbean	whole	Carbonised	-	x
Vicia sativa L.	Common vetch	whole	Carbonised	-	x
Vicia sp.	Bean	whole	Carbonised	-	x
Grasses					
Panicoid	Grass-like		Carbonised	-	x
Weeds					
Ornithopus perpusillus/sativus	bird's foot	pod	Carbonised	-	x
Trifolium L.	Clover	whole	Carbonised	-	x
Rubus	Blackberry	whole	Carbonised	-	x
Umbelliferae		whole	Mineralised	-	x

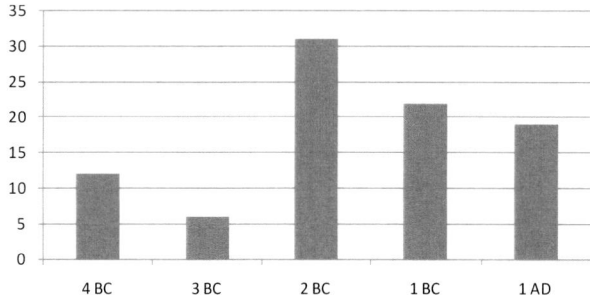

FIGURE 127: COUNT OF CARBONISED TAXA RECOVERED FROM THE HOUSE OF THE VESTALS (VI.I.VII) (ADDITIONAL DATA FROM CIARALDI 2001 PHD THESIS)

The number [n=22] of carbonised taxa from the House of the Vestals (VI.I.vii) suggests that the highest number of taxa [n=31] occurred during the 2nd century BC. The lowest number of carbonised taxa are from the 4th and 3rd centuries BC [n=6], although few deposits could be securely dated to this time period. There is a drop in number of taxa from the 2nd century BC to the 1st century BC [n=22] and again a slight drop in the 1st century AD [n=19]. The decrease in number of carbonised taxa from the 1st century AD [from n=31 2nd century BC to n=19 in 1st century AD] may be influenced by taphonomic factors, such as looting and tramping. It is also possibile that less trade activity was occurring after the earthquake of AD 62.

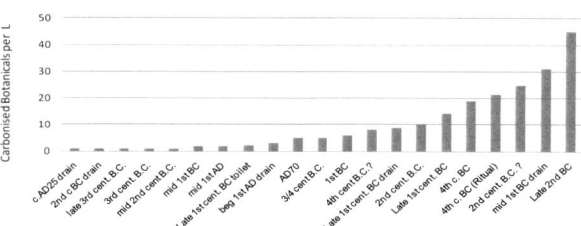

FIGURE 128: CARBONISED BOTANICAL REMAINS PER L FROM SPECIFIC CONTEXTS FROM THE HOUSE OF THE VESTALS (VI.I.VII) (ADDITIONAL DATA FROM CIARALDI (2001) PHD THESIS)

TABLE 5: Presence and Absence table for Domestic properties within Insula VI.I

	Fruit		Preservation	House of the Surgeon	House of the Vestals
Cerasus avium	Cherry	half	Mineralised	-	x
Cucumis melo L.	Melon	tip	Carbonised	x	-
Cucumis L.	Melon	seed	Carbonised	x	-
Ficus carica L.	Fig	achene	Mineralised	x	x
Malus domestica Borkh.	Apple	seed	Mineralised	-	-
Malus domestica Borkh.	Apple	seed	Carbonised	-	x
Olea europea L.	Olive	endocarp	Carbonised	x	x
Punica granatum L.	Pomegranate	seed	Carbonised	-	x
Prunus armenicara L.	Apricot	pit	Carbonised	-	-
Prunus persica (L.) Batsch.	Peach	whole	Carbonised	-	x
Prunus sp.			Carbonised	-	x
Vitis vinifera L.	Grape	fruit	Carbonised	x	x
Vitis vinifera L.	Vascular tissue indeterminate	fragment	Carbonised	x	x
Vitis vinifera L.	Fruit exocarp	fragment	Carbonised	x	x
Vitis vinifera L.	Fruit mesocarp	fragment	Carbonised	-	-
	Nuts				
Amygdalus communis L.	Almond	shell fragment	Carbonised	-	-
Corylus avellana L.	Hazelnut	shell fragment	Carbonised	-	x
Juglans regia L.	Walnut	shell fragment	Carbonised	x	x
Pinus pinea L.	Pinenut	shell fragment	Carbonised	x	x
Nut shell	Nuts (generic)	shell fragment	Carbonised	-	x
	Cereals				
T.aestivum/T.dicoccum L.	Free-threshing wheat	caryopsis	Carbonised	x	x
T . cf. aestivum/T.dicoccum L.	Free-threshing wheat	caryopsis	Carbonised	x	-
cf. *T.aestivum/dicoccum* L.	Free-threshing wheat	caryopsis	Carbonised	-	-
T.dicoccum Schübl.	Emmer wheat	caryopsis	Carbonised	x	-
T.dicoccum Schübl.	Emmer wheat	spikelet fork	Carbonised	-	-
T . cf. dicoccum Schübl.	Emmer wheat	caryopsis	Carbonised	x	-
cf. *T. dicoccum* Schübl.	Emmer wheat	caryopsis	Carbonised	-	-
T.durum	Bread wheat	spikelet fork	Carbonised	-	-
T . cf. monoccocum	Einkorn	caryopsis	Carbonised	x	-
Triticum sp.	Wheat	caryopsis	Carbonised	x	x
Hordeum vulgare L.	Barley	caryopsis	Carbonised	-	x
cf. *Hordeum vulgare* L.	Barley	caryopsis	Carbonised	-	-
Chenopodium/S. italica	Chenopodium/Italian millet	whole	Carbonised	-	-
Panicum miliaceum L.	Common millet	whole	Carbonised	x	x
Panicum miliaceum L.	Common millet	whole	Mineralised	-	-
Panicum cf. *miliaceum* L.	Common millet	whole	Carbonised	x	-
Panicum sp.	Panicum	whole	Carbonised	x	-
Setaria italica Beauv.	Italian millet	whole	Carbonised	x	-
Setaria cf. *italica* Beauv.	Italian millet	whole	Carbonised	-	-
cf. *Setaria italica*	Italian millet	whole	Carbonised	x	-
	Pulses				
Cicer arientium	Chickpea	whole	Carbonised	-	-
Lens culinaris Medik.	Lentil	whole	Carbonised	x	x
cf. *Lens culinaris* Medik.	Lentil	whole	Carbonised	x	-
cf. *Lens culinaris* Medik.	Lentil	fragment	Mineralised	-	-
Lens culinaris Medik.	Lentil	whole	Mineralised	-	-
Pisum sativum L.	Pea	cotyledon	Carbonised	-	x
cf. *Pisum sativum* L.	Pea	cotyledon	Carbonised	-	-
cf. *Vicia ervilia* (L.) Willd.	bitter vetch	whole	Carbonised	-	-
Vicia faba L.	Broadbean	whole	Carbonised	-	-
cf. *Vicia faba* L.	Broadbean	whole	Carbonised	-	x
Vicia sativa L.	Common vetch	whole	Carbonised	x	-
Vicia cf. *sativa* L.	Common vetch	whole	Carbonised	x	-
cf. *Vicia sativa* L.	Common vetch	whole	Carbonised	x	-
Vicia sp.	Bean	whole	Carbonised	x	x
	Grasses				
Panicoid	Grass-like		Carbonised	x	x
Pinus pinea L.	Umbrella pine cones				
	Weeds				
Apiaceae			Carbonised	x	-

TABLE 6: PRESENCE AND ABSENCE TABLE FOR COMMERCIAL PROPERTIES WITHIN INSULA VI.I

			Preservation	Vestals Bar	Soap Factory	Inn	Inn Bar	Bar of Acisculus	Bar of Pheobus
	Fruit								
Ficus carica L.	Fig		Mineralised	x	x	x	x	x	x
Olea europea L.	Olive	endocarp	Carbonised	x	x	x	x	x	x
Punica granatum L.	Pomegranate	seed	Carbonised			x			
Prunus persica (L.) Batsch.	Peach	whole	Carbonised			x			
Prunus sp.			Carbonised	x					
Vitis vinifera L.	Grape	fruit	Carbonised	x	x	x	x	x	
Vitis vinifera L.	Vasicular tissue indeterminate	fragment	Carbonised	x		x	x	x	
Vitis vinifera L.	Fruit exocarp	fragment	Carbonised	x				x	x
Vitis vinifera L.	**Nuts**								
Amygdalus communis L.	Almond	shell fragment	Carbonised						x
Corylus avellana L.	Hazelnut	shell fragment	Carbonised	x			x		
Juglans regia L.	Walnut	shell fragment	Carbonised	x			x		
Pinus pinea L.	Pinenut	shell fragment	Carbonised	x		x	x	x	x
Nut shell	Nuts (generic)	shell fragment	Carbonised	x	x	x	x	x	
	Cereals								
cf. *T. aestivum/durum* L.	free-threshing wheat	caryopsis	Carbonised			x			
T. dicoccum Schlübl.	Emmer wheat	spikelet fork	Carbonised	x					
Triticum sp.	Wheat	caryopsis	Carbonised	x	x	x	x	x	
Hordeum vulgare L.	Barley	caryopsis	Carbonised		x	x	x		
Chenopodium/S. italica	Italian millet/Chenopodium	whole	Carbonised		x	x	x	x	
Panicum miliaceum L.	common millet	whole	Mineralised					x	
Panicum sp.	Panicum	whole	Carbonised			x			
Setaria italica Beauv.	Italian millet	whole	Carbonised		x				
	Pulses								
Lens culinaris Medik.	Lentil	whole	Carbonised		x	x		x	
cf. *Lens culinaris* Medik.	Lentil	whole	Carbonised	x					
Pisum sativum L.	Pea	cotyledon	Carbonised			x			
cf. *Pisum sativum* L.	Pea		Carbonised	x					
cf. *Vicia ervilia* (L.) Willd.	bitter vetch	whole	Carbonised			x			
Vicia faba L.	Broadbean	whole	Carbonised					x	
Vicia sativa L.	common vetch	whole	Carbonised	x		x		x	
Vicia cf. *sativa* L.	common vetch	whole	Carbonised		x	x			
Vicia sp.	bean	whole	Carbonised	x	x	x	x	x	
	Grasses								
Panicoid	Grass-like		Carbonised	x	x	x		x	x
Pinus pinea L.	Umbrella pine	cone			x				
	Weeds								
Agrostemma githago	Corncockle	whole	Mineralised			x			
Celtis L.	Hackberry	whole	Mineralised			x			
Galium aparine L.	Goosegrass	whole	Carbonised			x		x	
Echium vulgare	Viper's Bugloss	whole	Mineralised	x			x		
Illex aquifolium L.	Holly	whole	Mineralised			x			
Lithospermum/Buglossoides		whole	Mineralised	x		x	x		
Ornithopus perpusillus/sativa	Bird's Foot	pod	Mineralised	x					
Ornithopus perpusillus/sativa	Bird's Foot	pod	Carbonised	x					
Plantago L.		whole	Carbonised			x			
Persicaria		whole	Carbonised			x			
Polygonium		whole	Carbonised			x			
Sherardia arvensis L.	Field madder	whole	Carbonised			x			
Sherardia arvensis L.	Field madder	whole	Mineralised			x			
Solanum nigrum L.		whole	Carbonised			x			
Arrhenatherum elatius (L.) P. Beauv.	Tall oatgrass	fragment	Carbonised			x			
Rubus	Blackberry	whole	Carbonised	x		x			
Rubus	Blackberry	whole	Mineralised			x			
Rumex		whole	Carbonised	x					
Cypress	Cypress	seed	Mineralised				x		

No unique deposits for the House of the Vestals (VI.I.vii) were analysed in this study. Therefore, the archaeobotanical remains recovered from this property came from secondary fill. If the House of the Vestals owners' controlled nearby commercial areas, such as the Bar of Acisculus, then it is possible that some of the fill may have come from construction events occurring in these other properties.

4.12 Summary

Aside from the Via Consolare every property within Insula VI.I had charred olive endocarp fragments and mineralised fig achenes present. The ubiquity of these botanical remains may be due in part to their physical properties: in the case of mineralised fig achenes to the large number of seeds per fruit and from olives the large number of endocarp fragments.

The archaeobotanical background 'noise' observed by Ciaraldi (2001, 2007) from the House of the Vestals was mainly composed of fragments of charred olive endocarp. Further examination of the other properties within Insula VI in the present study found that the archaeobotanical background 'noise' can be expanded to include fragments of mineralised and carbonised grape pips, mineralised and sometimes carbonised fig achenes and often carbonised indeterminate fruit vasicular tissue along with olive endocarp fragments. It is likely that contexts which have a combination of the archaeobotanical remains list above in low quantities represent an archaeological matrix

Table 7: Presence and Absence table for Ritual contexts from Insula

			Preservation	Shrine 1AD	Well 1AD
	Fruit				
Cucumis L.	Melon	seed	Carbonised	x	
Ficus carica L.	Fig	achene	Mineralised	x	x
Malus domestica Borkh.	Apple	seed	Mineralised	x	
Olea europea L.	Olive	endocarp	Carbonised	x	x
Punica granatum L.	Pomegranate	seed	Carbonised	x	
Prunus persica (L.) Batsch.	Peach	whole	Carbonised	x	
Prunus sp.			Carbonised		
Vitis vinifera L.	Grape	fruit	Carbonised	x	x
Vitis vinifera L.	Vasicular tissue indeterminate	fragment	Carbonised	x	
Vitis vinifera L.	Fruit exocarp	fragment	Carbonised	x	
Vitis vinifera L.	Fruit mesocarp	fragment	Carbonised	x	
	Nuts				
Corylus avellana L.	Hazelnut	shell fragment	Carbonised	x	
Juglans regia L.	Walnut	shell fragment	Carbonised	x	
Pinus pinea L.	Pinenut	shell fragment	Carbonised	x	
Nut shell	Nut (generic)	shell fragment	Carbonised	x	
	Cereals				
T.aestivum/T.durum L.	Free-threshing wheat	caryopsis	Carbonised	x	
T . cf. aestivum/durum L.	Free-threshing wheat	caryopsis	Carbonised	x	
cf. *T.aestivum/durum* L.	Free-threshing wheat	caryopsis	Carbonised	x	
T.dcoccum Schübl.	Emmer Wheat	caryopsis	Carbonised	x	
T. cf. *dicoccum* Schlübl.	Emmer Wheat	caryopsis	Carbonised	x	
cf. *T.dicoccum* Schübl.	Emmer Wheat	caryopsis	Carbonised	x	
T.durum	Bread wheat	spikelet fork	Carbonised	x	
Triticum sp.	Wheat	caryopsis	Carbonised	x	x
Hordeum vulgare L.	Barley	caryopsis	Carbonised	x	
cf. *Hordeum vulgare* L.	Barley	caryopsis	Carbonised		x
Chenopodium/S.italica	Italian millet/Chenopodium	whole	Carbonised	x	
Panicum miliaceum L.	common millet	whole	Carbonised	x	
Panicum miliaceum L.	common millet	whole	Mineralised	x	
Panicum cf. *miliaceum* L.	Panicum	whole	Carbonised	x	
Panicum sp.	Panicum	whole	Carbonised	x	
Setaria italica Beauv.	Italian millet	whole	Carbonised	x	
Setaria cf. *italica* Beauv.	Italian millet	whole	Carbonised	x	
	Pulses				
Lens culinaris Medik.	Lentil	whole	Carbonised	x	
Lens culinaris Medik.	Lentil	whole	Mineralised	x	
Pisum sativum L.	Pea	cotyledon	Carbonised	x	
Vicia faba L.	Broadbean	whole	Carbonised	x	
cf. *Vicia faba* L.	Broadbean	whole	Carbonised	x	
Vicia sativa L.	common vetch	whole	Carbonised	x	
Vicia sp.	bean	whole	Carbonised	x	
	Grasses				
Panicoid	Grass-like			x	x
	Weeds				
Apiaceae		whole	Carbonised	x	
Agrostemma githago L.	Corncockle	whole	Mineralised	x	
Galium aparine L.	Cleavers	whole	Carbonised	x	
Echium vulgare L.	Viper's Bugloss	whole	Carbonised	x	
Echium vulgare L.	Viper's Bugloss	whole	Mineralised	x	
Lithospermum/Buglossoides		whole	Mineralised	x	
Ornithopus perpusillus/sativa	Bird's foot	pod	Mineralised	x	
Ornithopus perpusillus/sativa	Bird's foot	pod	Carbonised	x	
Arrhenatherum elatius (L.) P. Beau	Tall oatgrass	fragment	Carbonised	x	
Rubus	Blackberry	whole	Carbonised	x	
cf. *Rubus*	Blackberry	whole	Carbonised	x	

Chapter 5

Comparison of Archaeological Evidence

To situate the archaeobotanical results from Insula VI.I it is necessary to examine the paleoenvironmental data from excavated sites within the city of Pompeii, its surrounding hinterland, the Vesuvian region, Italy and finally the wider Roman world.

5.1 Initial observations on plants at Pompeii

The first attempts to study foodstuffs at Pompeii began in the 18th century with a number of botanists, Joakim Frederik Schouw, Orazio Comes, and Casella studying various art historical sources and artefacts recovered from Pompeii and the Vesuvian region and producing lists of taxa based upon these often unrealistic and idealised, illustrations of plants (Ciaraldi 2001, 28; Francissen 1987, 111; Meyer 1980, 401). Some of these observations have been supported by archaeobotanical evidence and others have not. It was not until the 20th century that the idea that the plants depicted in these works of art were largely based upon metaphorical and allegorical principles rather than scientific botanical study became accepted.

5.2 Archaeological research on Pompeii

Carbonised plants recovered from the sites destroyed by Mount Vesuvius in AD 79 offer valuable and often unique archaeological data on the many staple food plants utilized by the inhabitants of Campania. Over the past two centuries of excavations at Pompeii only a handful of published works deal with either the identifications of preserved plant remains located in museum stores (Borgongino 1993, 2006; Meyer 1980, 1988, and 1994; Wittmack 1904) or the plant assemblages recovered from recent excavations (Ciaraldi 2001; Jashemski 1968, 1973, 1977, 1979, 1986, 1987, 2002; Robinson 1999, 2002).

Earliest evidence from outside of Pompeii

The early Bronze Age village, *Nola-Croce del Papa*, buried by the *Pomici di Avellino* eruption in 3550 BP by Mount Vesuvius was discovered in May 2001. Three huts were uncovered along with the enclosed area of a threshing floor and an animal pen made of wattle and daub (Livadie 2002, 941; Swedish newspaper The Local 2005; Veal 2009, 12). The oldest evidence of macrobotanicals and impressions of botanical remains in ash included carbonised cereals, einkorn (*T.monococcum*), emmer (*T. dicoccum*), barley (*Hordeum vulgare*), almonds (*Amygdalus communis*), olive stones, acorns (*Quercus* sp.). Carbonised wood was represented by beech (*Fagus*), black hornbeam (*Carpinus*) and fig (*Ficus*).

Shops

Charred plant materials recovered from the shops on the north side of the macellum at Pompeii were identified as fruits such as figs, plums, grapes. Lentils, chestnuts, grain, and prepared baked items such as loaves of bread and cakes were also found (Mau 1902 in Ciaraldi 2001, 44). Carbonised figs, chestnuts, plums, grapes, fruit in glass vessels, lentils, and grain were discovered from shops on *Strada degli Augustali* at Pompeii (Jashemski, Meyer and Ricciardi 2002, 476).

Bars

According to Ellis (2005) there are 158 bars within the city of Pompeii, broadly identified by their masonry counters facing the street with embedded *dolia*. It is likely that hot items such as stews and soups and dried foodstuffs e.g. grains, dried fruit and vegetables, were commonly stored in the shop *dolia*. Nuts are recorded as found in Shop IV.XVII (Borgongino 2006; Jashemski, Meyer and Ricciardi 2002, 120). At Herculaneum carbonised remains of nuts were retrieved from the *dolium* in the bar of the House of Neptune and Amphitrite. The lower classes, who lacked kitchens within their apartments (approximately 90 to 95% of the population of Pompeii according to Ellis (2004, 381)), would have obtained their cooked food from among the numerous *thermopolia* (food and drink) and *cauponae* (drink only) along the street.

Domestic houses within Pompeii

Early discoveries of botanical remains in Pompeii which were easily recognisable and present in large quantities include *dolia* filled with olives and pine cones from the marine warehouse of *M. Cellius Africanus* near the ancient port of Pompeii (reported by Fienga in Meyer 1988, 184) and Stone pine nuts from the *Casa dell'Atrio Corintio* (V.XXX) (Meyer 1988, 194).

The House of the Vestals (VI.I.vii)

In her analysis, Ciaraldi (2001, 140) found that plant remains, recovered from a variety of contexts from the House of the Vestals (VI.I.vii) were distributed unevenly through the broadly defined phases. There was a noticeable paucity of archaeobotanical material from defined earlier deposits dating to the 4th to first half of the 2nd century BC, aside from common millet (*Panicum miliaceum* L.) and an olive found in one area. According to Ciaraldi (2007, 149), the few early deposits, dating to no earlier than the 4th century BC, had plant assemblages that were fairly homogeneous in composition, which included cereals such as emmer, free-threshing wheat, barley, both common millet and Italian millet (*Setaria italica*), and fruits such as fig and grape (Ciaraldi 2007, 149). The presence of cereal chaff within these early deposits suggests that cereals were cleaned daily prior to cooking.

The settlement patterns at this time hint that the city was less densely populated than before its destruction in the 1st century AD (de Caro 2007, 76). Ciaraldi (2007) inferred that at this time agricultural related activities took place within the House of the Vestals (VI.I.vii) and further suggests that each household within the city could have been a self-sufficient unit. Ciaraldi (2001, 217) found that in contexts dating from the second half of 2nd century BC to AD 79, no cereal chaff was found. Similarly, at the House of Amarantus (I.IX.xii) there is an absence of cereal chaff in later deposits. Rubbish deposits were also absent from this domestic property, which supports the inference that crop processing was no longer occurring within this property in the later Roman period (Robinson 2002, 96).

From the 4th to the first half of the 2nd century BC Ciaraldi (2001, 213) argues that there appears to be an increase in pulses within the House of the Vestals (VI.I.vii). Ciaraldi (2001, 212) suggests that lentils and chickpeas were imported from North Africa but they were also cultivated throughout Italy. TThe archaeobotanical assemblage from contexts representing the second half of the 2nd century BC to AD 79 is richer and more varied than from earlier phases of this property, with common millet, Italian millet, *Vicia ervilia* (bitter vetch), *Pisum sativum* (pea), *Lens culinaris* (lentil), walnut, and more fruit species including *Punica granatum* (pomegranate), *Pyrus/Malus* (pear/apple), peach, *Malus domestica* (apple), and *Citrullus* sp. (watermelon). This pattern may be partly due to new taphonomic factors such as the mineralisation of plant remains as well as possible changes in the provisioning of the city, and changing agricultural practices. Ciaraldi (2001, 216) argues that the increase in fruit species may be partly due to their cultivation as cash crops on farms in close proximity to Pompeii. The increase in 'exotic' plants including, peach, *Sesamum indicum* (sesame), *Phoenix dactylifera* (date), walnuts, *Pistacia vera* (pistachio) and *Ziziphus jujuba* (jujube), suggests an increase in trade at Pompeii, particularly in items originating within the East.

Ciaraldi (2001, 153) lists Cucurbitaceae, including probable *Citrullus* sp. (watermelon) and *Cucumis* spp. (cucumber/melon), as exotic trade items. Janick et al. (2007, 1441) using ancient agrarian texts, images from art historical sources and references in Rabbinic law, otherwise argue that there is no evidence for *Cucumis sativus* (cucumber) in the Mediterranean region at that time. The references to *cucumis* in Columella and Pliny were probably *Cucumis melo* subsp. *melo* not cucumber. Janick et al. (2007, 1456) found evidence for *Cucumis melo* (melon), *Lagenaria siceraria* (bottle gourd), with *Citrullus lanatus* (watermelon) and other round-fruited forms of *Cucumis melo* (melons) being less common and grown as vegetables in the Mediterranean region. *Lagenaria siceraria* (bottle gourd) was probably also used as a vessel. A number of cucurbit species may have been grown strictly for medicinal purposes, e.g. *Bryonia alba* (white bryony), *B. dioica* (red bryony), *Citrullus colocynthis* (wild gourd or bitter apple) and *Ecballium elaterium* (squirting cucumber).

Expanding the excavations from the House of the Vestals (VI.I.vii) into the bar area (VI.I.v), faunal analysis revealed little difference in the domestic animal bone assemblages between these two properties, aside from the age of the sheep being consumed. In the bar, animals were butchered at an older age (mutton), compared to the House of the Vestals in which higher quality cuts of meat were being consumed (lamb) (Richardson 2006). Limited studies regarding waste have been undertaken at Pompeii, therefore it is difficult to ascertain if this result has wider applications to the rest of the city.

The House of Amarantus (I.IX.xii)

Both pre-Roman and Roman assemblages contained cereal grain, fruits and nuts which differ from the early and later deposits (Robinson 2002, 93). During the pre-Roman phase large grained cereals such as free-threshing *Triticum* sp. (durum or bread wheat), legumes, in particular chick pea, walnut and grape dominate the assemblage in the burnt deposits. Significantly, pre-Roman domestic assemblages contained cereal processing products such as chaff and weed seeds along with cereal grains (Robinson 1999, 101-102). In contrast, in the Roman domestic assemblage there was a notable absence of evidence of cereal crop processing, suggesting that the inhabitants were not involved in the final processing stage of cereals. A greater range of fruit and nuts were found in burnt deposits from the Roman phase, particularly pine nuts, and fig. *Corylus avellana* (hazel nut), walnut and large cereal grains such as emmer wheat were also present in a number of these burnt deposits. The appearance of dates during the Roman period may indicate an increased trade with North Africa due to the expansion of the Roman Empire (Robinson 2002, 96).

The earliest charred stones of olive were recovered from the very latest pre-Roman deposits. Stone pine was found only in late deposits, as well as one poppy seed cake (Robinson 2002, 98). Robinson (2002) explains the relatively late appearance of the olive and stone pine to the Roman influence on the town after the establishment of the Sullan colony. Interestingly, 'olive stones were some of the most frequently excavated charred remains in the general domestic contexts of the 1st cent. A.D. but they were poorly represented amongst the burnt deposits' (Robinson 2002, 96). In a similar manner Ciaraldi noted '…an intensification in the use of olives and their by-products' in the House of the Vestals (VI.I.vii) (Ciaraldi 2007, 198). This increased presence of olive may be due to changes in their uses and/or their processed by-products, e.g. using the olive lees for fuel and/or fertiliser. They would have been abundantly available and inexpensive after pressing for oil. Olive dregs were often mixed into mortar for the construction of granary floors and walls, as a kind of insecticide, well into the 1600s (Ciarallo 2000, 38).

The House of Hercules' Wedding (VII.IX.xlvii)

The stratigraphy from the House of Hercules' Wedding (VII.IX.xlvii) spans approximately seven centuries. The palaeobotanical remains were preserved through both mineralisation and carbonisation, and were recovered from a variety of domestic contexts including kitchen areas, cesspits and a ritual deposit. Homogeneous archaeobotanical assemblages were found in the earliest layer, which included cereals e.g. emmer, free-threshing wheat, possibly einkorn, barley, common and Italian millets, and fruit, including figs and grapes. The detection of wheat chaff during the earliest phase is a strong indication that daily cleaning of the cereals was occurring on a small scale before being consumed (Ciaraldi 2001, 173).

The presence of cereal chaff from the 4th to the 2nd century BC contexts of several domestic properties suggests that some (limited) processing of wheat took place in Pompeii. It is possible that little or no distinction was made between agricultural and urban activities. Ciaraldi (2001, 34) proposed that because agricultural activities were not limited to non-urban areas it is perhaps technically incorrect to call 3rd century BC Pompeii an urban centre. It is possible that the occupants of the House of Amaranthus (I.IX.xii), the House of the Vestals (VI.I.vii), and Hercules' Wedding (VII.IX.xlvii) bought prepared flour or bread from among the 33 bakeries found in the city (Jashemski 1979b, 14; Robinson 1999, 101-102).

Similar results are seen from the House of Amarantus (I.IX.xii), in which the pre-Roman assemblages contained evidence of cereal processing, which was absent in later Roman context (Robinson 2002, 96). Therefore, it appears that by the later Roman period crop-processing of cereals was done outside the city, which suggests that a major shift in processing, transportation and storage strategies had taken place. Large-scale agriculture was probably pushed outside the city gates as the process of urbanisation continued from the 2nd century BC to the 1st century AD. Ciaraldi (2007, 197) suggests that the use of slaves as cheap labour on large agricultural estates favoured the removal of crop-processing activities from the city and the presence of threshing floors within the majority of the *villae rusticae* that surrounded Pompeii as support for this theory. In a similar manner, the predominance of fish scales and vertebrae within Insula VI.I suggests that cleaning was occurring elsewhere and it is likely that the inhabitants were now were buying their fish already cleaned (Jones 2006). It is therefore reasonable to suggest that during this period Pompeii was becoming an increasingly urban market in which more citizens were being removed from the primary production of their own food.

5.3 Villae

Villae rusticae

The most well-known example of a Roman *villa rustica* is the Villa of the Mysteries (Villa dei Misteri). This site was initially excavated by a private citizen and then by Italian archaeologist A. Maiuri between 1929 -1930. It has been dated to the 2nd century BC. It is thought to have been a working farm located just outside the Herculaneum Gate along the Via Consolare (Nappo 1998, 148).

A debatable figure of around one hundred and forty villas have been discovered in the countryside around Pompeii, in what are now the modern districts of Boscoreale and Boscotrecase (Moorman 2007, 440). Relatively little is known about these villas because they were never properly recorded due to a lack of interest at the time of their discovery and their location on private lands. Modern field survey of the area surrounding Pompeii is impossible as the ancient remains are now deeply buried by successive eruptions of Mount Vesuvius and the overlying modern housing. However, an expanding number of agricultural settlements have recently come to light with the increasing number of archaeological excavations (Guzzo 2007, 5).

The majority of the villas around the Bay of Naples were working agricultural complexes rather than merely country retreats (Adams 2006, 69). From ruined Stabiae, the remains of 16 working farms were excavated from 1759-1782. John Day in 1932 argued that the majority of the villas (26 of the 39 known at the time) contained evidence of wine production. Three villas (villas 1, 2, and 6) 6 km north of Pompeii, adjacent to the road which ran from Pompeii to Nola, possessed wine presses and cellars. Based upon the number of villas and evidence of viticulture, it is likely that a surplus of wine was being produced for local markets and possibly also for export (Jongman 2007, 504).

A considerable distance from Pompeii, near Boscoreale, the Villa Regina was one of the first villas excavated and published to modern archaeological standards. Due to the modern housing surrounding the site it was only possible to investigate an area of 2000m^2. Using the techniques of creating casts of root-cavities, along with archaeobotanical and pollen analysis, Jashemski concluded that the majority of land surrounding this villa was dedicated to viticulture. Identification of tree roots based solely upon casts is questionable aside from a few very distinctive taxa and Jashemski's conclusions were mostly drawn from the regular gird layout of the root cavities. A range of fruit trees were also discovered among the grapevines including olive, almond, fig, apricot, and cherry or walnut. Based upon the pollen analysis, over 80 plant species were thought to have been grown on this estate (Ling 1996, 346).

Although animal husbandry within Pompeii has not been extensively studied, faunal and archaeological evidence indicate that it was practised. This data suggests that the pronounced division between the town and country are a modern construction that cannot be correctly applied to the ancient city. The fertility of the soil may have created a need to maintain some areas of the city for agricultural use instead of building upon it. It appears that the assigned pattern of productive land use in Pompeii was not based on function as there are no areas of the city specifically associated with craft workshops or productive gardens (Guzzo 2007, 4). Little separation between residential, small-scale cultivation, retail or production areas is evident within the city walls (Laurence 1994, 67).

5.4 Archaeobotanical evidence from outside Pompeii

Herculaneum

Herculaneum was another ancient city located in the Bay of Naples that was buried by Mount Vesuvius during the 79 AD eruption. Archaeobotanical remains have also been recovered from Herculaneum in a similar fashion to those at Pompeii, sporadically, mainly from large, identifiable deposits of food items, during poorly documented 18th and 19th century excavations. For example, Maiuri reported the remains of figs and legumes in *Casa d'Argo* at Insula II.II (Meyer 1988, 194). Like at Pompeii, a number of preserved archaeobotanical remains have been stored in the local deposito, *Decumanus Maximus*, at Herculaneum including 4kg of carbonised chickpeas (with *Vicia ervilia* as a weed contaminant), 1000 fig achenes, a few hundred well-preserved dates, carob, walnuts and almonds. A number of properties from this city yielded *Allium cepa* L., (onion) including the *deposito, Casa Graticcio* (III.13-15), *Casa dei Cervi* (IV.21), and *Casa del Bel Cortile* (V.8). 3.5 kg of oak galls were recovered from the main street in Herculaneum (Larew 1988, 145). From *Casa della Stoffa* (IV.19-20) 1kg of carbonised lentils were recovered and 0.5 kg of carbonised lentils from *Casa dei Due Atrii* on *Cardo* III. *Vicia faba* has been recovered from Shop II.13 and the *deposito*. The majority of cereal recorded from Herculaneum appears to be emmer. Specifically, emmer was recovered from Casa della Stoffa (IV.19-20) and V.iii. The *Casa del Bel Cortile* (V.8) produced a half bushel of well-preserved un-threshed spikelets and cereal caryopses along with oats and barley.

Oplontis

Oplontis was thought to have been a coastal suburb of Pompeii (Adams 2006, 57; Ward-Perkins 1984, 27-28). Ricciardi and Aprile (1978, 318) reported a large cache of well preserved carbonised plant remains believed to be from a 1st century AD ancient hay mow and interpreted as animal fodder. It was discovered during the 1961 excavations at the villa of *L. Crassus Tertius* at *Oplontis* or modern day Torre Annunziata (Meyer 1980, 402). In Ricciardi and Aprile (1979, 318) preliminary report, 111 taxonomic entities are listed, (mainly herbaceous dicotyledons and one true fern) representing 26 families and 69 genera, of which 88 are identified to species.

Ricciardi and Aprile (1988, 318) argue that the presence of woody species such as leaves, tendrils, and twig fragments of grape, leaves of olive, English oak (*Quercus robur*), and Prunus sp. suggest that the hay was cut from a neighbouring vineyard close to the villa (Meyer 1980, 402). Campanian vineyards often included various kinds of trees. These results suggest that hay in the past was floristically more diverse than today (Grieg 1981, 274). Assuming that the climate was similar to the present, Ricciardi and Aprile (1988, 318) noted similarities between the species found in the archaeological assemblage and vegetation found in modern Vesuvian hay meadows (Ciaraldi 2001, 137). This discovery was of such importance that 'sixty-seven species, thirty-one genera, and one family have been added to the list of 408 plants probably known in the first century BC' in the area (Jashemski 1987b, 34). This assemblage is botanically unique, representing the oldest documented record of some species and provides information on the hay meadows within the vicinity of Pompeii (Meyer 1980, 402).

5.5 Garden evidence

Garden evidence at Pompeii

Gardens and cultivated areas were an important part of life in ancient Campania, as testified by the approximately 626 gardens uncovered to date in Pompeii, Herculaneum and surrounding villas preserved by the eruption (Jashemski 2007, 487). A *hortus* could refer to a variety of known Roman garden types: from a small vegetable garden to large landscape gardens on palatial estates situated within the town or country. Originally, Roman gardens were probably perfunctory vegetable gardens, part of a 'kitchen-garden economy' rather than the elaborately decorated ones associated with the later Imperial period (Farrar 2000, xi). Gardens were important features attached to the Roman households and were used for a myriad of functions including enjoyment, pleasure, work, entertaining, and religious worship (Jashemski 2007, 487).

The Italians added peristyles to their atrium houses which transformed the Hellenistic style beaten clay, or paved with cobblestones, cement, or mosaic centre courts into a garden. The centre of the peristyle house was the garden, which provided light, fresh air circulation and communication with the rest of the house (Jashemski 1979b, 25; Farrar 2000, xii). Limited environmental data was recorded during the early excavations at Pompeii, nevertheless Spano, an archaeologist excavating in 1910, reported cavities in four peristyle gardens, although size, shape or location were not noted. Italian archaeologist Spinazzola was the first to modify Fiorelli's method of making plaster casts of human bodies for the purpose of making plaster casts of large tree-root cavities found in the back of the garden of the House of the Moralist (Regione III, insula II-III). The unique preservation conditions in which the lapilli and ash formed a protective layer over the organic plant material which subsequently decayed leaving behind the ancient root cavities (which sometimes in-filled with lapilli) allowing plaster casts to be made, and in some instances for the identification of certain casts (Jashemski 1970, 26). Unfortunately these casts have been damaged but those made by the Italian archaeologist Maiuri of tree-root cavities at Pompeii have survived (Cooley 2003, 99).

Garden archaeology is a relatively recent field in Classical archaeology. In the past there has been a tendency to concentrate on 'more tangible aspects' of ancient cities such as grand public buildings rather than on gardens. Until the pioneering work of Jashemski in the 1950s and 1960s, gardens were regarded as 'empty spaces' within the city. Jashemski's large scale environmental work provides a more complete picture of the spatial dynamics of the town and the placement of ornamental gardens, vegetable plots, market gardens, and vineyards within the city. Jashemski (1979, 24) calculated that buildings occupied 64.7%, gardens and cultivated land occupied 17.7%, and streets and other areas 17.6% of the total excavated areas within the city of Pompeii. Therefore, a large proportion of the excavated city comprised open green areas including gardens, parks, vineyards, and orchards. Jashemski (1979, 14) argues that this figure would be even larger if the entire city was exposed.

Jashemski further attempted to systematically identify the location and past vegetation in the gardens and green spaces of Pompeii, specifically agricultural plots in Regioni I and II to the south-east of the city (Cooley 2003, 99; Laurence 1994, 64-67). Her subsequent statistical reconstruction of these supposed 'open' spaces within the city and surrounding areas at Pompeii has altered previously held views on the economy of this city, strongly suggesting that it was a busy commercial centre, not the wealthy resort town as previously believed (Ciaraldi 2001, 44; Jashemski 1977, 217). Through limited subsoil excavations in 1966 and using the famous Pompeian plaster mould technique in every possible root cavity in Regione II, insula V, (located north of the amphitheatre in Pompeii) an area originally thought to been the *Foro Boario*, or Cattle Market, was identified as the largest vineyard in Pompeii (Ciarallo 2000, 57; Jashemski 1970, 26). Vineyards have also been found on the *Via di Sarno, Via de Nocera* and *Via di Castricio*, which all possessed winemaking equipment including presses, *dolia* and *amphorae*. Indeed, the majority of the discovered agricultural plots were in the south-eastern end of the city near the amphitheatre. This area, where Jashemski carried out most of her excavations, appears to have had a lower settlement density and therefore space for agricultural purposes. The insulae in this area are situated upon a south-facing slope, having greater sunlight, an advantage for agriculture, that other parts of the city did not possess (Laurence 1994, 67).

From the House of *Loreius Tiburtinus* (II.II.ii), excavated at the beginning of the 1930s, traces of pergolas were recovered. Plaster casts of the root cavities have been identified as ornamental plane trees, vines and fruit trees. Unfortunately this garden was damaged by bombs during WWII. Fruit orchards were identified from the garden of the Villa of *Julia Felix* (II.IV.ii) and the garden of the Fugitives (I.XXI.ii/vi) (Ciarallo 2000, 57).

Surprisingly, market gardening was also discovered in Regione VII, a densely populated part of the city, situated to the east of the Forum (VII.XI.i, VII.X.xiv and VII.XI.xi/xiv). The smallest garden found by Jashemski (1977, 227) was the shop-house garden. It was the first small scale garden of its kind to be excavated and studied and has yielded important insights into plebian Pompeii. It offers a poignant example of an 'informally, but intensively planted [garden] in trees, vines, and vegetables, [which] enabled the family to produce their own wine and fruit, with perhaps a surplus to sell' locally (Jashemski 1977, 227).

Ancient Pompeii was famous for its flower cultivation and production of garlands and perfume (Ciarallo 2000, 57; Cooley 2003, 109; Jashemski 2007, 496). The House of the Vetti (VI.XV.i/VI.XV.xxvii) contained the now famous frescoes of Cupids collecting, harvesting and transporting flowers for garlands and making perfume. However, it is likely that the majority of Pompeian gardens were planted with evergreens and flowers had only an incidental role (Jashemski 1979b, 54).

During the 1970s, influenced by Jashemski's environmental work, the *Giardini ripristinati* (restored gardens) was created at Pompeii (Ciarallo 2000; Francissen 1987, 118). The first peristyle garden, the old Samnite House of *C. Julius Polybius* (IX.XII.i-iii) on *via dell'Abbondanza*, was excavated in 1973 using newly developed scientific techniques (Jashemski 2007, 488). Ciarallo and Lippi (1993, 110) explain that the garden of the House of the Chaste Lovers (*Casa dei Casti Amanti*) (IX.XII.vi) was restored to its AD 79 appearance using the recovered environmental data; geometrically shaped beds, enclosed by trellises of common reed, *Phragmites australis* (Cav.) Trin. and giant reed *Arundo donax* L. were discovered; from the pollen record *Juniperus* and *Rosa, Artemisia* and other shrubs and herbs from *Caryophyllaceae, Asteraceae, Poaceae* and ferns appear to have grown along perimeter drains. Similar modern vegetation is found in the area today thus it is difficult to ascertain if the pollen truly dates to the time period under study.

Pollen evidence from gardens

Gardens present unique archaeological problems because features often vary considerably between gardens, and leave few visible traces of their presence. Meticulous methods of excavation are needed to produce adequate results (Farrar 2000, xvii). Ancient pollen and fern spores, recovered by Jashemski (2002) for the first time from a Roman site in the Mediterranean region, have brought to light a new species hitherto unidentified at Pompeii. Unlike in the previously mentioned pollen study by Veal (2009) at Pompeii which was unsuccessful Dimbleby, then of the Institute of Archaeology, University of London, analysed soil samples from the gardens of Hercules,

Polybius and other gardens buried beneath a protective layer of lapilli for pollen analysis. Comparing these pollen studies it would appear that the undisturbed lapilli layer is necessary for ancient pollen preservation. He was successful in identifying over 20 types of pollen including both woody and herbaceous plants. One sample contained large quantities of olive (*Olea europea*) pollen, 75%, which strongly suggests that olives were grown on the site (Dimbleby and Grüger 2002, 211), although Dimbleby states that due to the poor preservation conditions pollen data from Pompeii must be regarded with caution.

At the Villa of Poppaea at *Oplontis*, where the protective lapilli layer had been removed, little pollen was recovered. That which was recovered and identified by Dimbleby and Grüger (2002, 181) was poorly preserved due to the alkalinity of the soil (pH 7.8-8.2), in addition to microbial decay within the upper layer of soil caused by the warm Mediterranean climate. Irrigation, if present, would also have contributed to pollen decay. Another sample from a different part of the same orchard produced large amounts of fern spores including 48% *Polypodium* sp. (genus of true ferns) and 25% bracken, *Pteridium* sp.; *Polypodium* is rare in the Pompeii area today, while bracken remains common (Dimbleby and Grüger 2002, 182). Also identified from pollen were woody species including *Pinus* (Pine), *Juglans* (Walnut), and *Quercus* (Oak), and herbaceous plants, probably weeds, *Plantago lanceolata,* and plants identified to family only including *Chenopodiaceae*, *Compositae* (Liguliflorar), *Gramineae*, and *Ranunculaceae* (Jashemski, 1979; Meyer 1980, 403). Dimbleby and Grüger (2002, 182) observed that based on the pollen and spore data, floras do not differ substantially between the vineyard of Boscoreale, the gardens from *Oplontis* and the gardens of Hercules and *Polybius* from Pompeii.

Even with improved recovery methods, based on the comparative modern soil samples taken from similar areas, the majority of ancient pollen appears to have been lost and 'that necessarily makes any quantitative interpretation appear doubtful' (Dimbleby and Grüger 2002, 199). Recent examinations of sediments from several excavated *ollae perforatae* from the garden area of the House of the Surgeon (VI.I.x) showed no pollen present (Veal 2009, 96). From the garden of the *Casa delle Nozze di Ercole ed Ebe°d,* Pompeii Vitis and *Citrus* pollen were identified. A ground cover of herbs, similar to those found in the Vesuvian area today was also evident (Marta 2000, 205). The majority of the Campanian flora were wind-pollinated species, therefore it is difficult to reconstruct the past vegetation from pollen for such small defined areas as gardens, and often not possible to specify which plants were grown in individual garden plots (Dimbleby and Grüger 2002, 189, 211).

The plaster and excavation techniques made famous by Jashemski have been used on a variety of different gardens throughout Italy, including the *villa rustica* at Boscoreale, Villa Regina, in order to investigate the layout of the planting of the surrounding area. Although it was not possible to expose the whole farm-estate due to the modern housing surrounding the site today, an area of around 2000m^2 was excavated and palaeoenvironmental data collected, including pollen, plaster casts of tree roots and macrobotanical evidence. Pollen analysis revealed the presence of more than 80 species of plants in the vicinity of the villa. The analysis revealed that the majority of the land was used for viticulture with the vines, probably on a grid of poles (*Vitis compluviata*). Other trees were identified including olive, almond, fig, walnut, cherry, apricot, peach, willow and poplar. Pines and plane trees were possibly grown along public roads at the southern end of the site (Ling 1996, 346).

The accuracy of Jashemski's identifications and statements has been questioned by other scholars. In one instance identification of root casts were confirmed based upon a single carbonised olive fruit found near the excavation of tree roots (Ciaraldi 2001, 46; Jashemski 1979b, 211). Ciaraldi (2001, 46) notes that, aside from vines, identification of trees based solely upon root casts is uncertain. There has been criticism regarding Jashemski's reconstructions of the general structure of the gardens as unrealistic and the lack of a precise definition of the term 'garden', which is used for a variety of different types of planted areas such as 'courtyard gardens', 'nursery gardens' and 'market gardens' (Ciaraldi 2001, 46-47).

Garden evidence from the House of the Vestals (VI.I.vii)

A drain deposit in the House of the Vestals (VI.I.vii) produced a large number of garden species. Ciaraldi (2001, 173) argued that this was the first time in Pompeian archaeology that preserved plant macrobotanical remains were used as evidence for the presence of select garden species, which she views as more reliable than pollen, which can be transported over long distances, or the questionable identification of plaster root casts. The inclusion of what appears to have been garden plants could be due to their transport by rain water, as there is a channel which starts in one corner of the garden that flowed into the main drain.

The mineralised flower of the strawberry tree (*Arbutus unedo* L.) was recovered from the drain deposit (Ciaraldi 2001, 168). One of the charcoals from the ritual deposit from the House of the Vestals (VI.I.vii) came from this tree (Veal 2009, 159). Interestingly, the strawberry tree is one of the most frequently depicted garden plants on Pompeian wall paintings. At *Villa Vesuvio*, at *Scafati*, a large number of immature fruit from the strawberry tree were recovered from the waterlogged deposits. It is possible that a strawberry tree was planted in one of the house gardens or in the near vicinity. Plants that might have been used as ornamental flowers include mayweed/chamomile (*Anthemis* sp.), which was also recovered from the drain deposit (Ciaraldi 2001, 168). Ciaraldi (2001, 168) argues that the large numbers of mineralised weedy species such as seeds of bird's foot pods, commonly found in animal fodder, suggests the presence of manure. Roman authors advocated regular application of fertilisers including wood ash, animal manure, or green manure to fields (Dupouey et al. 2002 cited in Blondel 2006, 715).

Garden evidence from outside of Campania

Work on Campanian gardens was extended to investigate the gardens of the Roman Empire. Using analogies with the well-known and studied Campanian villas, it has been suggested that there was a similar kind of planting arrangement in the gardens of the Villa of Livia, outside Rome. In several cases planting ceramic vessels, *ollae perforatae*, were found within the areas thought to be gardens. Sediment samples were taken from the allae perforatae for macro-fossil and pollen analysis. Garden archaeology in the Lazio region is different than that of the Vesuvian region, due to differing preservation factors, i.e. the absence of pumice and lava to preserve the traces of ancient plants in Lazio (Klynne and Liljenstolpe 2000, 221). In 1943 Pierre Grimal's work drew together multiple lines of evidence on Roman gardens, including images of painted plants, the ancient authors and recent excavation data from gardens. In his conclusions he warns against labelling gardens fortuitously saved at Pompeii as type gardens for the rest of Italy and the Roman world (Cooley 2003, 110; Francissen 1987, 118).

Limitations

Limited archaeobotanical studies are available for the city of Pompeii and the surrounding region for the period of interest here (*circa* 300 BC to 79 AD). Thus, extrapolating these results from Pompeii at a regional level must be attempted with caution. Nevertheless, preliminary work begun on the macrobotanical remains by Borgongino (1999, 2006), Ciaraldi (2001, 2007), Ciaraldi and Richardson (2000), and Robinson (1999, 2002) provides a finer perspective in terms of the chronological and spatial patterns present in the archaeobotanical assemblage across the entire city of Pompeii.

In an attempt to integrate the environmental results, Dimbleby and Grüger's (2002) pollen results were compared with the plant macroremains reported by Meyer (1980) and the root identification results of Fideghelli (Jashemski 1979b). Ciaraldi (2001, 83) notes that inter-site comparison cannot be made from simple lists of environmental results without reference to stratigraphy and method of collection. Significantly, little is known of Pompeii's relationship with its surrounding hinterland, including the *villa rustica*e and trade on the River Sarno (Cooley 2003, 112).

5.6 Ritual botanical evidence

Pompeii

Ritual deposits, in which food items of particular significance are burnt in an offering to the gods, provide unique information about foods that are otherwise rarely preserved. Robinson (2002, 98) argues that the burnt offerings from the House of Amaranthus (I.IX.xii) are the first domestic ritual deposits to be archaeologically identified. Robinson (2002, 98) attributes these cremations from the House of Amaranthus (I.IX.xii) 'to the burnt offerings and sacrifices that formed part of private domestic worship'. Many of the ritual deposits were found to contain cones and seeds from stone pine, figs, grapes and nuts such as walnut and hazelnut. Interestingly, the archaeobotanical remains from the kitchen area were similar to the 1st century AD burnt offerings from this property (Robinson 2002, 93).

Robinson (2002, 96) found that the majority of the charred seeds from the burnt ritual deposit were Mediterranean crops such as cereals, field legumes, fruits and nuts, similar to crops grown in the region today. Dates, which were found in some of the 1st century AD burnt deposits, were the only species recovered here that does not grow in the region and was likely imported from North Africa. Stone pine, represented by cone bracts and nuts, was a major component of the 1st century AD burnt ritual deposits. "*Ficus carica* (fig), large cereal grain including wheat and barley and *Juglans regia* (walnut) were the species most frequently present. They comprised a wider range of taxa than were present in the other contexts on the sites such as refuse deposits" (Robinson 2002, 96). Non-food plant remains were present including the shoots, seeds and cone scales of *Cupressus sempervirens* (cypress), the immature fruit of *Myrtus communis* (myrtle). These plants were burnt during religious practices for their pungent fragrance. Both are believed to have been grown as ornamental plants in the gardens at Pompeii (Jashemski 1979b, 25). Many grass seeds and disturbance loving weed seeds were recovered from burnt deposits.

From the House of the Vestals (VI.I.vii), a large foundation ritual deposit was found in the atrium area (Bon et al. 1997; Ciaraldi 2001, 63). The deposit consisted of two layers, with different concentrations of charcoal. Ten miniature incense cups were discovered in the archaeological matrix along with foetal pig bones, fish bones, eggshells, sea urchin fragments, charred remains of cereal, fruits and nuts along with evidence of *in situ* burning (Bon et al. 1997; Ciaraldi 2001, 144-145; Ciaraldi and Richardson 2000, 75; Veal 2009, 58). This deposit was interpreted as a ritual foundation offering during the construction of the house, either before the establishment of the first walls, or just after their raising. Unfortunately, the pit was partly truncated due to a WWII bomb crater in 1943 (Bon et al. 1997). Obvious taphonomic issues are associated with this destruction event. Therefore, limited soil from this pit could be taken for analysis and the recovered plant materials were badly preserved.

Broad bean (*Vicia faba* L.) was the most numerous ecofact recovered as they were often found in ritual deposits of differing periods and used as food offerings in religious practices (Ciaraldi 2001, 145). Pliny (NH XVIII. 118) emphasised the sanctity of broad beans and their use as offerings to the gods. Fragments of hazelnut shells (*Corylus avellana* L.), almond (*Prunus amygdalus* L.) and *Prunus* spp. were also recovered from the deposit. There was a high concentration of cereals in the form of bread or a cake-like item. Present in the deposit were fragments of fig and grape fruit and pips. One seed of woundwort (*Stachys* sp.) was recovered, suggesting a possible floral offering due to the plant's attractive flowers. The ritual charcoal assemblagrecovered, suggesting a possible floral offering due to the plant's attractive flowers. The ritual charcoal assemblage included fragrant woods like myrtle (*Mrytus* spp.) and rose (*Rosaceae*). Symbolic wood like cypress associated with funerary rites was also found. Utilitarian woods, such as beech and hornbeam were also present and were likely to help ignite and sustain the burning of the offerings (Veal 2009, 157). This ritual wood use pattern differs from domestic and commercial contexts of Pompeii.

A ritual deposit was found in the north end of Insula VI.I which consisted of an egg placed between two plates holding other environmental remains (unfortunately, all evidence of this deposit was misplaced in the excavation storage area in Pompeii). In addition, small votive cups were recovered from the Shrine area in the south end of Insula VI.I (Veal 2009, 129). At the House of the Wedding of Hercules' (VII.IX.xlvii) a ritual deposit was discovered which showed similarities with the palaeobotanical assemblage from the defined ritual deposit of the same period from the House of the Vestals (VI.I.vii) in which there was an abundance of cereal grains (Ciaraldi 2001, 144; Ciaraldi and Richardson 2000, 75). The palaeobotanical remains from both of these ritual contexts are also similar to the House of Amaranthus (I.IX.xi) deposit where a wide range of cereals, pulses, fruits and nuts were found (Robinson 2002, 98). Rectangular ritual pits near the large altar discovered by Mau during the 1765-1766 excavation of the Temple of Isis in Pompeii, contained carbonised figs, dates, chestnuts, walnuts, hazelnuts, and stone pine kernels and cones (Ciaraldi and Richardson 2000, 79; Overbeck and Mau 1884 cited in Robinson 2002, 98).

'In Pompeii, food was an integral part of both everyday and exceptional ritual activity' (Ciaraldi and Richardson 2000, 81), therefore it is perhaps anachronistic to try and separate food into different categories. A symbolic offering of every meal consumed was ritually offered to the household *Lares*, as with observed by fragments of carbonised pine cone and four pine nuts on the altar of the House of *N. Popidius Priscus* (VII.II.xx) in Pompeii (Meyer 1988, 184). These burnt offerings represent a unique opportunity to witness private domestic worship, which is rarely preserved in the archaeological record outside of Pompeii.

Recent excavations at the *Porta Nocera (Nuceria* Gate) *necropolis* in Pompeii allowed for the analysis of the archaeobotanical remains offered to the dead from the 1st century AD by Matterne and Derreumaux (2008, 105). Archaeobotanical remains recovered from ground surfaces consisted mainly of fruit residues, particularly fig and grape, while a much wider range of species was observed in the tombs, including cereals, pulses, other kinds of fruits, weeds and bread/pastry. Botanical remains were preserved through the intentional burning of food, as the rituals took place in the funerary enclosure to the ancestors; fig was found to be the most important taxon. Ritual offerings appeared to be selected from locally available produce. These ritual botanical offerings were composed of the staples of everyday life within the Mediterranean world and these results accord well with van der Veen's (2003, 407) terminology of funerary rite foods which meet instrumental needs rather than luxuries.

5.7 Medicinal drugs

Ciaraldi (2000) investigated the contents of a *dolium* or storage vat found in the cellar of the *villa rustica*, located near Pompeii, during the 1996 excavation. This deposit, from the bottom of the storage vat, yielded a high percentage of seeds from plants with known medicinal properties, including *Papaver somniferum* (opium poppy). Also found in this residue were the bones of reptiles and amphibians. It is known that ancient medicine often utilised both animals and plants (Ciarallo 2000, 10). Ciaraldi (2000, 9) suggested that this preserved assemblage is evidence of medicinal preparation.

Villa Vesuvio, dated to the AD 79 destruction layer, has a number of architectural elements that firmly situate it as a modest rural house. Ciaraldi (2000, 91) estimated that 70% of the *dolium* was filled with peach stones and 30% with walnuts. Preservation was defined as 'excellent' by the author and the likely presence of entire peach fruits and 'green' walnuts coincides with the harvesting of these plants in late summer and provides further evidence of the date of the eruption on August 24th (Ciaraldi 2000, 93). However, it is also possible that these fruits could have been preserved in their green state for over a year.

Many of the plant remains recovered from *dolia* 2 could have been used for medicinal purposes in ancient times. Ciaraldi (2000) states that 58% of the plants identified to species level can be grouped as medicinal plants, the majority of which ancient sources mention as agents against poisons. *Cupressus sempervirens* L. (Cypress) was recovered from pre-AD 79 excavations in Pompeii and appears frequently at the site in the present day. Cypress is known from ancient sources to have had both medicinal properties in the curing of snake bites and economic value as a dyeing agent and for recovering boggy areas; results must be interpreted cautiously as it is unknown if these results are unique as there is a 'scarcity of archaeobotanical studies in Italy and particularly in southern Italy' with which to compare this assemblage (Ciaraldi 2000, 97). In addition, in the ancient world, plants often served a variety of functions, including medicinal uses, so it is difficult to assess from one assemblage whether taphonomic factors or small sample size bias the interpretation as solely medicinal.

5.8 Charcoal analysis from Pompeii

Veal's (2009) diachronic anthracological research from several locations within the city of Pompeii, including

Insula VI.I, the *Porta Stabia* Project (VIII.VII), Temple of Venus garden (VIII.I) and two test trenches within the House of the Silver Wedding (V.II.g) revealed that the dominant taxon (representing between 50-75% of the charcoal) from each of these sites was the montane beech (*Fagus sylvatica*). Beech is believed to have grown in the lower level managed montane deciduous forests (*circa* 500m in southern Italy) or the higher altitude natural forests surrounding the city of Pompeii (*circa* 1200-1600m). The nearest source of beech would have been 15- 25 km from the city, approximately a day's travel in ancient times. The transport of raw timber or charcoal down the Sarno River, originating at the foot of the Apennines, was likely the major trade route linking Pompeii to the inland cities (Veal and Thompson 2008, 10).

Important cultural and economic implications can be inferred from the dominance of beech (*Fagus sylvatica*) within every assemblage studied (Veal 2009, 131). Significantly, beech is a high calorific fuel. This is useful for producing a high heat for industrial or domestic purposes. Veal (2009, 168) notes that beech's scant mention in the ancient sources is contradictory to its ubiquity within the Pompeian assemblages examined.

Veal (2009, 228) analysed the charcoal from the 3rd century BC to the 1st century AD. Beech, although dominant in each century, diminishes over time in tandem with secondary fuel woods, i.e. oaks, maple and hornbeam. There is a slight increase in diversity of woods, mainly of fruit and nut types from the 2nd century BC to the 1st century BC. Interestingly, there is a marked increase in the diversity of woods from the 1st century BC to the 1st century AD, with large amounts of grapevine, olive, ash, elm chestnut and alder observed in small quantities for the first time within the assemblages, which coincides with the establishment of the Sullan colony founded in 80 BC. Veal (2009) posits that this diversity may represent the increasing influence of Roman tastes through the interaction of the installed Roman veterans with the local Samnite population or just a natural increase in these types as occurred elsewhere in the Roman Empire with population increase.

5.9 Trade

Transoceanic trade between India, the Arabian Peninsula and East Africa began at least by the early Holocene, and became especially intense during the Greco-Roman period when many cultivated plants were introduced into new areas via established land and sea routes. However, imports that can be unambiguously linked to the Mediterranean area include largely edible fruits and include the stone pine, the almond, the walnut, the apricot (*Armeniaca vulgaris*), and the juniper (*Juniperus*). In particular, the stone pine and almond are well represented in both the early and late habitation period at Berenike, which was an important harbour along the Red Sea coast (Cappers 2006, 1).

Commodities imported into Berenike were most likely intended for further transport into the interior of the Roman Empire, at smaller ports such as Pompeii. However, many of the exotic plant products that entered Berenike may have been available to the local inhabitants. In particular the archaeobotanical evidence shows that foodstuffs that were delivered in bulk quantities, like black pepper, were probably consumed locally. Luxury is a relative and flexible notion and is primarily created by supply and demand. Exotic foodstuffs are construed as luxury items based upon their distant origins, storage life, irregular supply and any additional expenses incurred in cultivating and transporting such items (van der Veen 2003, 405). Trade centres such as Berenike would have presented unique examples because they engaged in the transfer and redistribution of large quantities of so-called exotic produce. Such commodities could have been temporarily available in even greater quantities than regular staple food items, which possibly depreciated the status of these food items amongst the local inhabitants (Cappers 2006, 165).

5.10 Broad trends in archaeobotanical evidence from the Roman world

Long-distance trade within the Mediterranean Basin goes as far back as the Bronze Age (Grove and Rackham 2001, 80). The diet of the Bronze and Iron Age largely consisted of cereals, pulses, wild fruits and nuts (Bakels and Jacomet 2003, 542). Naked barley disappeared from Western Italy after the early Bronze Age according to published archaeobotanical reports (e.g. Ciaraldi 2001; Hopf 1991). During the first phase of the Bronze Age there appears to be a shift from naked towards hulled barley in the western Mediterranean similar to developments north of the Alps (Bakels 2002, 7).

Rome

Little environmental analysis is available for the period of 9th to 6th centuries BC (Motta 2002, 71). Carbonised plant remains from pre-urban centers of Rome, in *circa* the 8-7th centuries BC, Forum Romanum tombs excavated in 1954, consisted of cereals, mainly einkorn, emmer and spelt, and leguminous seeds with fruit added to this recovered meal (Helbaek 1956, 288-289, 294).Helbaek (1956, 292) argued that there is little evidence for horsebean/ broad bean/faba bean (*Vicia faba*) in Italy until the Iron Age. Although faba bean cultivation appears rather late in the 3rd millennium BC in Europe, more recently large quantities of charred broad bean seeds have been discovered in Neolithic/Bronze Age sites (Zohary and Hopf 2000, 114). Grass Pea, (*Lathyrus sativus* L.) became widely distributed in 3rd millennium BC in Italy.

From on-going investigations at Iron Age sites in and around Rome, Motta (2002, 72) produced palaeobotanical assemblages from key sequences of Rome and its surrounding hinterland. Motta (2002, 72) argues that a shift in agriculture, from a wide diversity of crop processing strategies to increasing homogenisation, reflects the change from small-scale domestic farming towards a more centralised food distribution system and increasing urbanisation. Aside from providing palaeobotanical evidence from this underrepresented time period, the present study offers broader theoretical implications for multi-period archaeobotanical analyses and interpretation as it attempts to move beyond the traditional 'before and after' phase approach when investigating issues in urbanization and processes of state formation (Motta 2002, 76).

The density of recovered plant remains was low; 50 samples were found to contain more than 25 items of grain, chaff or weeds. A limited range of taxa were recovered in nearly every context. Cereals dominated all analysed assemblages, with legumes and fruit represented by only a few remains. Five cereals including einkorn, emmer, spelt (*T. spelta* L.), free-threshing wheat (*T.durum/turgidum/ aestivum*) and barley were represented by grain and chaff in every phase. Relative proportions of grain/chaff/ weeds were examined to assess changes in crop processing patterns during this period (Motta 2002, 73).

The dominant wheat recovered from all contexts was emmer. This corresponds to the results of other small assemblages from early Rome (Constantini and Giorgi 2001 cited in Motta 2002, 73; Helbaek 1956, 289). Einkorn was scarce, having a ubiquity of only 21% from all palaeobotanical remains from all contexts. Although spelt is usually recovered in low frequencies from prehistoric and Iron Age contexts in peninsular Italy, it was found in significant quantities (ubiquity of 33%) from all contexts, while free-threshing species are poorly represented from only 10% of all contexts (Motta 2002, 73).

Several species of cultivated legumes were present, including horsebean, bitter vetch, common vetch, pea, *Vicia/Lathyrus* spp. and indeterminate *Fabaceae*. A limited range of fleshy fruits was recovered from the site. Among the cultivars, grape pips were frequent and occurred in each context. Fragments of olive stone and fig seeds were also recovered. Wild plants included strawberry and plum. *Sambucus* cf. *ebulus* L., a non-edible elder, can occur as a ruderal, may have been intentionally gathered or represent the natural vegetation of the site.. This plant is known to have cultural purposes such as medicinal uses or dyeing of fabric and may have been intentionally gathered (Motta 2002, 73).

Wild species known to inhabit arable fields and weeds of cereal crops were also recovered, predominately Lolium (*L.temulentum* L., *L. multiflorum* Lam., *L. perenne* L.). Plants from wetland environments such as *Carex* spp., *Eleocharis palustris* (L.) Roem & Schult. and *Scirpus lacustris* L. were also found in the recovered assemblages. Interesting interpretations based on the charred assemblage containing species from both grassland and wetland habitats are possible. One explanation is that these contexts represent mixed materials such as the burning of dung or weeds (Motta 2002, 73). Another explanation is that these assemblages of cultivated cereals and mixed wild grassland and wetland species represent cultivated crops and arable weeds and that the arable weed communities were different than in the present day; or that cultivated fields were in close proximity to marshes or ditches. On the other hand, this type of mixed charred assemblage may be the result of crop processing activities and therefore include crop husbandry products and by-products (Motta 2002, 73).

Motta (2002, 74) uses the standard archaeobotanical concept that a higher proportion of cereal grain represents centralised crop processing and can be an indication of urbanisation. Motta (2002, 74) argues that if the archaeobotanical remains analysed are representative, then the increase seen over time in the average proportions of grain, chaff and weeds, in particular during the 7th and the 6th century BC contexts, could support the view that the urbanisation process was completed between the 7th and the 6th century BC at Rome. Distinction between consumers, who were not directly involved in agriculture e.g. soldiers, and those who were engaged directly in agriculture would be evident.

Using this same framework one could argue that the uniform absence of cereal crop processing remains from the properties of Insula VI.I supports the theory of an urbanised centre in which cereal grains are being cleaned in the countryside and brought into the city to bakeries for processing and baking into bread where it is then purchased by consumers. This uniformity in cereal processing, as Motta (2002, 76) notes based upon anthropological models, suggests a fully urban centre with a regulated and controlled supply of cereal coming into the city from the countryside, along with commercial businesses in place to process the cereal into bread and sell the bread to consumers.

5.11 Summary

A range of surviving sources including art historical, literary texts and environmental evidence (pollen, charcoal, and macrobotanical remains) were surveyed to attempt to understand the past environment, agricultural activities and diet in the Vesuvian area. This review encompassed a wide range of environmental research within the Vesuvian area, investigating different property types including commercial and domestic properties, villae, and special deposits of interest from Pompeii e.g. of ritual and pharmacological interest, together with garden evidence. Environmental data continues to be the most conclusive line of evidence for understanding the past diet and environment of Pompeii.

Early discoveries of carbonised botanical remains tended to be of easily recognisable food types (i.e. fruits and bread) found in large quantities among a range of property types including shops, bars and domestic properties. More recent purposeful and intensive environmental collection methods have increased the quantity of archaeobotanical material available allowing for more detailed analyses to be made. Archaeobotanical analysis has elucidated aspects of urbanisation in the ancient city. The presence of chaff within early pre-Roman deposits and the disappearance of cereal processing by-products during the Roman period, is a pattern seen earlier in Rome (7-6th century BC) and later in several recently stratigraphically excavated domestic properties from Pompeii, including the House of the Vestals (VI.I.vii), the House of Amarantus (I.IX.xii), and the House of Hercules' Wedding (VII.IX.xlvii). The lack of crop-processing remains from nearly all the properties from Insula VI.I support this view that Pompeii was an urbanised centre by the late 1st century BC/1st century AD. This implies that in the above listed cases agricultural activities were being confined to the countryside and that both Rome and Pompeii were becoming urbanised centres with clear separation between producers in the countryside and consumers within the city.

Chapter 6

Food and Roman Culture

6.1 Roman agriculture

Throughout Roman history agriculture was the dominant economic activity, affecting multiple aspects of Roman life. The majority of Romans would have had a direct connection with agriculture as small farmers, wealthy landowners or elite consumers (MacKinnon 2004, 11; White 1967, 1). During the Republican and Imperial periods the Pompeian economy was thought to have been a hard-driven traditional agricultural economy (Jongman 2007, 503; Moorman 2007, 435, 445). This is evident from the intensive agrarian methods of inter-cultivation and combination cropping practices with vines, olives and grain and crop rotation with legumes and fodder crops (Jashemski 1979b, 618; Kron 2000, 277-278; Spurr 1986, 1; White 1965, 107). Thus, it is inferred that the stress in place at the time of Imperial Rome on the vegetation of Italy was at least as great as any the Mediterranean experienced before the 19th century industrial age (Grove and Rackman 2001).

6.2 Food, ideology and Roman culture

The study of food systems and eating provides scholars (historians, anthropologists, sociologists and archaeologists) with a means of investigating the structure and fabric of a societies' collective cultural traits, attitudes and social institutions (Bourdieu 1979; Douglas 1997; Dunbabin 2003b; Garnsey 1999; Sahlins 1972 cited in Hastorf 1995, 132). Food is a particularly relevant cultural element with which to study Pompeii, because as in all pre-industrial societies, food permeated nearly every aspect of Roman daily life (Garnsey 1999, xii; MacKinnon 2004, 11). Food is simultaneously a necessity of the physical realm and a constructed cultural ideology which is largely governed by social and cultural traditions (Garnsey 1999, 142). Food is 'a highly condensed social fact' (Appadurai 1981, 494 cited in van der Veen 2003, 405). What is consumed is the combination of individual decision-making within a larger complex socio-cultural framework (Smith 2006, 480). Food ways illustrate the division of labour, control of resources and social symbolism and cultural dynamics within a society (Hastorf 1995, 135). Food, particularly in ancient societies, acted as vehicle for communicating status, for maintaining familial, social and political relationships and served to reinforce the rigid social hierarchy present within Roman society (Garnsey 1999, xi).

Dietler (1996, 8, 88) argues that research needs to move beyond the traditional approach of focusing on diet and instead begin to regard food, with its multiple cultural dimensions, as a critical element in the formation and manipulation of social relations within society. In regards to palaeodietary information, 'one needs to understand the cultural context that ultimately envelops and shapes the data before one can draw reliable conclusions' (MacKinnon 2004, 12). Past biases in the literature were largely created by archaeologists and archaeobotanists, who were predisposed to understand ancient agricultural practices and food choices within the framework of rational optimizing resource strategy focusing on procurement and production rather than the existing cultural milieu and consumption (Hamilakis 1999; MacKinnon 2004, 11). Therefore, this research will attempt to undertake a holistic approach in which food and food systems are examined within their cultural framework.

6.3 Cuisine

Culinary culture is composed of the ensemble of attitudes and tastes that people bring to cooking and eating and thus different societies develop different tastes and attitudes over the course of the development of their respective societies (Mennell, Murcott and van Otterloo 1992, 20). Practically, cuisine is confined by the production and distribution of food and is daily reinforced by the social and cultural traditional practices of that society and includes pragmatic issues such as: how and where the food is obtained, who is allowed to or expected to prepare it, where it is prepared, when it is prepared and who is allowed to consume it and with whom (Garnsey 1999, 141-142; Goody 1982, 98).

On a philosophical level, cuisine and the food it encompasses, can express fundamental human values and attitudes with their meanings expressed in an understood social code. According to socio-cultural anthropologist Lévi-Strauss (1958), who pioneered this early structural approach, to decipher this code is to penetrate the 'deep structures' of a civilization as '[t]he cooking of a society is a language into which it unconsciously translates its structure- or else resigns itself, still unconsciously, to revealing its contradictions' (Lévi-Strauss cited in Garnsey 1999, 7).

Following this structural theoretical framework anthropologist Mary Douglas argued that

> '[i]f food is treated as a code, the messages it encodes will be found in the pattern of social relations being expressed. The message is about different degrees of hierarchy, inclusion and exclusion, boundaries and transactions across the boundaries' (Douglas 1997, 36).

The main criticism of Levi-Strauss's classic structuralist analysis, involving his famous triangle of the raw, uncooked and cooked, is that it oversimplifies the complexities found within the code of food of a culture for the sake of academic clarity. A particular case study is presented as an overarching model and anachronistically assumes that one culture has a single uniform mentality. This model largely negates individual agency within the different social groups present within a society (Gowers 1992, 6).

Food as a social code is one entrance into the mental attitudes of a society (Lévi-Strauss 1958 cited in Garnsey 1999, 35). Although this structural theory was designed for anthropological research into the foodways of living societies, its usefulness can be extended to the discipline of archaeology. Admittedly, without informants the majority of the symbolic and social meaning behind food will have been lost along with the civilization that created it. However, using this theoretical framework it may be possible to tease out remnants of the social and cultural meanings behind certain foods through archaeological analysis by investigating what types of food are utilised and which are ignored or forbidden, how the food is being prepared, where the food and food-processing remains are recovered and where they are not found. These are all questions that the physical remains of food can potentially answer. By using this anthropological theory one can begin to gain a deeper insight and understanding behind the foodstuffs used by the Romans and move beyond optimum resource strategy models and appreciate the 'deeper' meanings attributed to certain foods within Roman society.

6.4 Symbolism

Relying upon the literary sources, Dupont (2000, 116) points out that from Cato's *De agricultura* to the *Historia Augusta* the symbolic significance of a variety of foods including bread, meat, and vegetables within Roman society remained static for nearly a millennium and therefore should be analysed as a largely enduring framework. A classic example from antiquity are the symbolic associations of eggs, apples, and pomegranates which were used to represent life and fertility and are often depicted at marriages or even funeral ceremonies (Garnsey 1999, 8). It was the French sociologist Bourdieu (1990) who argued that '[t]he daily routine of food consumption reflects and recreates the social and symbolic codes of a society' (cited in van der Veen 2003, 415). A meal is composed of a set of food types which conveys to both viewer and consumer a set of established meanings that have been internalised through repetition (Hastorf 1995, 135). Dupont (2000, 114-116) provides the persuasive argument that food symbolism and dietary norms were tied up with Roman identity and built into political institutions and social practices which were reinforced daily and therefore only a fundamental shift in cultural traditions, such as witnessed with the adoption of Christianity, could bring about changes in food customs (Smith 2006, 480).

Scholarship has tended to focus on symbolic 'meaning' within specific cultural contexts and this has resulted in the lack of attention to long term change within symbolic meaning. Goody (1982, 36-37) argues that the domain of cooking itself is highly conservative and it is the introduction of foreign foods and cooking techniques from outside a culture that result in changes in its cuisine. Goody proposes that symbolic structures cannot be treated as immutable but rather must be analyzed as embedded within a world economy, as seen in Sidney Mintz's work on the trans-Atlantic trade in sugar, cane and rum during the 19th century. He theorises that since food is connected to the mode of production of material items 'the analysis of cooking has to be related to the distribution of power and authority in the economic sphere, that is, to the system of class or stratification and to its political ramifications' (Goody 1982, 37). Thus, growing, allocating, cooking and eating food all represent the production, distribution, preparation and consumption of food within society.

There has been a growing awareness within academic circles that traditionally ignored 'feminised' concepts of cooking and cuisine reflect important economic, social and political changes within a society than previously believed. Although the Roman perception of traditional Roman food items may have been relatively static it seems likely that they would have been influenced by the nearby Greek colonies, local Samnite populations and increasing contact with foreign peoples through trade. Not only is it important to understand the symbolic and cultural meaning behind the foods utilised but the historical factors and economic undercurrents which also influence food choice and usage in Roman society during the period under examination. Thus, change in Roman food reflects change in Roman society.

6.5 Cultural identity

Within Roman society an individual's cultural identity was regarded as more important than their social identity. A strict dichotomy was established between a citizen who lived in a civilized society and the uncivilized barbarian who lived in the wilds. Within Roman ideology the key factor separating these two distinctions was the capacity to domesticate and have control over nature and therefore over one's food (Montanari 2000, 70). Roman attitudes towards their natural environment were extremely complex (Hughes 1994, 70). Ironically, the need to distinguish the concepts of wild and tamed within the Roman psyche may be attributed to the fact that although sedentary farming was established within the Mediterranean before the Classical period hunting and gathering was still occurring on accessible uncultivated land (Garnsey 1999, 36-37). During the late Roman period categories of wild and domesticated were practically broader than the Roman construction of these terms would suggest (Montanari 2000, 72).

Lévi-Strauss's (1975) famous dichotomy of the raw and cooked is homologous to the Roman dichotomy between nature and culture. The geographer Strabo stated that the division between civilized and barbarian rested on the concepts of settled agriculture and urban living, a statement that reveals why little value is given to the produce of uncultivated land, e.g. gathered foodstuffs such as mushrooms, asparagus, laurel, wild fruits, and snails (Lomas 1995, 5). Excluded from this perception were medicinal plants such as nettles and some elite foods such as wild asparagus and mushrooms (Corbier 2000, 129). All parts of wild plants (*herbae*) were considered fodder for livestock and therefore regarded as inappropriate for human consumption. Such types of uncultivated food items were marginalized and largely excluded from the literary sources. Although they possess different preservation biases no wild food stuff were recovered from Insula VI.I and this may offer an insight into the Roman's distain for wild versus cultivated plants.

The Roman attitude towards cereal grains can be explained by Levi-Strauss's framework of raw, cooked and rotten (Levi-Strauss 1975 cited in Garnsey 1999, 29). Roman food ideology perceived harvested grain seeds as 'raw' because they had the potential to produce new plants, thus they were considered 'alive'. The roasting and milling of glume wheats was thought to destroy this generative power thus the resulting flour was considered 'uncooked' and so likely to decay and therefore its use was subject to food taboos similar to those associated with foods that rot, e.g. fish and meat. For this reason the Romans stored wheat as spikelets which were milled as needed and immediately baked the flour into bread or boiled it as porridge. Practically, the glumes help protect the wheat from disease and insects (Bruan 1995, 35). Plants within the Roman category of *Legumina* were regarded in a similar manner to cereals because once their seeds were 'killed' they began to decay and more rapidly than wheat. This ideology about food decay explains why Jupiter's priests (*flamines*) were forbidden to eat broad beans, flour or flesh (Dupont 2000, 119).

Grain was the most extensively cultivated crop throughout Roman Italy. A variety of cereals were cultivated but wheat and barley were the most common (Spurr 1983, 1986, 1). The diet of middle and lower class Romans was heavily concentrated on cereals (Brothwell and Brothwell 1998; Bruan 1995, 25; White 1995). The poorest Romans likely subsisted on wheat, either as bread or crushed and boiled to make porridge or puls. Boiling was probably the most common method of preparation because few had access to an oven (Shelton 1998). Archaeobotanical evidence shows that the cultivation and consumption of hulled barley and naked wheat continued throughout the 2nd century BC to the 11th century AD in the western Mediterranean. Wheat, barley, leguminous crops, grapes and figs were the basis of the Mediterranean diet (Martinez 2005, 341).

Wheats

Wheat was an important buffer against famine in Roman society. The availability of wheat prevented peasants from the degrading activity of scavenging for wild plants such as acorns, nettles, and vetches because wheat was perceived as an acceptable substitute. In Roman society, famine threatened not only one's health but one's civic liberties (Dupont 2000, 127). Indeed, beginning in the Republican period a grain dole was instituted to distribute among the Roman poor to prevent the shameful foraging of wild foods and allow, as the Plebeians demanded, access to the diet of civilized free men or of a Roman soldier.

The results from all the properties within Insula VI.I showed no evidence of uncultivated food items and indeed the archaeobotanical assemblage was dominated by fruits, including olives and figs. Cereals also composed a smaller portion of the recovered archaeobotanical material. Although wild or uncultivated food items are less likely to become preserved in the archaeological record than processed or prepared food items their absence lends credence to the idea that Romans did not eat, sacrifice or bring into their homes wild plants.

The people who transgressed these cultural boundaries were defined as being outside Roman civilization and therefore classified as the barbarous 'other' (Dupont 2000; Montanari 2000, 70). Indeed, it was considered scandalous for Roman citizens to forage like herbivores and only in dire need would individuals do so. According to Dupont (2000, 113-121) shepherds, hunters, slaves and extremely poor peasants were the only classes of individuals within Roman society for whom it was considered acceptable to do so; Germanic tribes who consumed meat and animal by-products rather than domesticated plants were by definition, also barbarians. Also categorized as uncivilized were foreign peoples encountered along the Roman borders such as the Numidians, tyrants and other social usurpers, and groups within Italy outside the boundaries of civilized Roman society such as philosophers, wealthy freedmen, and gladiators.

6.6 Late Republican to Early Imperial Period Dietary shifts in meat preference

The Roman diet was divided into *fruges*, products of the soil and *percudes*, foods derived from animals (Strong 2002, 18). Pigs were important animals to the ancient Romans in Italy. Pork has the advantage that its flesh preserves well in the meat safe (*carnarium*) in which butcher's products smoked over the kitchen hearth. Lamb appears to have been second in Roman meat preferences; possibly because ewe's milk was considered the most nutritious variety of milk and therefore nursing offspring were culled early (Bober 1999, 181). Italian pig and pork production centres were located in the regions of Campania, Samnium, Bruttium, and Lucania. Zooarchaeological data informs us that the Romans preferred a smaller, thinner, leaner, and long-legged breed of pig. According to ancient texts, this breed was kept in herds and allowed to, and was believed to be well suited to, foraging in nearby deciduous woodlands for its food (MacKinnon 2001, 664).

Interestingly a larger breed of pigs, which was raised in smaller numbers, and was stall fed is the breed that dominates Roman artistic depictions of pigs. A hypothesis has been suggested that this larger breed was preferred as a sacrificial victim in religious ceremonies. However, it is difficult to substantiate this claim as the majority of the zooarchaeological data was recovered from deposits believed to have been associated with food processing rather than with any religious activities (MacKinnon 2001, 649- 666).

The importance of pork in the Roman diet is so great that some scholars have even argued that its consumption and later presence in the archaeological record can be constructed as a characteristic feature of Romanisation; with the rather simplistic model of 'more pork signifying 'Roman', and less denoting 'native' cultures, depending on the region and time period in question' (MacKinnon 2001, 649). Varro (Rust. 2.4.3 cited in MacKinnon 2001, 649), the ancient agronomist, succinctly summed up this important economic relationship by asking the rhetorical question: 'who of our people runs a farm without keeping pigs?' MacKinnon's (2001, 649) work offers an example of how the integration of literary, visual, and zooarchaeological data sources, using specifically the example of pigs in Roman Italy, can provide a more complex and multidimensional aspect to Roman farming and dietary practices.

Faunal evidence from west central Italy, comprising the regions of Etruria, Latium and Campania, during the Late Republican and Imperial period shows a very clear pattern of high pig bone percentages. This differs from the faunal evidence of the northern part of Italy, including the Arno and Po valleys, which revealed a higher percentage of cattle bones and a corresponding low percentage of pig bones. Sites in the southern part of Italy, mostly in the former *Magna Graecia* region, tend to display a differing pattern from the rest of Italy and have a high sheep/goat pattern, which conforms to the known western Greek influence of this pattern. However, this pattern is not always present in west central Italy. At Greek, Etruscan, and other pre-Roman sites in Italy, including a number from the regions of Etruria and Campania, significantly fewer pig bones have been recovered and instead a higher proportion of sheep/goat and cattle are present. The faunal analysis at many of these sites reveals equal proportions of the main domesticates and provides evidence of a largely meat diet which was more reliant on beef than other kinds of meats (King 1999, 169).

During the late Republic a pattern of high pork consumption emerged as a feature of the Roman diet; this chronological distinction in meat consumption is most clearly seen in the Bay of Naples area. Thus, the cattle-dominated diet of the 1st century BC was replaced by the high pig pattern seen in the Imperial period (King 1999, 169). Indeed, preliminary research on the faunal assemblage from Regione VI, insula I at Pompeii by Richardson (2006, 174) revealed a similar high pig pattern indicating that dietary preferences shifted over time and class distinctions in differential meat consumption are noticeable in the archaeological record.

Present faunal data suggest that a similar pig-dominated pattern emerged earlier, in the 3rd to 2nd century BC, in the region of Latium, north of Campania (King 1999, 169).

At the *Settefinestre villa* this shift in diet is revealed on a micro-scale. The intra-site analysis revealed higher pig bone percentages recovered from what were interpreted as higher-status parts of the site as compared to parts of the site which were interpreted as lower status and possibly inhabited by slaves. This difference is interpreted as showing that pork, in particular suckling and young pork, was considered a high-status and expensive dietary element, possibly because the animal is conspicuously killed before reaching its economic reproductive value for its owner. Similar to other villas, this particular villa was thought to be a site of intensive production, supplying pork and other types of meat to the surrounding urban centers, such as cities like Rome and Ostia which tend to have observed pork-rich diets (King 1999, 169-171).

King (1999, 171) suggested that pig herding could have been added to the established agrarian system in which pigs, in particular the more numerous smaller variety, were allowed to graze amongst the groves of olives and grape vines, taking advantage of the shade provided by the cultivated trees, as well as possible woodlands along the edges of the orchards (Parsons 1962 cited in King 1999, 175). This form of agriculture would leave little available land for sheep farming and despite the possibility that sheep and goats could have been grazed amongst the hills of central Italy, it remains clear from the analyzed faunal assemblage that during the Roman period cultural selection of pork over mutton was occurring.

A clear pattern emerges from all early Imperial faunal assemblages that cattle bones are always ranked third in order of percentage, suggesting the lack of importance of beef in the diet at this time. The lack of cattle bones could be due to religious reasons, or more pragmatically large-scale cattle rearing would have needed the creation of large areas of low-level pasture, during dry Mediterranean summers few areas would have been available for good pasture land, which in turn would have resulted in fewer fields for growing cereals for human consumption (King 1999, 171).

This consistent pork-rich dietary pattern in the region surrounding Rome during late Republic and early-to-middle Empire may be explained by practical reasons related to the agricultural conditions, or to simply topographical factors. However, the faunal pattern, in which pig bones comprise almost half of faunal domesticate assemblages, could be due to cultural preferences, i.e. a clear 'Roman' dietary pattern intended to distinguish the Roman population from the surrounding Italic and Greek people within the expanding Roman Empire (King 1999, 171).

Very little of the 'Roman' pattern is evident in archaeological assemblages from Roman-occupied Provence, despite prolonged Roman occupation. However, local and Greek influences dated to the second half of the 1st millennium BC are evident. In addition, Greek faunal assemblages from the Aegean, from both before and during the Roman period show a preference for the consumption of sheep and goat. It thus appears that Hellenisation was more influential in terms of diet than the later Romanising presence (King 1999, 177).

By the late 2nd century AD the Romanising influence on dietary patterns and the intensification of agriculture had most likely reached its apex. Many urban centres in western central Italy, such as Naples, Ostia and Rome, continued the dietary pattern established in the early Imperial pattern into the 4th and 5th centuries AD. However, by the 6th century AD at Naples, the faunal record reveals a move away from this pattern as sheep and goat appear more frequently than pig (King 1999, 177).

6.7 Dietary shifts in cereal preferences

Motta (2002, 73) found that the dominant wheat recovered from all contexts from her study of early Rome was emmer. This corresponds to the results of other small assemblages from early Rome (Constantini and Giorgi 2001; Helbaek 1956, 289; Motta 2002, 73). Einkorn was found to be scarce, having a ubiquity of only 21% from all palaeobotanical remains from all contexts (Motta 2002, 73). Although spelt is usually recovered in low frequencies from prehistoric and Iron Age contexts in peninsular Italy, it was found in significant quantities (ubiquity of 33%) from all contexts, while free-threshing species are poorly represented from only 10% of all contexts (Motta 2002, 73). Motta (2002, 73) argues that if the archaeobotanical remains analysed are representative, then the increase seen over time in the average proportions of grain, chaff and weeds, in particular during the 7th and the 6th century BC contexts, could support the view that the urbanisation process was completed between the 7th and the 6th century BC at Rome. Distinction between consumers, who were not directly involved in agriculture, e.g. soldiers and the producers, those who were engaged directly in agriculture would be evident.

Archaeobotanical assemblages representing traditional Mediterranean fare were recovered from every property examined from Insula VI.I. Aside from a few unique contexts from the House of the Vestals (VI.I.vii) analysed by Ciaraldi (2001, 213) which yielded exotic items like pepper and spices the majority of the archaeobotanical assemblage from the properties of Insula VI.I can be considered standard Mediterranean food items used by the Romans on a daily basis. Hence the archaeobotanical evidence from the properties of Insula VI.I represents the food use of the everyday.

The homogeneity of the agricultural practices within the Roman Empire, based upon the traditional Mediterranean triad, has been noted despite the variations throughout this large geographical area in climate and geology. Uniformity in agricultural practices can possibly be attributed to the retired centurions taking up farming throughout Italy and the provinces and the subsequent development of the large scale productive villae.

The Roman army has always been made up of men from disparate backgrounds and ethnicities from either allied or subjected people (James 1998, 23). Through their military service they most likely would have came into contact with and interacted with other ethnic and social groups aside from the Roman people (Gardner 2001, 43). The military was one constructed social identity amongst many 'nested affiliations' which a Roman soldier possessed and negotiated in his daily life, both with other military officers and the local civilian population (Gardner 2007, 199). Ex-military men, particularly veterans who had served together, would be aware of their mutual shared status, as for example Sulla's veterans awarded for their service with land in Pompeii, likely formed what James (1998, 15) calls 'a self-conscious 'imaged community''.

Sullan's veterans, who settled in Pompeii in 80 BC, were likely a heterogeneous group of soldiers from different parts of the Roman Empire, including from different regions of Italy, with different ethnic, social, political, cultural, religious, and kinship relations who influenced their interaction with each other and the local populations they encountered in their military service and their subsequent retirement from military service in Pompeii. However, this perceived community of soldiers and their dependents also would have come into daily contact with and developed increasing ties with the local Samnite population of Pompeii through social, legal, commercial and financial interactions (James 1998, 24). 'These relationships [would have been] dynamic across time and space' (Gardner 2001, 43-4). Hence, the veteran's military status and identity was not static, rather they were influenced and modified by many factors which would have altered their understanding of their own ethnic, social, political and cultural identities including the veterans new status as land-owning Roman colonist in southern Italy within the former Samnite city of Pompeii (Haynes 1999, 11-12; Mattingly 2004, 15). Thus, for these military veterans their multiple culturally and socially constructed identity categories would have intersected on multiple levels, transforming their perceptions of their identities as ex-Roman soldiers (Gardner 2007, 208).

These culturally constructed aspects of the veteran's identity would require both 'maintenance and negotiation on the point of actors at the same time as having associated structural features and obligations' (Gardner 2007, 202). Mattingly (2004, 15) argues that the Roman military used material culture, e.g. armour, weapons, and clothing, to distinguish soldiers from civilians rather than uniting them with the local populous. Therefore, it is not unreasonable to assume that other aspects of social identity could be expressed through different media such as diet (Millet 2007, 71). Both food and the technology and labour that are used to prepare it are part of everyday life and therefore have both implicit and explicit cultural significance

(Gardner 2007, 90; Gumerman 1997, 105). Eating is an integral part of daily practice and reinforces cultural and socially constructed identities (Gardner 2007, 128-9); 'people doing things in a particular way' (Gardner 2007, 199). Hence, '[i]n complex societies individuals from distinct social, economic, gender, or age groups often consume different foods because of various economic, political, and ideological factors' (Gumerman 1997, 105).

Although these retired Roman soldiers would have been from various parts of the Roman Empire and exposed to difference agrarian traditions throughout their childhoods and military careers it appears that, based upon the archaeobotanical evidence, within the Mediterranean region a standard set of crops were grown consistently. It is likely that through these agrarian, economic and political actions the process of Roman civilization was transferred, in particular the diet, throughout the Empire (Longo 2000, 160-161). Indeed, it appears that it was the capital of the Roman Empire that forced its own diet onto the people they encountered.

Dining

Within Roman ideology there were several aspects of food which were regarded as lifting it from the level of the uncivilized barbarian masses and making it acceptable within civilized Roman society including the social nature of food consumption, conviviality, the types of foods selected and how these foods were prepared. Within Roman society bread, along with wine and oil, became the classic symbols of civilization and therefore separated humans and beasts. Hence, through these processes the physical act of eating, which even barbarians did, was regarded as being transformed and elevated into a civilized social and communicative event with other members of the same social class (Montanari 2000, 69).

Elite feasts were a social mechanism by which membership of elite status was confirmed and social competition occurred within culturally defined boundaries (Dietler 1998, 98). Through examining the systems of redistribution in Roman society one can begin to understand how the divisive qualities that Roman food possessed could be used to naturalise, confirm, and enforce inequality in wealth, power, social status, and gender relations (Bourdieu 1984 cited in Dietler 1998, 98; Corbier 2000; Garnsey 1999; Goody 1982; Hastorf 1995). Goody (1982, 97-98) has put forward the theory that the presence of a strongly hierarchical cuisine, as seen in Roman society, reflects a society that is strongly stratified both socially and politically, characterized by sharp divisions between elites and common people, as expected in a relatively advanced agricultural system, involving the use of the plough and the exploitation of slave labour.

Breakfast or *Jentaculum* was eaten first thing upon rising and consisted of bread and fruit. The midday snack or *prandium*, was a cold vegetarian meal consisting of bread, olives, wine, and vegetables dressed in olive oil with figs (Dupont 2000, 126; Strong 2002, 18). Both were simply meals taken standing up with no utensils. Practically, the one proper meal of the day for Romans was dinner, or *cena*, taken in the evening or the grander form *convivium*, a substantial meal eaten reclined along with guests (Sassatelli 2000, 107; Strong 2002, 19). The *cena* comprised three courses. The first course was the *gustatio*, which consisted of two symbolic foods, eggs and olives, and was accompanied by bread and honeyed wine. The second course, *secundae mensae* normally consisted of fruit dishes, which having been cultivated were considered the most 'civilized' aspect of the dining experience, and included walnuts, dried figs, and grape preserves (Dupont 2000, 125).

Eating was one means by which Romans were able to situate themselves within 'time, space, and society' (Dupont 2000, 115). Latin literature is replete with references to dining and cookery as analogies to the complex social obligations of Roman society: 'entertainment demanding reciprocity; sharing and hospitality that may highlight either friendship or subjugation; extravagant spectacle countered by virtuous frugality and abstemiousness' (Bober 1999, 169, 176). The *cena* forced diners into a hierarchical and competitive environment in which the coded nuances and understood symbolic meaning were apparent to the largely male audience in attendance and was conducted for the enhancement of the head of the household's reputation amongst the invited male guests (Bradley 1998, 37, 52). To the Romans one could ask the well-known maxim: 'Tell me what you eat and I will tell you who you are' and equally 'Tell me with whom you eat' (Dupont 2000, 114).

Early Roman period meals were served in the *atrium*, used as a formal anteroom for the reception of clients and then subsequently in the room labelled the *cenaculum*. Initially the *tablinum* doubled as a dining room. Later, when reclining became fashionable, *triclinia* were added, which were small rooms with three couches for Greek style dining, often with excellent views for the guests' enjoyment. Wealthy Romans would have had larger banqueting halls (Strong 2002, 28). Wallace-Hadrill's hierarchy of domestic space based upon the public to private spectrum reflects Roman social structure of the Late Republic and Early Empire. Hence the *atrium* and *tablinum* were considered public areas of the house in which uninvited guests or clients might enter whereas the dining rooms, baths and bedrooms were reserved for invited guests (Clarke 1991, 12).

This division between public and private, as defined by Wallace-Hadrill, is difficult to see in the recovered distribution of mineralised or carbonised archaeobotanical remains from Insula VI.I. Looking at aspects of dining within the House of the Vestals (VI.I.vii) (Figure 84, Figure 85), the areas that are normally classified as dining areas, i.e around the peristyle and towards the back or more private areas of the House, cannot be distinguish from the public areas of the House, based upon the archaeobotanical assemblage distribution by room. In the 1st century AD there are relatively few carbonised remains recovered from the back area of the House of the Vestals. It should be acknowledged that this may also be due to taphonomic or preservational issue.

Looking at the other main domestic residence within Insula VI.I, the House of the Surgeon (VI.I.x), no discernable patterns of public **versus** private can be seen based upon the archaeobotanical assemblages as displayed by room (Figure 79, Figure 80). According to traditional ideas of Roman dining the garden area would be a favoured dining spot in the warmer months. However, as mentioned previously only a few specimens of archaeobotanical remains were recovered from this area, Room 16/20, located at the back of the House of the Surgeon. This, again, may be due to the damaging preservation or taphonomic issues created by garden areas.

Interestingly, the areas within Insula VI.I that do show a slight difference between their front and back rooms or public and private sections, respectively, are the tiny commerical buildings such as the Bar of Pheobus and Bar of Acisculus. In both bar properties the back rooms fronting onto the Vicolo di Narciso have recovered carbonised and mineralised archaeobotanical remains from the 1st century AD and not from the earlier 1st century BC time period. It may be that this is simply a taphonomic issue as both bar properties were converted into and used in their present form in the 1st century AD and it may be that the ecofacts present from the 1st century BC have either been destroyed or removed in the construction process.

The Inn and the Soap Factory, as commercial properties, did not reveal any patterns in the archaeobotanical assemblages as displayed per room in terms of separation between commercial and more private activities within these properties. Although there are shifts in the archaeobotanical assemblages between rooms and time perids, as will be discussed in Chapter 7, there are no visible differences in room usage or patterning in the archaeobotanical assemblages in these two commercial properties.

Wallace-Hadrill's theory regarding public and private divisions within the Roman household were not applied to the Triclinium, which was essentially one area in the 1st century AD, located within the Inn. In addition, although the spatial distribution of the recovered archaeobotanical assemblage was plotted and examined for the Well and the roads, the Via Consolare and the Vicolo di Narciso, from Insula VI.I, the same dimensions of analysis of private and public that have been applied to domestic residences were not used on these areas of Insula VI.I.

Luxury

van der Veen (2003, 405) defines luxury foods as those that offer refinement in taste, texture, fat content or another desirable quality (either as a stimulant or inebriant) and distinction, either because of their quantity or quality or because they offer pleasure and enjoyment based upon the qualitative refinement of a basic good. The emphasis on quality, style, labour-intensity, rarity and 'foreignness' is characteristic of hierarchical societies with institutionalized forms of social ranking such as is found within Roman society (see Hayden 1996). According to van der Veen (2003, 406), fundamentally they represent an indulgence because luxury foods always contain an element of exclusivity and can eliminate or create social distance. For example, Cicero regarded oysters, fish, and home-baked bread as markers of high status within the Roman world as only aristocratic households could afford to bake their own bread (Corbier 2000, 132, 134).

The elites frequently developed and cultivated symbolic meanings for select foods and would limit the access of these food items from their social inferiors (Garnsey 1999, 6). Emulation by the lower classes was thwarted by use of exotic foods, acquired through exclusive networks of acquisition, and the proper consumption paraphernalia surrounding these food items (van der Veen 2003, 405). If a strict monopoly was not maintained than the symbolic force of the luxury food item would become 'devalued' and could cause new shifts in style to arise in reaction to the attempts of emulation (Dietler 1998, 98). Wallace-Hadrill (1994, 8) comments upon this paradox of luxury as the elites' attempt at exclusion also makes them socially penetrable.

A recurrent theme within the Classical world was the corrupting qualities of luxuries and their associated negative connotations (van der Veen 2003, 405). Ancient Roman literary sources describe the powerful role of food in projecting an individual's moral and cultural values (Gowers 1992, 4). Gastronomy became a socio-political metaphor within Roman society, reflecting certain tensions in Roman society that are documented from the early 2nd century BC onwards. Diametrically opposed views arose from the conflict between the newly created folk myth during the reign of Augustus of traditional Roman values of archaic frugality and conservative pride in native Italic agrarian roots and emerging new attitudes open towards foreign modes of living and thought which took hold during the expansion and duration of Roman power (Garnsey 1999, 10; Strong 2002, 18). This presented for the Roman elite a dichotomy between conservative ancestral values and the increasing reality of the multicultural ways of life within Roman society; it could be compared to the difference between high culture and the vernacular (Bober 1999, 176).

Although relatively unacknowledged within the constructed dietary model of Greco-Roman ideology with its triad of traditional Mediterranean food, bread, wine and oil, pulses were widely used in Italy and were likely an important protein component of the poor person's meal (Montanari 2000, 73). The most important for human consumption were broad beans (*Vicia faba*), chickpeas (*Cicer arietinum*), lentils (*Lens culinaris*) and peas (*Pisum sativum*) (Garnsey 1999, 15); beans appear to have been more favoured than peas (Brothwell and Brothwell 1998, 106). Unsurprisingly, it is believed that the majority of the protein found in the Etruscan diet came from pulses (Sassatelli 2000, 108). There is a relative neglect of pulses in modern accounts of ancient diets, based upon the frequent references made to them in a variety of ancient sources, it is known from both Greek and Roman contexts that both the rich and poor included lentils, beans and chickpeas as well as cereals in their diets (Garnsey 1999, 15, 19, 121).

6.8 Historical perspective on food

From the 5th and 4th centuries BC the surrounding Etruscan and Greek civilizations' cuisine became increasingly more sophisticated (Dupont 2000, 123; Strong 2002, 21). During the 3rd century BC Pliny (NH XVIII. 83-84) notes that emmer was the staple food-grain of the Romans, as he mentions that at one time the Romans were regarded as 'gruel (*puls*) eaters' as grain was traditionally pounded in a wooden mortar mixed with water to produce gruel or baked to form an unleavened flatcake (Corbier 2000, 132; White 1995, 39-40). Archaeobotanical evidence supports this claim; emmer was the main crop of the Mediterranean basin and was grown throughout Roman Italy (Bruan 1995, 34). Although only a few grains of emmer were recovered from Insula VI.I.

Although the Roman citizen farmer was the ideal citizen throughout this period, an individual who owned and cultivated his land to meet his family's dietary needs, grain production was becoming an increasingly unprofitable venture. Roman law required grain to be sold at cost or distributed free of charge while the raising of livestock was a more profitable activity. Contrary to their generic name Dupont (2000, 127) proposes the unlikely theory that the Romans preferred vegetables to wheat which was largely eaten during food shortages. Dupont argues that only those of considerable wealth and belonging to the property hierarchy were able to raise wheat; with the vegetable and fruit garden being the 'poor man's farm'. However, this theory contradicts the archaeobotanical evidence to date which shows a strong and consistent presence of cereals including barley, emmer, free-threshing wheats and millets throughout the Roman period, including throughout the city of Pompeii (Table 9, Table 10) and Insula VI.I. (Table 8).

The use of culinary practices to mark social distinction increased in the 3rd century BC in tandem with a decline in subsistence farming. Influences from the Greek colonies in *Magna Graecia* from southern Italy, now under tighter Roman control, increased. This tension was illustrated by the general approval by the Roman people of the sack of Syracuse in 212 BC, as both Sicily and southern Italy represented to the Romans the gastronomic flesh-pots of Hellenistic culture and the all corrupting influences associated with this metaphor (Bober 1999, 169). These cultural and gastronomic pressures surged with the continued absorption and increasing contact with other foreign cultures over the course of the Imperial conquest and domination of other people during the late Republic and Roman Empire. Roman society was being transformed by these processes including the growing cash economy and the increasing wealth and expenditure of privileged groups. Romans generally regarded the growing disparity in terms of their traditional status hierarchies, even if, paradoxically they were aware that these newly created hierarchies of consumption were destroying not only the rules on which their civilization was based but also their own self-image as a Roman people (Corbier 2000, 138).

Wealthy Romans often demonstrated their social status with extravagant displays of consumption of exotic foodstuffs; 'much wealth was spent for these purposes rather than engaging in cautious saving or shrewd investment' (MacKinnon 2004, 11). Grand dinner parties were the defining event of 1st century AD Roman society. Overindulgence in food and drink became a standard political weapon in late Republican and early Imperial Rome (Garnsey 1999, 10). This over-indulgence by elite Romans in luxury food items led to a series of senatorial Republican sumptuary laws passed in an attempt to rein in displays of conspicuous consumption including a decree in 161 BC limiting the amount that could be spent on a single dinner, the number of guests invited to a dinner party and banning eating fattened hens and in 115 BC the infamous ban on the eating of fattened dormice, shellfish and imported birds (Strong 2002, 21). Although it appears to have had little effect upon these lavish lifestyles, it does highlight the contradiction between elite indulgence and the undercurrent of conservative values which still respected the idea of frugality in regards to food consumption (Cordier 2000, 137; Goody 1982, 103). Like most elites throughout history, wealthy Romans had relatively free access to every kind of food available in their society, standard or prestigious, in larger quantity, at a higher frequency throughout the year and of a better quality (Garnsey 1999, 127; Mennell, Murcott and van Otterloo 1994, 54).

Aside from (biased) literary and artistic ideological representations, the so-called 'food of the others' was not an issue for the Romans, whose diet was largely free of taboos. Interestingly, Pliny the Elder advised the avoidance of beans as a food as they were thought to contain the souls of the dead. In a passage in Athenaeus, eating beans is thought to 'amount to the same thing as eating the heads of one's parents" (Garnsey 1999, 88). These ancient references would offer one explanation for the use of beans in memorial services for dead relatives (Garnsey 1999, 19). With the Roman ancestor cult of worship the feeding of the departed would have maintained the domestic relationships which existed among the surviving ancestors (Goody 1982, 191).

Throughout the empire, the tables of elite Romans displayed varieties of local and foreign food items (Longo 2000, 160). The transformation of the regional food of peasants with the addition of exotic items characterises high cuisine throughout a number of cultures. Roman elites do not appear to have distained plebeian meals. Rather, they created distinct dishes with the same ingredients as the peasants but with different methods of preparation and rules of eating the dish. 'When they ate broad beans, for example, they added a costly, refined sauce that completely disguised the beans' taste' (Corbier 2000, 135). This may be one reason why exotic and/or expensive food items were not found within the elite properties from Insula VI.I, including the House of the Surgeon and the House of the Vestals. Hence, Roman high cuisine was largely based upon the subtle art of complexity and transformation of regional foods with exotic luxury foods. Therefore, a strong image of social superiority was projected to their social and economic inferiors through the combination of consuming these basic food staples along with a variety of other more inaccessible foods (Garnsey 1999, 19; Goody 1982, 191).

Typically, the higher the status of the individual, the wider the range of contacts with the outside world and incorporation of a greater range of foreign items into their diet (Goody 1982, 105). This trend is well illustrated by the spice trade in antiquity, which expanded after Alexander the Great's conquest of the east. Romans adopted Greek cuisine by the end of the 3rd century BC as witnessed by the great increase in exotic spices and herbs (Strong 2002, 23). By the 1st century AD it is believed that the Romans enjoyed very sweet and highly spiced foods (Edwards 1985, xxix). The etymology of the word spices comes from the Latin *species*, referring to an item of rare value as opposed to items of ordinary commerce. The Roman spice trade drew on products from China, South-East Asia, India, Persia, Arabia and East Africa and the Spice Quarter became a well-known part of the city of Rome (Cappers 2006, 165; van der Veen 2008, 26). Spices were traded for precious metals, demonstrating the wealth of the Roman Empire at this time. Like other rare commodities, spices became associated with the social hierarchy of Roman society (Goody 1982, 105).

Cookery

In Greece and Rome recipes were written to provide a base upon which set practices could be further elaborated in the kitchens of the very wealthy. This is illustrated by *De re conquinaria* (On Cooking) by *Apicius*, the earliest cookery book to survive, which was only directed at a 'favoured few' (Goody 1982, 103; Strong 2002, 22). Within Roman society the virtues associated with foods were not constant or intrinsic to that food but rather could be altered by artificial means, including its associated environment, preparation, and culinary transformations. The huge variety of food items mentioned by classical writers could even be extended by dividing these items into finer distinctions including geographic provenance, climate, and cultivation methods (Mazzini 2000, 145).

In reality food had no place in the moral discourses of the majority of the poorer Greek and Italian populous, people who did not see the great quantities of luxury foods described by ancient authors. On the other hand, no detailed description of a 'standard' Roman meal is recorded in the surviving ancient sources (Gowers 1992, 2). The diet of the Roman middling and lower classes, aside from local variations, was most likely heavily concentrated upon cereals, along with olives and grapes and other less important crops such as millets, oats, and rye (Bruan 1995, 25; Garnsey 1999, 13, 15; White 1995, 38-39). Supplementing this diet would have been dried or fresh vegetables and fruits, honey, a variety of nuts including walnuts, almonds, hazelnuts, pinenuts, and chestnuts, and animal protein from milk, cheese, meat, and fish (Corbier 2000, 129).

Gardens

Similar to other Mediterranean cultures, the Romans relied heavily upon vegetables for their diet. The ideal was for every Roman family to have their own garden to cultivate their own produce, even within urban areas (Jaskemski 1979, 89). Vegetables from the garden, fruits and grapes were considered the most civilized food, pure and uncorrupt, edible raw or in the form of salad or wine (Strong 2002, 20). The Romans regarded vegetables and grains as being partially or wholly cooked naturally by the sun's rays while growing in the field and therefore after harvesting there was little worry of decay. During peacetime, the prevailing folk notion of a meal composed of 'civilized' garden vegetables were all a Roman living in the country needed.

Roman gardens or *horti* encompassed vineyards and orchards, which were regarded as the most civilized type of land taken into cultivation. Roman gardens were expected to produce vegetables all year round through a variety of techniques for irrigation and fertilization. The Roman category of *holera*, meaning a variety of different green vegetables from the garden, included colza, tubers, and edible bulbs, specifically this term included 'various kinds of cabbage, varieties of cardoon (*Cynara cardunculus*, a Mediterranean plant related to the artichoke and cultivated for its edible leaves and roots), salad greens, leeks, herbs, turnips, carrots, parsnips, garlic, and onions' (Dupont 2000, 118). What was not included under the term *holera* fitted under the term *legumina* which included beans, peas, and other pod plants.

Jashemski (1979, 24) estimated that productive gardens account for 9.7% of the urban area while ornamental gardens account for 5.4% of the known space within the city of Pompeii. Relatively recent archaeological excavation alongside environmental analysis including pollen and macrobotanical analysis and examining plaster casts of root cavities has revealed the presence of vineyards, fruit orchards, ornamental plants and flower and vegetable beds within the confines of the city. Based upon these calculations small-scale cultivation was an important feature of the urban landscape of Pompeii. It is likely that the produce from these agricultural plots was for local consumption rather than export to other cities (Laurence 1994). At Rome, writes Pliny (XIX.51f) a 'garden was in itself a poor man's farm' (White 1970b, 49).

	Well	Bar of Pheobus	Bar of Acisculus	Soap Factory	Shrine	House of the Surgeon	House of the Vestals	Vestals Bar	Inn	Inn Bar	Triclinium	Via Consolare	Vicolo di Narciso
Barley				X	X		X		X	X	X		X
Einkorn													
Einkorn/Emmer							X						
Emmer					X	X	X	X	X				X
Free-threshing wheat			X		X	X	X		X				
Oats							X						
Rye									X				
Millet													
Common millet			X	X	X	X	X	X	X	X	X	X	X
Italian millet				X	X	X	X				X		
Fruit													
Apple					X		X				X		X
Citron							X						
Date													
Fig	X	X		X	X	X	X	X	X	X	X		X
Grape	X		X	X	X	X	X	X	X	X	X	X	X
Peach				X	X		X		X				X
Pomegranate				X	X		X		X		X		X
Olive	X	X	X	X	X	X	X	X	X	X	X	X	X
Almond		X		X	X						X		
Hazelnut				X	X		X		X				X
Stone pine		X	X	X	X		X		X	X	X		X
Walnut				X	X	X		X	X	X	X		X
Bitter vetch				X	X	X			X				
Broadbean			X	X	X		X				X		
Chickpea													
Lentil			X	X	X	X	X	X	X		X		X
Pea				X	X		X	X	X				
Cypress					X		X			X	X		

TABLE 8: INSULA VI.I PRESENCE AND ABSENCE

	House of Hercules' Wedding	House of Amaranthus	House of the Chaste Lovers (fodder)	House of the Ship of Europa	Temple of Isis	V.III	I.IX.8	I.IX.15	II.ii	IV.7	II.9 (shop)	IV.17 (Herculaneum)	Vineyard II.v	Humble house on Via Nocera (I.XIV.2)	Villa Vesuvio	Villa rustica	Villa poppaea	Shop in Insula Orientalis II.13	House of C. Julius Polybius IX.XIII.1-3	Casa del Bel Cortile
Cereals																				
Barley		X	X	X	X															
Einkorn	X																			
Einkorn/Emmer	X																			
Emmer	X	X	X																	
Free-threshing wheat																				
Oats	X		X																	
Rye																				
Millet																				
Common millet	X	X																		
Italian millet	X	X				X	X													
Fruit																				
Apple																				
Citron	X																			
Date	X	X		X																
Fig	X	X	X	X										X	X				X	
Grape	X	X																		
Peach		X								X					X					
Pomegranate	X	X																		
Olive	X	X																		
Nuts																				
Almond	X	X	X							X					X			X	X	
Hazelnut	X	X	X																	
Stone pine	X																	X	X	
Walnut	X	X								X	X									
Pulses																				
Bitter vetch			X															X		
Broadbean	X		X	X									X	X			X	X		
Chickpea	X	X																		
Lentil																				
Pea	X																			
Tree																				
Cypress	X	X																		

TABLE 9: POMPEII SITES PRESENCE AND ABSENCE

	vineyard II.v	humble house on Via Nocera (I.XIV.2)	Villa Vesuvio	Villa rustica	Villa Poppaea	Shop in Insula Orientalis II.13	House of C. Julius Polybius IX.XIII.1-3	Casa del Bel Cortile	Casa dell Stoffa	Casa del Due Atrii	Casa d'Argo Insula II.2	Marine warehouse of M. Cellius Africanus	Porta Nocera Necropolis	Shops from Strada degli Augustali	Carbonised Hay from Oplontis	Naples Museum	Pompeii Museum
Cereals																	
Barley											X					X	
Einkorn								X									
Einkorn/Emmer																	
Emmer																	
Free-threshing wheat															X		
Oats								X									
Rye																	
Millet																	
Common millet																	
Italian millet																	
Fruit																	
Apple																	
Citron																	
Date																	
Fig													X	X			
Grape		X	X			X				X							
Peach																	
Pomegranate			X														X
Olive											X						
Nuts																	
Almond																	
Hazelnut				X			X	X								X	X
Stone pine											X						
Walnut				X	X						X						
Pulses																	
Bitter vetch											X		X				
Broadbean							X									X	X
Chickpea	X	X				X	X									X	X
Lentil											X						
Pea									X	X	X		X			X	X
Tree																	
Cypress											X						

TABLE 10: POMPEII SITES CONTINUED PRESENCE AND ABSENCE

Cereals

Romans classified cereals into a hierarchy in which barley was replaced with hulled wheats in terms of preference. Hulled wheats were in turn replaced in preference with naked/free-threshing wheats. The literary and archaeobotanical evidence imply that during the Roman period various hulled grains, such as emmer and einkorn (*Triticum monococcum* L.), and millet were less fashionable but persisted alongside other more valued varieties (Garnsey 1999, 1-11; Spurr 1983, 4).

Einkorn

There was only one firm identification of einkorn from the Vesuvian sites, from the House of Hercules' Wedding (VII.IX.xlvii) recovered from a pit dating to the 4th to the early 2nd century BC. From Insula VI.I the only instance of possible einkorn/emmer identification came from the House of the Surgeon (VI.I.x) from the 1st century AD. Ciaraldi (2001) identified possible einkorn from the House of Hercules' Wedding (VII.IX.xlvii) from the cesspit and pit dating to the 4th to the 2nd century BC and the House of the Vestals (VI.I.vii) dating to the 4th century early deposits and to the late 1st century BC from the toilet feature.

Emmer

Although farmers planted a wide range of cereals, based upon regional variations suitable for certain areas, two varieties of wheat, naked wheat and husked emmer, have received the majority of attention from historians (White 1995, 38-39). Emmer (*Triticum dicoccum* L.) was the most widely cultivated husked wheat in Roman Italy for over 300 years, into the Imperial period, until at least the 5th and 4th century AD (Bruan 1995, 34-6; Meyer 1988, 215; Spurr 1985, 13). Ancient sources (Columella and Pliny) report that Campanian fields with their fertile volcanic soils were sown twice a year, in the autumn and spring with emmer and once with millet (Spurr 1986, 11). According to the ancient author Varro the best emmer wheat came from Campania (White 1970b, 66).

Emmer, the hardiest of the wheat grains, was utilized by the Romans because it tolerates a wide range of soil conditions and poorly prepared land, and can be sown earlier than other cereals. Moreover, emmer is more nutritious than naked wheats (Meyer 1988, 215). Ancient Romans are known to have used emmer pottage (Pliny XVIII. 84) or emmer grain and salt in their sacrifices (Bruan 1995, 34). Emmer could be fed green, in a husked state, in a mixture with barley, vetch and other legumes known as *farrago* as fodder to livestock. Pliny writes that emmer was mixed with rye to make it more palatable; he further implied that the practice of polyculture was occurring (Spurr 1986, 13). Like oats (*Avena sativa*), rye (*Secale cereale* L.) does not thrive well in the dry Italian climate (Hughes 1994). The fact that rye was found archaeologically in *San Giovanni in* contexts dated to 30 BC- AD 200 and from the Inn in Insula VI.I dating to the 1st century AD suggests that Pliny's delimitation of rye to the north-west of Italy was perhaps not entirely accurate (Spurr 1983, 3).

This apparent disappearance of emmer (hulled wheat) during the Roman period may be due to an archaeological sampling bias. Many Roman sites have been excavated without an appropriate archaeobotanical research design. Emmer has been recovered in limited frequencies from other sites in the Mediterranean into the 1st and 2nd century AD (Spurr 1983, 4) including from Insula VI.I.

The archaeobotanical evidence from Pompeii reveals that emmer wheat had a continued presence at Pompeii (Table 8, Table 9, Table 10, Figure 129). From Insula VI.I emmer wheat was found in contexts dating to the 1st century BC in the Inn, Vicolo di Narciso and from the 1st century AD from Vestals Bar, the Shrine and from both time periods from the House of the Surgeon (VI.I.x). Emmer wheat has been recovered from the House of Amarantus (I.IX.xii), from the garden area dating to the 1st century AD (Robinson 2002, 96). It was also identified by Ciaraldi (2001, 151) from the 4th to 2nd century BC from the House of Hercules' Wedding (VII.IX.xlvii), the House of the Vestals (VI.I.vii) from the 4th century BC and late 1st century BC. In addition it was also found in the fodder from the House of the Chaste Lovers (IX.XII.vi) and the Naples Museum.

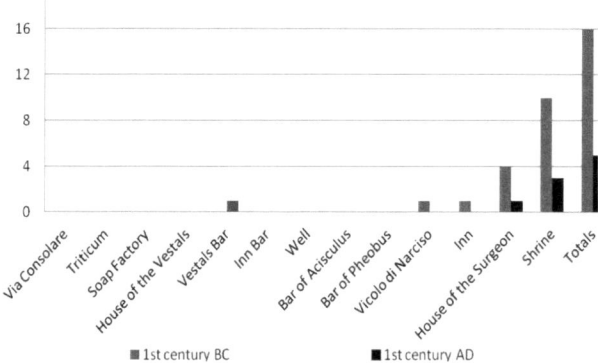

FIGURE 129: COUNT OF *T. DICOCCUM* FROM INSULA VI.I

Barley

According to Pliny (NH XVIII. 74) barley (*Hordeum vulgare* L.) a crop of minor importance was the first cereal grown in Italy but declined in status to animal fodder (Bruan 1995, 32-33; Garnsey 1999, 1-11). Barley cultivation has been largely underestimated despite the fact that it was probably grown throughout Italy (Spurr 1983, 6). Indeed, hulled barley was the dominant cereal from the Imperial period in the south of France, at sites such as Ambrussum (Martinez 2005, 347).

Barley is suited to the Mediterranean environment because it is somewhat drought resistant and can grow in soils of low fertility (Jongman 1988, 81). Ferrio et al. (2005, 215-216), in their controlled agricultural trials, found that in a Mediterranean climate barley has a shorter vegetation cycle than wheat, thus reaching maturity roughly two weeks earlier. Mediterranean-grown barley would thus form under moister conditions than wheat grains, even when grown in the same area. This growing habit is significant because the probability of drought increases greatly during the last weeks of the crop cycle, through May and June. Thus, as barley is a less drought-sensitive cereal crop than wheat, it is common practice in dry areas to keep barley as a rain-fed crop and reserve additional water or moister soils for wheat. Barley is an excellent backup resource for small farmers, although it is heavier and bulkier to transport (Jongman 1988, 81; Spurr 1983, 14).

Roman period texts (Pliny, Columella, Varro) mention the low status of barley yet note that it continued to be consumed in rural areas during the late 2nd and early 3rd centuries AD (Ciarallo 2000, 55; Garnsey 1999, 119). Fluctuations in the food supply brought about competition between human and animal consumers for various categories of foodstuffs including barley; it is questionable whether slave populations ate mostly barely. Ancient sources make numerous references to the use of low-grade cereals and legumes in times of scarcity (White 1970b, 39-40). However, the Greeks, who considered themselves 'bread eaters' ate barley, distinguishing themselves from the surrounding Italic people who ate spelt (*Triticum* spelta L.) (Longo 2000, 155).

Barley rations were served instead of wheat as a collective punishment for the normally well-fed Roman army as early as 214 BC (Bruan 1995, 33). Ulpian (Digest 33.9.3.8) wrote that barley should be stored for pack-animals and slaves (Columella RR 2.9.14 cited in Bruan 1995, 34). When food was scarce a six-rowed variety of barley called cantherinum (horse-barley) was regarded as better for human consumption than bad wheat. Barley was roasted before being pounded in a wooden mortar and made into gruel/porridge with highly nutritious barley-water or made into barley cakes or flat bread (Bruan 1995, 29; Ciarallo 2000, 53; White 1995, 39-40). Medicinally, a traditional potion called tisane containing barley was thought to have been in wide usage (Spurr 1983, 15). Sacrificial cakes made of barley for religious rites provide a clear indication of barley's traditional significance (Hughes 1994, 70).

Archaeobotanical evidence indicates a general decrease in the importance of barley and the increase in the prevalence of naked or hard/common wheats (Martinez 2005, 347). The evidence of barley from Pompeii suggests that barley continued to be in use up until the destruction of the city (Table 8, Table 9, Table 10, Figure 130). From Insula VI.I barley has been found from the 1st century BC from the Soap Factory and from the 1st century AD from the Triclinium, the Inn, Inn Bar, and from both time periods from the Shrine and Vicolo di Narciso. Barley grains have also been recovered from the Pompeii Museum, the fodder from the House of the Chaste Lovers (IX.XII.vi), the House of the Ship Europa (I.XV.iii), V.III, the House of Amarantus (I.IX.xii) from the 1st century AD, the House of the Vestals (VI.I.vii) from the 4th century BC, and 1st century BC and AD, and the *Porta Nocera, Necropolis*. As barley has been recovered from a wide range of properties and areas within Pompeii and Insula VI.I it is likely that barley was in common usage at Pompeii.

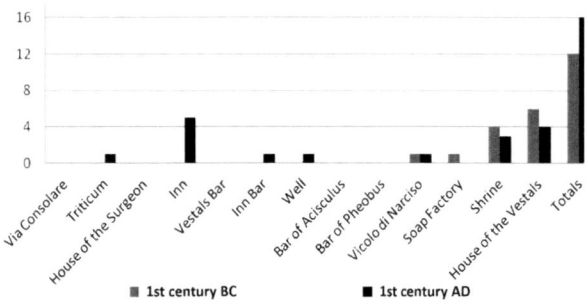

FIGURE 130: COUNT OF HORDEUM FROM INSULA VI.I

Free threshing/soft bread wheat

The cultivation of soft bread wheat (*T. aestivum*) is limited to the north of Italy due to the plentiful annual rainfall of 500-700mm. The lack of spring rain and excessive rain in the winter months presents an obstacle to growing 'soft wheat' in the south of Italy. 'Hard wheat' or 'macaroni wheat' (*T. durum*), grown in the south of Italy, is more drought resistant than bread wheat, and can survive with as little as 300 mm of rain. It is possible that both species were grown in overlapping zones in the centre of the country (Spurr 1983, 1986). *T. aestivum* and *T. durum* grains are impossible to distinguish unless their corresponding rachis segments are present. However, ecological factors would suggest that most naked wheat recovered from Insula VI.I is *T. durum*, recovered from the Shrine, the House of the Surgeon (VI.I.x), the Inn and the Bar of Acisculus. Figure 131 shows a decrease in free-threshing wheat within Insula VI. However, it is the large number of cereal grains [n=26] from the Shrine which are skewing the results towards the 1st century BC. Free-threshing cereal grains were discovered from the ritual deposit from the House of the Vestals (VI.I.vii) dating to the 4th century BC by Ciaraldi (2001, 63) and the *Porta Nocera, Necropolis* by Matterne and Derreumaux (2008, 105).

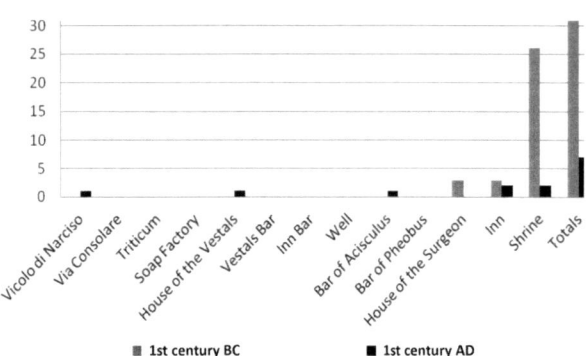

FIGURE 131: COUNT OF *T.AESTIVUM*/*T.DURUM* FROM INSULA VI.I

The apparent increase in the cultivation of naked wheat from the late Republic onwards, was probably due to upper-class urban sophisticated tastes. Porridge was replaced by bread and *Triticum aestivum* and *T. durum* became the most highly valued wheats due to their superior bread-making qualities (Garnsey 1999, 12-21; Spurr 1986). By the late Republic, bread was the preferred form of food and was consumed by large portions of the population (Jongman 1988, 82). Based upon the list of types of breads on the wall of XI.VII.xxiv/xxv at Pompeii it would appear that d

Different grades of bread were available for purchase. These included pane puero, bread for a slave or boy and pane cibar, an inferior bread probably made from roughly dehusked emmer wheat or barley (Mayeske 1979).

Ciarallo (2000, 53) argues that archaeobotanical evidence from 1st century AD Pompeii demonstrates a **preference** for naked wheat. Figure 131 shows the results from Insula VI.I for cereal grains identified as *Triticum* sp., a category that encompasses both *T.aestivum*/*durum* and *T.dicoccum* (i.e both naked and hulled cereal grains). There is a large number of *Triticum* sp. cereal grains [n=197 in total] from the Shrine. Unfortunately the results from this study are not quantitatively substantial enough to see a definite trend in this regard as relatively few identified *T. aestivum*/*durum* cereal grains [n=39] were recovered from Insula VI.I in this study.

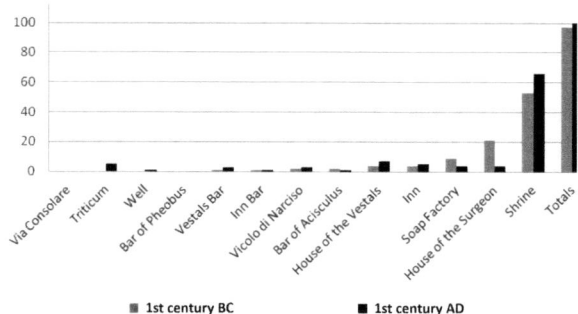

FIGURE 132: COUNT OF *TRITICUM* SP. FROM INSULA VI.I

Bread became the symbolic food of the citizen Roman soldier. Soldiers, whose *prandium* focused on wheat, were regarded as 'bread eaters' in Roman society. From a practical position, wheat is easily transported. Bread and dried fruit were eaten by travelers and soldiers. Bread was regarded as a compact cooked food which hardened rather than rotted over time. This imagery of bread was extended to the soldier's bodies who consumed it, as it was believed to have been heavily armored with a crust, hard and compact (Dupont 2000, 126).

Oats

Avena is the classical Latin name for oat (Meyers 1988, 206). Oats were not part of the Roman diet. None were recovered from the present study but a few were identified in the early deposits from the House of the Vestals (VI.I.vii) and the House of Hercules' Wedding (VII.ix.xlvii) in a pit dating to the 4th to the 2nd century BC by Ciaraldi (2001, 129). Although there is some evidence of cultivated oats it is relatively clear that in the Roman period oats were better known in their wild form as a weed (Spurr 1986, 61). Thus, oats were only cultivated as a distinct crop from the beginning of the 1st century AD and most likely as a replacement for millet (White 1970b, 38).

Oats have the practical advantage that they can thrive in the poorest soils but require more water than other cereals and do not thrive as well in the dry climate of Greece and Italy (Hughes 1994, 5; Spurr 1986, 14). The Romans regarded oats as an important high energy supplement for farm animals in their fodder and it was also later considered a staple in the human diet (Zohary and Hopf 2001, 77). Oats were identified in the fodder from the House of the Chaste Lovers (IX.XII.vi); and thus it is likely that Pompeians were aware of its nutrient value as cattle feed (Ciaraldi 2001, 129). Columella (RR II. 10.32) mentions the use of oats as fodder crop which could be used either green or dried. As oats were likely cut green for fodder only a sufficient amount of grain for next year's sowing were allowed to mature, which may be one factor accounting for the rarity of oats in archaeological finds (Spurr 1983, 1). No oats were recovered from Insula VI.I again reinforcing the theory of a fully urbanised centre by the late 1st century BC.

Millet

A large quantity of common millet was found dating to the 4th to mid 2nd century BC from the cesspit from the House of the Wedding of Hercules by Ciaraldi (2001, 152). Pliny (NH XVIII. 117) wrote of millet's low status as a poverty crop. In contrast, there is almost a complete absence of millet from similar deposits dating to a later time period, from the mid 2nd century BC to AD 79, from the House of the Vestals (VI.I.vii). Ciaraldi and Richardson (2000, 75) suggest that this may indicate disparate levels of wealth between these two domestic properties at different times in their development and perhaps reflect broader economic issues for the city of Pompeii. This shift in the presence of millet could also be interpreted as differences in household tastes or a general decline in the economic importance of common millet over time within Pompeii.

Millet has been recovered from every property within Insula VI.I aside from the Bar of Pheobus (Figure 133). It has also been discovered from the House of Amarantus (I.IX.xii) from the 1st century, the House of Hercules' Wedding (VII.IX.xlvii) and from the city of Herculaneum (Table 9, Table 10). A *dolium* filled with millet was discovered in *Pisanella Villa* near Pompeii; millet was also recovered from the Roman villa site at *Matrice* in *Molise*, dating to 200BC-AD 400 (Spurr 1983, 94).

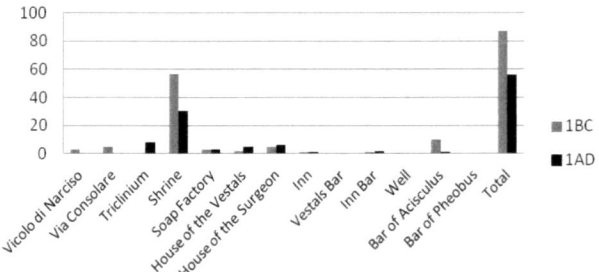

FIGURE 133: TOTAL COUNT OF MILLET FROM INSULA VI.I

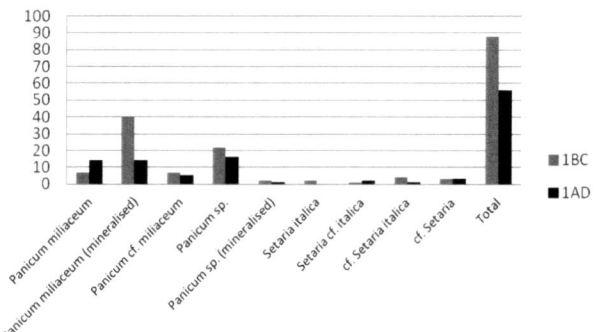

FIGURE 134: TOTAL COUNT OF MILLET FROM INSULA VI.I BY TIME PERIOD

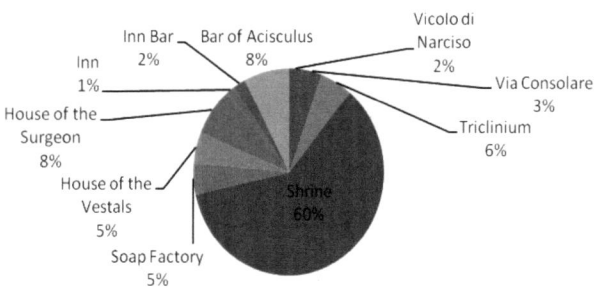

FIGURE 135: TOTAL COUNT OF MILLET PER PROPERTY FROM INSULA VI.I

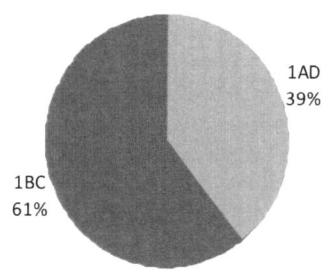

FIGURE 136: PERCENTAGE OF MILLET FROM INSULA VI.I BY CENTURY

More than half (61%) of the millet recovered from Insula VI.I came from the 1st century BC.

The majority of millet from Insula VI.I came from the Shrine and was composed of mineralised *Panicum miliaceum* L from the 1st century BC. Millets are often recovered in a mineralised state due to the fact that during dehusking some grains of millet may retain part or their entire seed coat. Being so tiny millet grains, with their intact pericarp, can pass through the human gastrointestinal tract and facilitate the mineralisation process. *Setaria italica* L. (or Italian millet) was also present in the archaeobotanical assemblage from Insula VI.I but in smaller quantities (Figure 137) including from the Soap Factory, the Triclinium, the House of the Surgeon (VI.I.x) and the Shrine (Table 8). Italian millet has been identified from the House of Amarantus (I.IX.xii), the House of the Vestals (VI.I.vii) and the House of Hercules' Wedding (VII.IX.xlvii) by Ciaraldi (2001), I.IX.viii, I.IX.xv and from the Pompeii Museum (Table 9, Table 10).

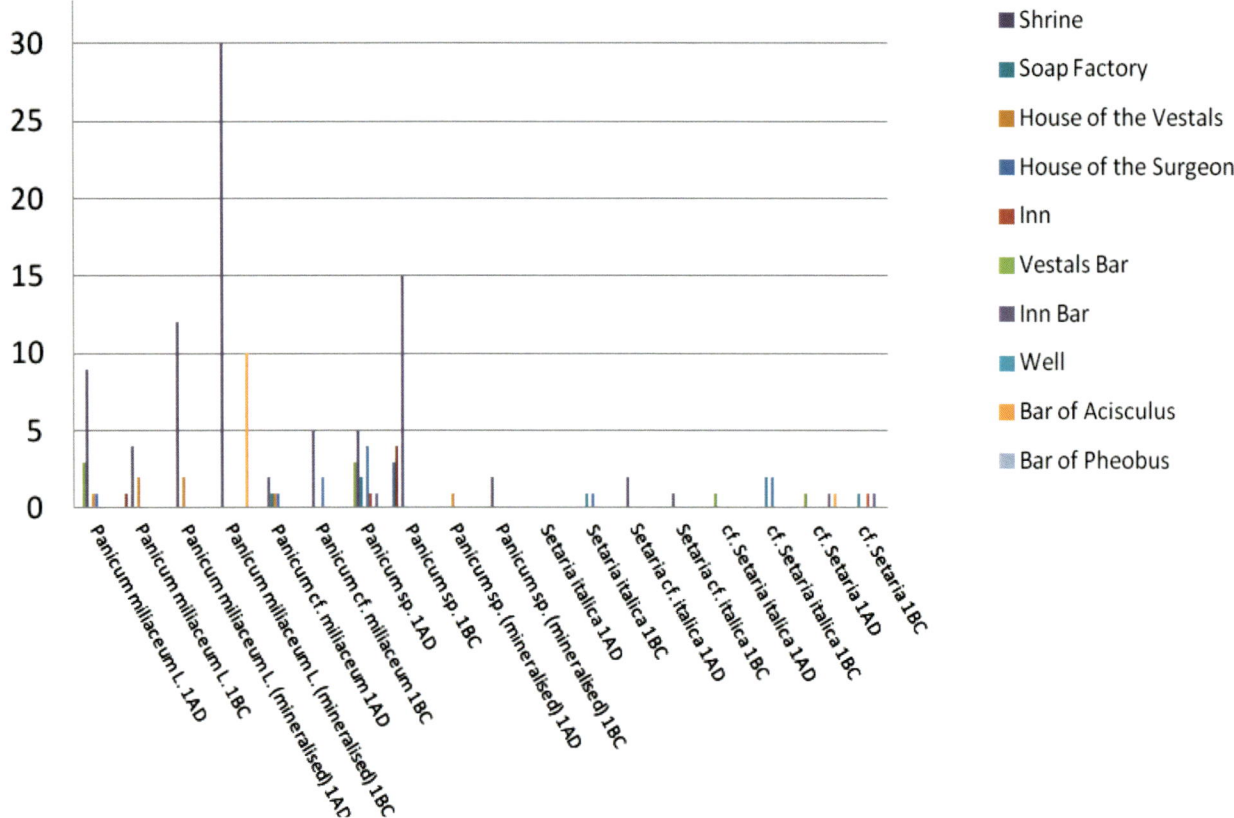

FIGURE 137: TOTAL COUNT OF MILLET PER PROPERTY FROM INSULA VI.I

Milium was the ancient Roman name for common millet (*Panicum miliaceum* L.) (Meyers 1988, 205). It is likely that millet was cultivated in the Po Valley and the region of Campania (Spurr 1983, 1). Indeed, Pliny notes that millet grew particularly well in Campania (Meyers 1988, 205). During the Roman period millet, which possesses a short vegetation cycle of three to four months, sown in the spring or summer, could act as a supplement to the diet if the autumn cereals failed. Pragmatically, millet is very resistant to drought conditions and grows in areas of poor fertility and requires less rainfall than other cereals (Spurr 1986, 89). Ancient sources advocated growing millet in areas not suitable for wheat, sandy or wet soil and foggy areas. This may reflect the Romans' understanding of millet's ecological adaptability. Millet is an important crop in times of scarcity as millet is richer in carbohydrates but poorer in digestible proteins than other cereals making it an excellent appetite satisfier to fend off hunger. Millet was probably a large component of the diet of slave households (Spurr 1983, 99).

The Romans used millet to make a white porridge and sweet bread and ate a combination of beans and Italian millet (*Setaria italica* L.). Millet, which is non-glutinous, makes a heavy flat bread. Mixed with wine-must millet was used as leaven for bread. Bread made from millet was more highly valued than bread made from barley. Millet was also used as fodder. Columella notes that millet was the most highly esteemed chaff for oxen. It was used as birdseed by farmers and excesses were sold at market (Spurr 1983, 102). Despite its many uses and the fact that it was grown throughout Italy, the importance of millet (and panic) is often overlooked (Spurr 1986, 14).

Millet is typically recovered from archaeological sites in smaller quantities than other cereals. This is likely due to the relatively small grain size. Archaeobotanists have simply classified the grains as 'millet'. Certainly, more sophisticated methods of recovery and identification are needed to more accurately assess the types of millets used in Roman Italy.

6.9 Spices

Opium

Opium poppy (*Papaver somniferum* L.) was grown as an ornamental garden plant. It was also recovered as a weed at Pompeii. Its presence is also known at Pompeii from wall paintings (Jashemski 1979b,). Dioscorides (Mat. Med. IV.64) mentions that the seeds were sprinkled on breads and cakes and, as today, used for its oil. Even in classical times its more famous usage was as a powerful narcotic and its use as a pain killer was well known (Jashemski et al. 2002, 139; Zohary and Hopf 2001, 136). No evidence of opium was recovered from insula VI.I although it is likely that the Romans were well-versed with its medicinal properties due to its appearance on other Roman sites.

Pepper

Several fragments of mineralised black pepper (*Piper nigrum* L.) were identified by Ciaraldi (2001, 157) from the House of the Vestals (VI.I.vii) which strongly suggests that the inhabitants had aspects of a high status diet, as black pepper must have been imported from the Indian sub-continent, and may have arrived via the Port of Puteoli (Ciaraldi 2001, 213). In contrast, thousands of peppercorns have been recovered from dump deposits and religious contexts at the important port trading centre of Berenike (founded approximately 275 BC) in the south of Egypt on the Red Sea coast, before being traded on to the rest of the Mediterranean (Capper 2006, 165-166).

Romans used the native myrtle berry as their prime seasoning. The discovery of pepper (*Piper nigrum* L.) supplanted this milder berry. Apicius (cited in Edwards 1985, xx) recommends using pepper in the preparation of most meats, vegetables, and fish, sprinkled over cooked dishes before they were served. No firm evidence of black pepper has been recovered from Rome to date and little trace within the Roman Empire.

Sesame

Theophrastos (Gallant 1985 cited in Zohary and Hopf 2001, 140) noted that sesame was widely cultivated in the Greco-Roman world, largely more for its edible seeds than its pure oil. Along with millet sesame was one of the main summer crops. It is possible that the sesame recovered and identified by Ciaraldi (2001, 158) from the House of the Vestals (VI.I.vii) was grown in nearby farms. No evidence of sesame was recovered from the other properties within insula VI.I.

6.10 Fruit

Olive

Margaritis and Jones (2006, 400) proposed that deposits containing large carbonised fragments of olive endocarp may represent the post-processing remains of olive oil by-products, as well as the remains of olive consumption. Olive pressing experiments undertaken by Warnock (2007, 1) found that Roman mills, with their roller and basin methods, produced a higher proportion of smaller fragments of olive endocarps in comparison with other forms of olive oil milling from the Near East. Experiments of olive oil processing using traditional style Roman mills by Tyree and Stefanoudaki (1996, 175) reached the same conclusion, with approximately 50-57% of medium and large sized olive stones crushed in the process. Modern and traditional methods of olive oil production crush the entire olive fruit including the stone to increase oil production as the broken edges of the pit continue to grind the olive fruit releasing more oil and facilitating the flow of the oil.

Pliny (NH XVII. 49) advocated the use of olive ashes as a fertilizer. Cato (On Agric. XXXVII. 2-3 cited in Ciaraldi 2001, 147) advises farmers to 'separate part of the olive seeds and throw them into a pit, add water and mix them thoroughly with a shovel. Make trenches around olive trees and apply this mixture, adding also burnt seeds'.

Pliny (NH XV. 33) complains that Cato prescribed olive lees, the black fluid left after the oil was separated, for all manner of uses. However, this does not appear to be the case as Varro (1.55.7) decried the fact that lees uses are generally not appreciated. It is known that olive lees were used to coat and preserve a variety of organic items including: heaps of grain were covered with a mixture of lees and chaff on which pure lees were then poured to protect it from infestations and rodents (Cato 92); dried figs were coated in lees and preserved in jars (Cato 99); as was oil (Cato 69, 100); and Myrtle twigs and berries, and fig branches with leaves and fruits were also preserved in lees (Cato 101 cited in Frankel 1999, 45). Cato (130) stated that 'soaking firewood in lees improved it' (cited in Frankel 1999, 45). Lees were also used as a preservative element in plaster, in which chopped straw, chalky or red earth and lees was combined (Cato 128). Grain threshing floors were often prepared by adding lees to earth and then tamping it down as the lees prevented insects from infesting the grain and the earthen floor from becoming muddy (Cato 91 cited in Frankel 1999, 46).

Aside from the Bar of Pheobus, evidence of grapes has been recovered from all properties within insula VI.I (Table 8). Ciaraldi (2001, 147) found a high concentration of burnt olive stones in one of the pits from the garden area (Area 41) from the House of the Vestals (VI.I.vii). Neef (1990 cited in Ciaraldi 2001, 147) has suggested that the round edges of the carbonised fragments of olive endocarps may indicate that they may have been broken before charring occurred and this could be an indication of burning lees (the residue from pressing). Indeed, Pliny (NH XV.22) recommends making use of these olive stones rather than wood as a fuel because olive stones generate a good heat with little ash produced and would have made ideal kindling, or lighting material within the household (Foxhall 1998; Rossiter 1981; Smith 1998 cited in Ciaraldi 2001, 171). By contrast Robinson (1999 cited in Ciaraldi 2001, 171) interprets the charred olive stones, dating to the same period from the House of Amarantus (I.IX.xii), as evidence of the culinary practice of preparing pickled olives.

Specific artefactual evidence from Pompeii of olives includes: an amphora found in the peristyle of the House of *M. Gavius Rufus* (VII.II.xvi) with the label *OLIVA* (CIL IV 5598b); a fragment of another amphora had the label *OLIVAS* (CIL IV 5762); a silver cup found at Pompeii, part of the silver hoard in the House of Menander (I.X.iv), has olives with leaves sculptured in high relief and realistic enough to be a botanical model (Jashemski et al. 2002, 135). Like the Greeks, the Romans used olive wreaths to decorate cavalry squadrons and to celebrate a minor victory. In ancient times the olive branch was a symbol of peace, more so than the laurel (Pliny NH XV. 134). Olives also had a number of medicinal uses.

Recent charcoal analysis of the House of the Vestals (VI.I.vii) revealed an increase in diversity in the wood charcoal assemblage after the 89 BC colonisation by

Sullan troops, which included orchard wood from fruit trees like *Prunus* spp. and *Vitis vinifera* (grapevines). No *Olea* sp. (olive) charcoal evidence was recovered from the House of the Vestals (VI.I.vii) (Veal 2009, 175; Veal and Thompson 2008, 7). In her analysis of the House of the Vestals (VI.I.vii), Ciaraldi (2001, 171) noted that charred fragmented olive stones comprised the majority of finds. Olive stones are fairly ubiquitous and usually recovered from most Mediterranean archaeological sites (Finley 1973, 31; Terral et al. 2004, 63-64). Indeed, olives were recovered from all the properties within Insula VI.I. Taphnomically, despite their compact and hard stone pits, olives are frequently fragmented so that the number of intact samples is often limited. Technically, the olive stone 'is a fusiform, uni-integumented and sclerified endocarp, composed of two asymmetric valves [separated by a longitudinal suture line] protecting one seed' (Terral et al. 2004, 64). The surface is covered in longitudinally aligned furrows. Due to the abortion of one of the ovules the sterile valve is usually flatter than the other fertile valve.

Olive is first present in deposits from the 2nd century BC onwards at the House of the Vestals (VI.I.vii), the House of Hercules' Wedding (VII.IX.xlvii) and the House of Amarantus (I.IX.xii) at Pompeii (Robinson 1995-6 cited in Ciaraldi 2001, 171). Olive deposits become more numerous in all the above listed houses in later deposits. From a few specific stratigraphic units from the House of the Vestals (VI.I.vii) olives were very abundant and were recovered as both fragments and entire stones (Ciaraldi 2001, 171). As the same pattern occurs in all three residential houses from different areas of the city of Pompeii it suggests that a real change in olive usage was occurring over time. This increased frequency of olive usage could be attributed to the culinary practice of preparing pickling olives (Robinson 1999) or the use of olive press residues for fuel or fertiliser for the gardens (Ciaraldi 2001, 171).

Columella wrote that the 'olive is first among all the trees' (*olea quae prima omnium arborum est*) (*RR De Arboribus* 5.8.1 cited in Warnock 2007, 1). Indeed, the olive was one of the most valuable and versatile crop of ancient Italy and a staple of the Roman diet (Jashemski et al. 2002, 135; Warnock 2007, 1). Most parts of the plants were used including the leaves, flowers, fruits, bark of the root, ash of the tree, and the oil, especially the green oil (Meyer 1980, 405; Pliny NH XXIII. 69-75).

Olives have been present in Greece at least since the Minoan era (3000 BC) as they have been recovered from the graves and the palaces of Knossos and Phaestos in Crete. Greeks of this period were using olives as both food and the olive oil as fuel for their small clay lamps. From ancient Greece the olive tree spread throughout the Mediterranean through human intervention. The olive tree is more easily propagated than other fruit trees, through both sexual and asexual modes, which is thought to have facilitated its relatively quick dispersal throughout the Mediterranean (Manousis and Moore 1988, 8) following both land and maritime routes to Italy, Spain, North Africa and France (Angiolill et al. 1999, 411).

During Roman times there were about twenty different kinds of olive being cultivated in Italy, Spain, and North Africa, differing in the shape, size, and colour of the olive fruits, from nearly white to green, violet and deep purple and in the shape and size of the leaves, from small linear-laceolate to large ovate (Newberry 1937, 5). By Pliny's day, in the 1st century AD, olives were common and Columella (RR 5.8.3), a contemporary of Pliny, lists nine known varieties of olive in the 1st century AD (Meyer 1980, 405). Interestingly, imported olives from Egypt and Syria were preferred for the table to those grown in Italy (Jashemski et al. 2002, 135). Cato (Agr. Orig. 58) notes that olives were a good source of protein for one's slaves (Sassatelli 2000, 108). Today over eighty different kinds are grown in the Mediterranean Basin (Newberry 1937, 6).

Low frequencies of olive pollen have been detected from 7th-6th century BC levels and low levels of olive pollen have also been recorded in a few palynological sequences undertaken for southern Italy (Allen et al. 1999; Carter and Costantini 1994 cited in Ciaraldi 2001, 172; Drescher-Schneider 2007, 279; Grüger 2002; Mercuri et al. 2002, 263; Watts 1985). Arboreal olive pollen shows a steep increase in pollen diagrams around the 1st century AD, which could suggest that intensification was occurring in southern Italy at this time (Mercuri et al. 2002, 263). It should be noted that the olive plant is insect-pollinated and therefore its pollen would not be expected to travel as far as wind-dispersed pollinated plants. It is not possible based upon palynological data to distinguish between wild and domesticated olives (Ciaraldi 2001, 172; Galili et al. 1997, 1141). However, recent genetic work has created genetic markers of *O. europaea* which would allow for the germplasm identification of olives, even with the use of poorly preserved material from herbarium specimens or archaeobotanical material (Besnard et al. 2003, 651).

The relatively late cultivation of olive trees at Pompeii is surprising given that olive stones are present in prehistoric sites in southern Italy (Castelletti et al. 1987; Costantini 1981; Costantini and Stancanelli 1994; Lacroix-Phippen 1975 cited in Ciaraldi 2001, 172). The wreckage of the Etruscan ship *Giglio* yielded olives preserved in brine dated to 600 BC. Olive pits were also discovered from the so-called Tomb of the Olives in Caere, 575-550 BC, as it is believed that olive pits were considered an offering to the dead (Sassatelli 2000). It is possible that olive cultivation occurred much earlier in other regions of Italy. Also, this does not rule out the exploitation and use of wild olives before domestication is visible in the archaeological record around Pompeii. Archaeological research at Lepine Hills, in southern Latium, has unearthed several platforms dating to the Late Republican period, from the late 4th century BC, constructed on hill slopes and interpreted by the researchers as being related to olive cultivation. Hence, archaeological evidence for the intensification of olive cultivation in central Italy is present despite their absence in the earlier deposits in Pompeii (Ciaraldi 2001, 171).

Olive oil

Olives were the primary source of edible fat in the Mediterranean. The oil was used in cooking, soap manufacturing, ointment, perfume, fuel for illumination, and medicinally (Costantini and Giorgi 2001, 246; Finley 1973, 31; Meyer 1980, 405). The genus *Olea*, belonging to the family *Oleaceae*, consists of more than 20 species found in tropical and subtropical areas over five continents. However, *Olea europea* (European olive tree) is the only species that produces edible fruit (Manousis and Moore 1988, 7). The olive is an oleaginous, one-seeded drupe with the fruit yielding up to fifty percent of its weight in valuable edible oil.

The olive fruit takes six to seven months to develop fully. The fruit grows slowly over the spring and the summer months. A faster increase in size takes place in the autumn due to the water content. Restriction of fruit development can occur if exposed to prolonged droughts. The olive weight can vary between 1 and 15 grams. Oil deposition within the fruit starts at the beginning of the month of August, increases throughout the autumn, and then reaches its peak during November-January when the fruit turns completely black. Harvesting takes place in the autumn after the dramatic increase in size has occurred and the fruit starts to change colour. Oil producing olives harvesting takes place when the olives are completely ripe, during the months of December or January, to permit the olives to obtain their maximum size and oil content. The commonest method of extraction of the oil from the pulp involves putting the fruit through mills of the edgerunner type (Manousis and Moore 1988, 11).

Grapes

Vitis vinifera (grape) is the sole Mediterranean representative of the genus *Vitis* (Zohary and Hopf 2001, 155). *Vitis vinifera* is hermaphrodite and occasionally with the escape of cultivated vines into the wild will revert back to *Vitis sylvestris* in all characteristics except for the fact that they remain hermaphrodite and the seeds of these escapees remain much closer to those of the cultivated form (Logothetis 1962 cited in Renfrew 1973, 125). In terms of elevation the plant can establish itself at a height of 400 m above sea level. Rarely is it found as high as 800 m although *Vitis vinifera* has a more restricted distribution than its wild variety (Renfrew 1973, 125).

The fresh fruits of grapes are rich in sugar, containing 15-25% sugar. Practically grapes can be easily stored as dried raisins in winter store-rooms and the juice can be fermented and made into wine, which was an extremely important trade item around the Mediterranean Basin (Brothwell and Brothwell 1998, 147; Zohary and Hopf 2001, 151).

Within the palynological record Vitis is noted in several locations in northern Greece with an increasing presence, albeit patchy, both spatially and temporally, since the beginning of the Holocene. Around 3500 BP (1550 BC) onwards there is a rise in the pollen record that is believed to reflect human manipulation and cultivation of the grape vine (Bottema 1982 cited in Valamoti et al. 2007, 56). Recent archaeobiological analysis has confirmed this complex interaction and spread of the grapevine through Europe and the Mediterranean region by humans (Terral et al. 2010, 443).

Peaches

As peaches are not native to the Mediterranean area they most likely reached the Near East and Europe from the East only during Greek and Roman times. As evidence, several large peach stones have been found in several sites in the Mediterranean basin from classical times onwards (Zohary and Hopf 2001, 182). Peaches were found in the Shrine, the Inn and Inn Bar, the Soap Factory and Vicolo di Narciso from Insula VI.I (Table 8). The fragment of preserved peach endocarp recovered from the House of the Vestals (VI.I.vii), according to Ciaraldi (2001), pre-dates by a century those recovered from the AD 79 eruption layer (Wittmack 1904 cited in Meyer 1980, 402).

Pliny described peaches, called *malum persicum* by the Romans, as a very expensive fruit and lists the different known types (Jashemski 1974). However, Varro nor Cato mention peaches in their writings which could provide support for their late introduction and cultivation in Italy (Ciaraldi 2001, 172). It is thought that the Romans did not cultivate peach trees until the 1st century AD and soon after this point peaches became an established and valued fruit crop found all over Italy and the Mediterranean basin (Pignatti 1982, 615; Zohary and Hopf 2001, 182). Peaches were an imported cultivated fruit known to the inhabitants of the Vesuvian area based upon the artistic images of peaches represented on wall paintings from both the cities of Pompeii and Herculaneum (Jashemski 1979b) and endocarp fragments recovered from archaeological excavations.

Fig

Every property within Insula VI.I had evidence of fig achenes aside from the main road, the Via Consolare (Table 8). Carbonized figs were found *in situ* in three gardens at Pompeii: the House of *C. Julius Polybius* (IX. XIII.i-iii), the humble house on the *Via Nocera* (I.XIV. ii), and the House of the Ship of Europa (I.XV.iii). From Herculaneum around 1,000 well-preserved whole carbonized figs with characteristic pyriform shape (inv. No 2319) were discovered on the *Decumanus Maximus* (Jashemski et al. 2002) (Table 9, Table 10). Usually, it is the small carbonized or mineralised seeds which are recovered during excavation. Despite their ubiquitous presence on excavations carrying out proper archaeobotanical analysis, the palaeobotanical record of figs is still rare and is likely biased against the small seeds, which can be easily overlooked by excavators (Cool 2006, 119; Zohary and Hopf 2001, 160).

Fig, along with the pomegranate and sycamore fig (*Ficus sycomorus* L.), is one of the many-seeded fruit trees associated with the beginning of *horti*culture in the Mediterranean basin (Simoons 1998, 279; Zohary and Spiegel-Roy 1975, 319). The fig is easily domesticated with fruit production normally starting three to four years after planting. Figs are believed to have been present in the area of Campania since at least the Early Bronze Age. Early cultivation of figs in the Mediterranean environment would have occurred in close association with the olive and grape vines (Zohary and Hopf 2001, 159, 164).

The fruit from the fig would provide fresh fruit in the summers and dried figs are easily stored all year round with their high sugar content (Matterne and Derreumaux 2008, 114). Columella states that it was dried apples, pears and especially figs which comprised the majority of the diet of country people in the winter (Brothwell and Brothwell 1998, 146). Archaeobotanical analyses from the Roman province of Britain hint at the presence of fig as a regular element of the military diet (Cool 2006, 121).

Figs were one of most highly esteemed fruits of ancient Campania and popular among all classes (Brothwell and Brothwell 1998, 146). Pliny (NH XV. 68) makes a number of references to the fig and describes 29 named varieties, even one called the Herculaneum fig. Both caprifigs and parthenocarpic figs were grown in ancient Italy. 'In the Mediterranean world, depending on the variety, figs were both a poor person's staple food and a delicacy for the elite' (Cool 2006, 121). The parthenogenic fig variety 'Dottata' or the Kadota fig was grown for centuries in Italy as the principal drying fig. Condit (1947) believes this may be the same variety that Pliny (NH XV. 83) said was brought to Italy by *Lucius Vitellius* from Syria to his country place at Alba during the reign of the emperor Tiberius. If this is true then the 'Dottata' fig could be the oldest continuously grown cultivar known to date (cited in Meyer 1980, 403).

Dates

Since antiquity date palms have been planted on the northern shore of the Mediterranean. However, they only bear fruit in one small area in southeastern Spain. Despite this fact, dates were well known to the Romans from sources outside of Italy. Palma is the Latin name used by the Romans for the date palm, and was the general term for all palms; whereas poma was sometimes used for the date fruit specifically (Meyer 1980, 435). Few archaeological sites have yielded evidence for date palm. This is likely due in part to ecological constraints, as the Near East and the Mediterranean basin are too cold for date cultivation (Zohary and Hopf 2001, 169). This appears to have been the case at Pompeii where dates appear to have been imported from Africa (Jashemski et al. 2002, 141). Date palm was cultivated in the Vesuvian plain but probably as a decorative plant because this region is not warm enough to permit fruiting (Ciarallo 2000, 12). No dates were recovered from Insula VI.I although archaeobotanical evidence for carbonised dates has been found from the House of Amarantus (I.IX.xii), the House of Hercules' Wedding (VII.IX.xlvii), the Temple of Isis, Herculaneum deposito, and the *Porta Nocera, Necropolis* (Table 9, Table 10).

Pomegranates

The Romans called Pomegranates (*Punica granatum* L.) '*malum punicum*', meaning the Punic apple, referring to its Carthagian origins (Brothwell and Brothwell 1998, 134). It is believed that the Greek colonies in southern Italy imported Pomegranate which had been cultivated principally for medicinal use (Ciarallo 2000, 18). Pomegranates are mentioned by Cato and Pliny cites nine varieties (Brothwell and Brothwell 1998, 134). Due to its multiple seeds Pomegranates have long been regarded by the ancients as a powerful fertility symbol. Pomegranates are one of the most frequently represented fruits from wall paintings and mosaics in the Vesuvian area. However, few carbonized remains have been recovered aside from the exceptional find of over a ton of pomegranates discovered stored between layers of straw from the Pompeii area (Jashemski 1987a, 64; Meyers 1988, 208). From Insula VI.I pomegranate seeds have been recovered from the Shrine, the House of the Vestals, the Inn, the Triclinium, and the Vicolo di Narciso (Table 8). Carbonised pomegranate seeds were also discovered in the House of Amarantus (I.IX.xii), the House of Hercules' Wedding (VII.IX.xlvii) and the *Porta Nocera, Necropolis* (Table 9, Table 10).

Apples

Romans were well aware of the apple (*Malus domestica* L.). Apples were cultivated throughout the western Mediterranean during the Roman period. Thirty two varieties have been recorded from Roman times (Martinez 2005, 352). Mineralised apple seeds have been recovered dating from the 1st century AD from the Triclinium, Vicolo di Narciso and carbonised and mineralised apple seeds from the House of the Vestals (VI.I.vii) dating to the 1st century BC and 1st century AD. Both carbonised and mineralised apple seeds were recovered from the 1st century BC from the Shrine (Table 8).

Citron

A single seed from the genus *Citrus* was recovered from the House of the Vestals (VI.I.vii) by Ciaraldi (2001, 175). It was difficult to identify, and as Ciaraldi (2001, 175) noted, it was probably an undeveloped seed. Another single *citrus* seed was recovered and identified from a 2nd century deposit at the House of Hercules' Wedding (VII.IX.xlvii). Based upon the artistic representations of what appear to be citrons in the wall paintings and mosaics at Pompeii it appears that Pompeians were familiar with the fruit. The spread and growth of *citrus* fruits to the Mediterranean region in Roman times may have been facilitated by their apomictic seeds (van der Veen et al. 2009, 269). The presence of these exotic food items, identified by Ciaraldi (2001) at the House of the Vestals (VI.I.vii) and the House of Hercules' Wedding (VII.IX.xlvii) hints at the possibility that these were elite houses whose inhabitants had access to expensive foods.

6.11 Nuts

From all properties within Insula VI.I a few carbonised nutshell fragments were found. The commonest carbonised nutshells recovered were walnuts and hazelnuts with a few identifications of almond (Table 8). Aside from being a nutritious and storable foodstuff it is possible that nut shells were used as tinder, which would explain why they are normally recovered in a charred state (Ciaraldi 2001, 151).

Hazelnut

Hazelnut (*Corylus avellana* L.) (also less commonly known as filberts) is indigenous and widespread throughout most of Europe as a shrub or small tree. It is known that hazelnut trees were being planted by the Romans (White 1970b, 259). The botanical name derives from an Italian town north of Mount Vesuvius, *Abella*, which Pliny (NH XV. 88-9) claims was famous for its fruit trees and nuts. The Romans also called hazelnuts *Corylus*, which Pliny sometimes uses. Hazelnut is indigenous and was well known in Campania during ancient times and still grows in the Pompeii area today, particularly on the lower slopes of Mount Vesuvius (Jashemski et al. 2002, 104; Meyers 1988, 186).

From Insula VI.I carbonised hazelnut shell fragments were recovered from the Vicolo di Narciso, the Soap Factory, the Inn, the House of the Vestals (VI.I.vii), Vestals Bar, and the Shrine (Table 8). Three hazelnuts were identified from an image on a small fragment of wall painting from the Museum at *Castellammare di Stabia*, believed to have been removed from the south wall of the peristyle of the House of the Cervi (IV.XXI) at Herculaneum. Several small pieces of carbonized hazelnut wood were recovered from a number of gardens in Pompeii and it is thought that they were most likely the remains of fertilizer. Imprints of hazelnuts (*Corylus avellana* L.) have been discovered from the villa of *L. Crassius Tertius* (Jashemski 1987b). Charred hazelnut shell fragments have been found in the House of Amarantus (I.IX.xii), the House of Hercules' Wedding (VII.IX.xlvii), the House of the Ship of Europa (I.XV.iii), and the *Porta Nocera, Necropolis* (Table 9, Table 10). Practically, Pliny notes that hazelnut wood was used for torches (NH XVI. 75) (Jashemski et al. 2002, 103). Pliny (NH XV. 86) warns that roasted hazelnuts 'put more fat on the body than one would think at all likely'. Parched hazelnuts are believed to have been used medicinally to cure cataracs (Meyer 1980, 407).

Walnut

Similarly to olives, walnuts are first recovered in deposits dating from the 2nd century BC onwards at Pompeii. From Insula VI.I charred walnut shell fragments were found in the Shrine, the House of the Surgeon, Vestals Bar, the House of the Vestals, Inn Bar, Inn, Triclinium and Vicolo di Narciso (Table 8). Remains of several varieties of walnut were discovered in the Temple of Isis at Pompeii and have long been speculated to have formed part of the last meal of one of the priests before the destruction of the city (Brothwell and Brothwell 1998, 150). Ciaraldi (2001, 175) argues that the walnuts recovered from the House of the Vestals (VI.I.vii) would have only been recently introduced into the area. From the literary sources Pliny (NH XV. 90) mentions that the walnut was likely unknown to Cato (234-149 BC). Walnuts were also found in the House of Amarantus (I.IX.xii), the House of Hercules' Wedding (VII.IX.xlvii), the *Porta Nocera, Necropolis*, the hay from Oplonits, IV.VII (Pompeii), and IV.XVII from the city of Herculaneum suggesting that this food was already known and possibly used by the 1st century BC (Table 9, Table 10). Walnut charcoal was also found in gardens in Pompeii and Jashemski et al. (2002, 117) speculates that it was most likely from the remains of wood-ash used as fertilizer.

Walnuts were not highly esteemed by the Romans. Pliny (NH XXIII. 147) writes that walnuts were considered difficult to digest and thought to cause headaches, stomachaches and were bad for a cough. However, other sources on early Rome note that walnuts were thought to have been regarded as a delicacy and were perhaps eaten as a dessert with fruit (Brothwell and Brothwell 1998, 151). However, their practical usages listed by Pliny (NH XV. 87) included using the shells of the walnut for the dyeing of wool and the young nuts were used for dyeing hair red (Meyer 1980, 414). Despite their reputation walnuts became more common from Roman times onwards and literary sources indicate that walnuts were widely grown in the Roman Empire (Zohary and Hopf 2001, 189). However, the pollen data indicates a late introduction of walnut in Italy at around 50 BC in the *lago di Ganna* sequence (Drescher-Schneider 1994; Magri 1999; Magri and Sadori 1999; Watts 1985; Watts et al. 1996 cited in Ciaraldi 2001, 172).

Almond

The almond (*Amygdalus communis* L./ *Prunus dulcis* (Mill.) D.A. Webb), an introduced tree, has been known in the Vesuvian area since antiquity (Jashemski et al. 2002, 118). The presence of almonds, based upon palaeobotanical evidence, has been well attested to in the Vesuvian area at the time of the eruption (Ciaraldi 2001, 175; Meyer 1980, 401; Wittmack 1904). Indeed, almonds were the most widely cultivated nut in the Mediterranean basin and most likely one of the early fruit trees domesticated along with Old World agriculture. Unsurprisingly, almonds thrive best in relatively warm Mediterranean-type climate and in comparison with olives and grapes can endure drier conditions (Zohary and Hopf 2001, 185).

Carbonised almond nut shells have been recovered from a number of sites in Pompeii. From Insula VI.I they have been recovered from the Bar of Pheobus, Soap Factory, Triclinium and Shrine (Table 8). They have also been recorded from the Pompeii and Naples museums, the House of Amarantus (I.IX.xii), the House of Hercules' Wedding (VII.IX.xlvii), House of Polybius (both the garden and house), the House of the Ship of Europa (I.XV. iii), shop II.IX, *Casa del Bel Cortile* in Herculaneum, and *villa rustica*.

6.12 Pules

Pliny assigns beans the highest place among legumes since they were used in crushed form (*lomentum*) in the baking of bread (Spurr 1983). A small amphora found at Pompeii labelled *lomentum* (CIL IV 2597) contained a paste-like mixture. Pliny (NH XVIII. 117) remarks that this bean paste, which was sometimes combined with Panicum, was used in bread for sale to increase its weight and also used by Roman ladies to preserve the smoothness of their skin. A vessel found in the peristyle of VIII.II.xiv at Pompeii was labelled *LOMENTVM FLOS EX LACTE ASININO VITCENSE* (CIL IV 5738) (Jashemski et al. 2002, 170).

Cultivated *Fabaceae* include broad beans (*Faba*), lentils (*Vicia*) and chick peas (*Cicer*). The pollen from this family is rarely identified. Beans are easily cultivated. The seeds can be scattered on broken soil and ploughed under, are adaptable to various soil types and climates and grow wild throughout Roman Italy. Beans also have a shorter growing cycle than wheat and therefore can be harvested earlier, in late June or before the late summer droughts occur (Spurr 1986, 115-116).,

Beans were likely part of a common vegetable-mix eaten by humans in the Mediterranean area. Beans were also used for animal fodder. The visible increase in wealth during the period from *circa* 200 BC to AD 100 may have resulted in a corresponding decline in the use of dried legumes (Spurr 1986, 107). However, during the Roman period a greater variety of legumes appear to have been cultivated than in earlier periods (Spurr 1983, 105).

Peas

Pea (*Pisum sativum* L.) is believed to be of Mediterranean origin. Peas were likely cultivated throughout the region due to their known climatic resistance and soil-enriching properties. Wittmack (1904 cited in Meyers (1988, 186) noted the presence of carbonised remains of the common pea (*Pisum sativum*) from Pompeii but they were not observed when re-examined by Meyers (1988, 186). Peas in Apicius's recipes were boiled before eating and this may be one reason for their lack of preservation (Spurr 1986, 113). From Insula VI.I peas were recovered from the Shrine, the House of the Vestals and Vestals Bar, the Inn, and the Soap Factory (Table 8). Peas were also recovered from the House of Hercules' Wedding (VII.IX.xlvii) and from the *Porta Nocera, Necropolis*.

Lentils

An amphora discovered at Pompeii (VIII.V-VI.xv) containing lentils bore a label in Greek (CIL IV 6580) stating 'lentils' (Jashemski et al. 2002, 121).

Lentil (*Lens culinaris* Medikus) is a nutritious legume which has adapted to subtropical or Mediterranean-type climate and many different soil types and now grows all over Italy (Pignatti 1982, 685; Spurr 1986, 115). Lentils became a valued legume in the Mediterranean basin, western Asia, as well as India and Ethiopia (Zohary and Hopf 2001, 108). The Romans named lentils lens or lenticula. They believed that a diet of lentils resulted in an even temperament. Lentils were made into soups by the Romans by roasting the lentil seeds and then pounding them with a mortar into bran (Meyer 1980, 401). Lentils were recovered from the Vicolo di Narciso, the Bar of Acisculus, Soap Factory, Triclinium, Inn, Shrine, the House of the Vestals, Vestals Bar, the House of the Surgeon, and the Shrine from Insula VI.I (Table 8). Lentils have also been noted from both the Naples and Pompeii museums, the House of Amarantus (I.IX.xii), House of Hercules' Wedding (VII.IX.xlvii), the fodder from the House of the Chaste Lovers (IX.XII.vi), *Porta Nocera,* Necropolis, the shop on the *Strad degli Augustali, Casa della Stoffa*, and the *Casa dei Due Atrii* on *Cardo* III (Table 9, Table 10).

Chickpeas

At Pompeii the word cicer was found lettered on two amphorae that had once held chickpeas (CIL IV 5728, 5729) (Jashemski et al. 2002, 101; Meyer 1980, 403). None were recovered from Insula VI.I. Chickpeas have been found in the House of Amarantus (I.IX.xii), the House of Hercules' Wedding (VII.IX.xlvii), and *Porta Nocera, Necropolis* (Table 9, Table 10).

Chickpeas were widely grown throughout Italy during Roman times (Jashemski et al. 2002, 101; Meyer 1980, 410; Meyer 1988, 190; Pignatti 1982, 670). Chickpeas were recovered from Iron Age Rome at the site of *Monte Irsi*. Lentils and chickpeas were also shipped in sizeable quantities from Alexandria (Spurr 1986, 116). Pliny (NH XVIII. 124-5) lists several varieties of chickpeas and their medicinal qualities. *Cicer* is the classical Latin name for chickpea and was also used as the cognomen for the famous Roman senator and orator *Marcus Tullius Cicero* (Jashemski et al. 2002, 101; Meyer 1980, 410).

Broad bean

The most numerous remains recovered from the ritual deposit from the House of the Vestals (VI.I.vii) were broad beans (*Vicia faba* L.), which are often found in ritual deposits of differing periods and food offerings in religious practices (Ciaraldi 2001, 205). Broad beans were recovered from Insula VI.I from the Vicolo di Narciso, the Bar of Acisculus, the Soap Factory, the Triclinium, the House of the Vestals and the Shrine (Table 8).They have been recovered from both the Pompeii and Naples museum, the House of Hercules' Wedding (VII.IX.xlvii), the burnt fodder from the House of the Chaste lovers (IX.XII.vi), the house on Via Nocera (I.XIV.ii), the House of the Ship of Europa (I.XV.iii) (garden and house), vineyard (II.V), the Villa of Poppaea, and the shop in Insula Orientalis (II.XIII) (Table 9, Table 10).

Pythagoras, a 6th century BC Greek philosopher, established a ban on faba beans. Followers of Pythagoras, called Pythagoreans, believed that faba beans contained and harboured the souls of the dead, considered the source of life, based upon the Artistotlian observation that faba beans possess un-jointed hollow stems which they believed served as a vehicle for spirits from Hades to enter the living world. It is often claimed that this ban was functional based upon what recent medical knowledge has identified as the disease, favism. Favism can cause haemolytic anaemic, jaundice and fever, in a small portion of the population, by eating raw faba beans. This disease still exists today in Mediterranean areas; however, there is no evidence that this ban was connected with knowledge of favism (Simoons 1998, 296; van der Veen et al. 2009, 269).

Pythagoreans considered faba beans symbolic of death and decay and therefore impure. Therefore based upon these beliefs eating faba beans was considered the most debased form of cannibalism in an ancestral society as 'eating beans is like eating the heads of one's father or parents' (Simoons 1998, 200). However, due to their connection with death faba beans were acceptable offerings to the dead to consume during rites to honour the dead. Pliny (NH XVIII. 118) also emphases the sanctity of broad beans and their use as offerings to the gods.

Thus, interestingly, aside from the dietary taboo and their association with religious contexts the majority of faba beans were recovered from contexts with no obvious religious connonations. Therefore, based upon their presence from a range of properties within Insula VI.I and other properties within the city of Pompeii and Herculaneum one can assume that faba beans were perhaps both eaten and offered as a ritual food to the dead.

Helbaek (1956, 292) argued that there is little evidence for horse bean, seeds of *Vicia faba* L. in Italy until the Iron Age, after this point horse bean was grown throughout Italy. Faba beans appear to have been a major food source from the Iron Age through to classical times based upon the abundance of archaeobotanical data of this bean from both Europe and West Asia (Zohary and Hopf 2001, 115). In both Europe and Asia the seed is regarded as valuable animal feed, hence the term 'horse bean'.

The *Fabii*, a prominent noble Roman family in Rome, took the name of this bean plant as their cognomen. (*Faba* being the classical Latin name for the broad bean). Interestingly, *Faber* means 'workers' in Latin. Romans had a solemn feast called *Fabaria*, in which beans were offered in honour of Carna, wife of Janus. Pliny (NH 18. 117) notes that amongst all leguminous plants 'the highest place of honour belongs to the bean (faba).... Beans are used in a variety of ways for all kinds of beasts and especially for man' (cited in Corbier 2000, 133). Faba beans were commonly used in the Mediterranean diet and were considered a nourishing delicacy, normally prepared as a purée (Corbier 2000, 128; Simoons 1998, 193). Jashemski (1974) suggests based upon the plaster casts and grid pattern revealed from the market-garden orchard at the House of the Ship Europa (I.XV.iii) broad beans were inter-cultivated between vines and trees. During the Roman period there is evidence of a greater variety of legumes cultivated than in previous periods with broad beans being the most widespread legume in Italy (Spurr 1986, 105). Faba beans were recovered from the villa of *Settefinestre* near *Orbetello* in Tuscany (late Republic to late 2nd century AD), the villa of *San Giovanni* near *Potenza* in *Basilicata* (early to late Imperial period), from Imperial Rome and at *Aquileia* (Spurr 1983, 105).

Epigraphic evidence, in the form of graffiti scratched on the *villa rustica* in the fundo Juliano at Boscoreale, has the words 'broad beans raised there' (*FABA M DLXXXVII*) (CIL IV 5430). However, faba beans were also believed to dull the senses and cause insomnia. A number of ancient medicinal uses were ascribed to this bean by Pliny (NH XXII. 140) including being 'roasted whole and thrown hot into strong vinegar it heals colic. The meal, boiled in vinegar, ripens tumours and heals contusions and burns' (Meyer 1980, 408). Significantly, broad beans are found more frequently, with twenty six to twenty-nine separate samples, than any other carbonized food plant in the sites destroyed by Vesuvius in AD 79 (Jashemski et al. 2002, 169; Spurr 1986, 105).

Bitter vetch

From Insula VI.I bitter vetch has been found at the Shrine, the House of the Surgeon, the Inn and the Soap Factory (Table 8). Amongst the archaeobotanical assemblages from Pompeii held in the museum storerooms bitter vetch was found as a weed contaminant in an assemblage of broad beans and chickpeas (Meyers 1988, 189). It has also been recorded at the Pompeii Museum, the House of the Chaste Lovers (IX.XII.vi), and at Herculaneum, (XI.X) mixed with lentils and from the shop in Insula Orientalis (II.XIII) (Table 9, Table 10).

Bitter vetch is found as a weed in grain fields and along roads throughout the Mediterranean and the Near Eastern regions (Jashemski et al. 2002, 169; Zohary and Hopf 1973 cited in Meyer 1980, 409). Meyer's (1980, 409) noted that the triangular to spherical shape of carbonized archaeological specimens of bitter vetch are comparable in size with modern material. Bitter vetch was well known to the Romans in Italy and it is believed that bitter vetch had been utilized primarily as an animal feed, as throughout its history it has never been used regularly as a food source for humans. Indeed, it was most likely used in times of need in antiquity by the very poor as a famine food as it is regarded as inferior for human consumption (Garnsey 1999, 38; Meyers 1988, 189; Zohary and Hopf 2001, 116). The medical historian Galen writing about the food crisis in the Thracian city of Ainos, observed symptoms of lathyrism, caused by the consumption of Lathyrus sativa and other related vetches and pulses e.g. muscular weakness and paralysis (Garnsey 1999, 38).

The bitter taste is caused by a cyanogenetic glycoside, which occurs in all parts of the plant, especially the seeds, which is highly toxic to pigs, horses and poultry, but ruminants and humans are highly resistant to it. Soaking the seeds in water for several days removes the toxic substance and makes the plant more palatable for human, cattle and other beasts of burden (Meyer 1980, 409; van Zeist 1988 cited in Zohary and Hopf 2001, 116). Bittervetch was believed to cause vomiting, intestinal trouble, heaviness in the head and stomach and enfeebled the knees. Despite these side effects Pliny (NH XXII. 151) lists a number of medicinal uses for bittervetch. Pliny (NH XVIII. 103) further notes that barley was leavened with the flour of either bittervetch or chickeling when making bread (Meyer 1980, 409).

6.13 Trees

cypress

From the House of the Vestals (VI.I.vii), Ciaraldi (2001, 167) discovered charred and mineralised seeds and cone scales of cypress (*Cupressus sempervirens* L.). The presence of cypress at the House of the Vestals (VI.I.vii) could have been from the nearby garden. Ciaraldi (2001, 167) suggests the possibility of a cypress hedge in the garden and leaves becoming charred during burning of garden or household waste. From Insula VI.I evidence of cypress has also been recovered from the Triclinium, Inn Bar and the Shrine (Table 8). Small fragments of cypress charcoal were identified in the sculpture garden of the Villa of Poppaea at *Oplontis* (Jashemski 1979b, 314; Jashemski 1992, 299; Jashemski et al. 2002, 104). Evidence of cypress has been found at the House of Amarantus (I.IX. xii), House of Hercules' Wedding (VII.IX.xlvii), and *Porta Nocera, Necropolis* (Table 9, Table 10).

Cypress (*Cupressus sempervirens* L.) is one of the most easily recognized arboreal species in frescoes of landscapes. It is indigenous to the Aegean but it is now found wild throughout the Mediterranean area (Pignatti 1982, 82). Cypress was and still is a favorite funerary tomb tree in Italy. Pliny (NH XVI. 142) ventured that Cypress was non-native to Italy, having been first imported into Tarentum, in the south of Italy from Crete. He cites the many uses of cypress, e.g. props for vines and general decoration, because it was resistant to decay and ornamental for clipped hedges as topiaria (Jashemski 1979b, 314; Jashemski 1992, 299; Jashemski et al. 2002, 104). Varro (RR 1.15) recommends using cypress trees on the boundaries of an estate, as he did at his place on Mount Vesuvius, to make his property more secure. In addition, Pliny (NH XXIV. 15) notes that cypress had medicinal properties for the treatment of fresh wounds and sunstroke (Ciaraldi 2001, 167).

Stone pine

Mediterranean Stone pine (*Pinus pinea* L.), or more commonly known as the umbrella pine, due to its distinctive characteristic umbrella-shaped crown, is indigenous to many regions along the north shore of the Mediterranean and is commonly found along the Bay of Naples and in the vicinity of Pompeii (Meyer 1980, 419).

The relatively large cones from this evergreen tree became a symbol of rebirth, due to its dormant outer cone and the numerous seeds inside the cone. Pragmatically, Columella (RR 12.30.2) suggests that the pinecones could be used to clean the necks and brims of wine vats (Jashemski et al. 2002, 144). The stone pine was also associated with Dionysus, the god of vegetation and fruitfulness (Simoons 1998, 279). Within the Roman world stone pine nuts were used in public and domestic religious ritual including weddings, triumphs and feasts (Atchley 1909 cited in Bird 2007, 122). Pinecones were often depicted in funerary scenes (Bird 2007, 122).

Stone pine nut shells has been found in Insula VI.I from the Shrine, the House of the Surgeon, Vestals Bar, the House of the Vestals, the Inn Bar, the Inn, the Triclinium, the Soap Factory, the Bar of Acisculus, the Bar of Pheobus, and Vicolo di Narciso (Table 8). Other physical evidence of stone pine from the Vesuvian region has been recovered from the Museums of Naples, Pompeii and Herculaneum, the House of Hercules' Wedding (VII.IX.xlvii), the Garden of the House of *Polybius*, Marine warehouse of *M. Cellius Africanus*, and *Casa del Bel Cortile* (Table 9, Table 10).

When burnt the cone releases a pungent incense associated with the rituals of the god's birth, as mentioned in an Egyptian papri dated to the 2nd and 3rd centuries AD (Richmond and Gillam 1951 cited in Bird 2007, 129). It is thought that they may have been used in mithraic traditions. Archaeologically, burnt pine kernels were recovered among the remains of offerings from the Temple of Isis at Pompeii. There is no evidence from pre-Classical times that pine kernels were being consumed. The Romans and Greeks considered edible pine nuts a delicacy and an expensive condiment. To this day pine nuts are still used in cooking in the region (Brothwell and Brothwell 1998, 150-151; Jashemski et al. 2002, 144).

6.14 Summary

The production and consumption of food affected nearly every facet of Roman society. Food is a useful concept to examine within Roman society because it illuminates the more elusive but routine activities of Roman society including widespread cultural practices, norms, beliefs and attitudes regarding food and dining, social competition and hierarchies. Roman literary sources and their biases often expose the disjunction between the ancient sources and the physical evidence provided by archaeological and increasingly, paleoenvironmental research.

Situating the Mediterranean food assemblage within its ancient cultural milieu provides a means of investigating the ecological origins of the different species, trade and manipulation, and food preferences of their human consumers. Moreover, the investigation of these ancient foods provides a way to understand the beliefs associated with different food uses, and how each food was prepared and eaten, and by whom. Examining the diachronic patterns in the archaeological evidence, including dietary preferences and trade practices at Pompeii, provides new insights into the impact of historical events, new contacts with foreign peoples and ideas, and the overall relationship of the population of Pompeii with their hinterland and the rest of the Mediterranean world.

Chapter 7

Discussion: Rubbish, Floors and Food

7.1 Interpretation of reults from Insula VI.I

The roads

Via Consolare

Scarcely any archaeobotanical remains [n=11.26] were recovered from the Via Consolare. It had the lowest recovery for archaeobotanical remains from Insula VI.I. This result is not surprising. Via Consolare was one of the first streets cleared of volcanic material in Pompeii in the 18th century and has been in continuous usage since that time. Therefore, it has likely been exposed to the strongest taphonomic pressure of any property or area within Insula VI.I, including the effects of trampling and prolonged exposure to the elements.

The few fruit pits/stones recovered included cherry [n=1], apricot [n=2], and peach [n=3]. Despite their questionable preservation, they are likely to be modern, due to their intact condition and the fact that they were retrieved *in situ* from contexts near the modern surface layer and hence were excluded from analysis. From the 1st century AD 0.55 fragments of carbonised olive endocarp and 0.07 carbonised and 2.06 mineralised fragments of grape pips were recovered. From the 1st century BC one carbonised common millet, 0.01 carbonised olive endocarp fragment and 0.57 grape pip were recovered.

Species	Via Consolare		Preservation	1BC	1AD
Olea europea L.	Olive	endocarp	Carbonised	x	x
Prunus persica (L.) Batsch.	Peach	fragment	modern		x
Vitis vinifera L.	Grape pip	pip	Carbonised	x	x
Vitis vinifera L.	Grape pip	pip	Mineralised		x
Cereals					
Panicum miliaceum L.	Common millet	whole	Carbonised	x	
Grasses					
Panicoid	Grass-like		Carbonised		x

TABLE 11: VIA CONSOLARE

In the 1st century AD there is a greater concentration of carbonised archaeobotanical remains with a greater range of taxa than in the 1st century BC. There is also a visible shift in the concentration of carbonised archaeobotanical remains in the 1st century AD towards the south end of Insula VI.I as opposed to the 1st century BC, which appears to be focused upon the middle and northern end of Insula VI.I (Figure 138). This could be due to the presence of the commerical triangle area shops in the 1st century AD.

There were no recovered carbonised remains from the 1st century BC from the southern end of Insula VI.I. This could be due to the fact that the commercial 'triangle' was not present until the 1st century AD, having undergone construction/renonovations during the late 1st century BC, thus likely removing a large majority of the earlier environmental evidence that may have been present.

No mineralised remains were recovered from the 1st century BC from the Via Consolare. This could be due to the destructive taphonomic factors mentioned previously and the lack of conditions conducive to mineralisation. In the 1st century AD only mineralised fruit was recovered from the north end only of Insula VI.I.

Vicolo di Narciso

A noticeable difference is present between the two streets, the Via Consolare and the back street, the Vicolo di Narciso. Vicolo di Narciso is much richer in terms of archaeobotanical remains [n=680.31] and yielded a greater variety of botanical material [20 taxa]. This contrast between the main and back street of Insula VI.I may, in part, be attributed to the Via Consolare's early clearance in the 18th century and continuous use as a street into the current day; whereas Vicolo di Narciso was left relatively undisturbed and not accessible to the general public. In addition, the known presence of drains which emptied onto the Vicolo di Narciso likely contributed to the range of species found along this back street. Also, in the past this street was probably used for transporting goods up and down the Insula including food supplies and perhaps even disposing of waste.

One deposit of interest, AA316 SU236, located by the back entrance to the Soap Factory, had 130 mineralised grape pips which were collected *in situ*/dry sieved and dated to the 1st century AD. Also recovered from the Vicolo di Narciso was one of the few instances of cereal chaff from Insula VI.I, a carbonised emmer spikelet fork, dating to the 2nd to 1st century BC and a carbonised cucumis (melon) seed from the 1st century AD.

As with the Via Consolare, there is a greater concentration of carbonised archaeobotanical remains towards the south end of Insula VI.I in the 1st century AD as compared with the 1st century BC. Again this may be attributed to the activity of the commercial triangle area in the 1st century AD. From the 1st century BC there are visible traces of carbonised archaeobotanical material in the back street possibly originating from the House of the Vestals, perhaps rubbish thrown out the back door onto the Vicolo di Narciso.

Similar to the results from the Via Consolare, only a few mineralised fruit remains were recovered from the 1st century BC. There is a greater concentration of mineralised archaeobotanical remains in the southern end of Insula VI.I surrounding the 'commercial triangle' properties in the 1st century AD. These contexts were dominated by fruit but also showed evidence of cereals and pulses. Interestingly, there is also a continued presence of mineralised material by the back door of the House of the Vestals from the 1st century BC to the 1st century AD. It may be that these archaeobotanical remains were mineralised in local cesspits and periodically emptied and dumped onto the back street.

Species	Vicolo di Narciso		Preservation	1BC	1AD
Cucumis L.	Melon	seed	Carbonised		X
Ficus carica L.	Fig	achene	Mineralised	X	X
Ficus carica L.	Fig	achene	Carbonised	X	X
Ficus carica L.	Fig mesocarp	fragment	Carbonised		X
Malus domestica Borkh.	Apple	seed	Mineralised		X
Olea europea L.	Olive	endocarp	Carbonised	X	X
Prunus persica (L.) Batsch.	Peach	stone	Carbonised		X
Punica granatum L.	Pomegranate	seed	Carbonised	X	
Vitis vinifera L.	Grape pip	pip	Carbonised	X	X
Vitis vinifera L.	Grape pip	pip	Mineralised		X
Vitis vinifera L.	Grape mesocarp	fragment	Carbonised	X	
Vitis vinifera L.	Grape exocarp	fragment	Carbonised	X	X
	Vasicular tissue indeterminate	fragment	Carbonised	X	
Nuts					
Corylus avellana L.	Hazelnut	shell fragment	Carbonised		X
Juglans regia L.	Walnut	shell fragment	Carbonised		X
cf. *Juglans regia* L.	Walnut	shell fragment	Carbonised		X
Pinus pinea L.	Pinenut	shell fragment	Carbonised	X	X
	Nuts (generic)	shell fragment	Carbonised	X	X
Cereals					
T. cf. *aestivum/T.durum* L.	Free-threshing wheat	grain	Carbonised		X
T. dicoccum Schübl.	Emmer wheat	spikelet fork	Carbonised	X	
Triticum sp	Wheat	grain	Carbonised	X	X
cf. *Hordeum vulgare* L.	Barley	grain	Carbonised	X	X
cf. *Hordeum vulgare* L.	Barley	embryo tip	Carbonised		X
Panicum miliaceum L.	Common millet	whole	Mineralised		X
Pulses					
Lens culinaris Medik.	Lentil	whole	Carbonised	X	
Lens culinaris Medik.	Lentil	whole	Mineralised		X
cf. *Lens culinaris* Medik.	Lentil	whole	Carbonised		X
cf. *Vica faba* L.	Broadbean	fragment	Carbonised		X
Vicia sativa L.	Common vetch	whole	Carbonised		X
Vicia sp.	Bean	whole	Carbonised	X	X
Grasses					
Panicoid	Grass-like		Carbonised	X	X
Weeds					
Ornithopus perpusillus/sativus	Bird's foot	pod	Carbonised	X	
Trifolium L.	Clover	whole	Carbonised		X
Rubus	Blackberry	whole	Carbonised		X
Umbelliferae		whole	Mineralised		X
Plant Parts					
	Thorn	fragment	Mineralised		X
Unknown					
	Indeterminate	fragment	Carbonised	X	X

TABLE 12: VICOLO DI NARCISO

Commercial triangle

FIGURE 138: COMMERCIAL TRIANGLE, COMPOSED OF FROM LEFT TO RIGHT THE SOAP FACTORY/METAL WORKSHOP, THE BAR OF ACISCULUS, THE BAR OF PHEOBUS, THE WELL AND FOUNTAIN (PHOTOMOSAIC @ 2007 JENNIFER F. STEPHENS · ARTHUR E. STEPHENS)

Bar of Pheobus (VI.I.xviii)

FIGURE 139: BAR OF PHEOBUS (PHOTOMOSAIC @ 2007 JENNIFER F. STEPHENS · ARTHUR E. STEPHENS)

The Bar of Pheobus, like the Via Consolare, yielded few archaeobotanical remains [n=31.18]. From the 1st century BC only one carbonised weedy panicoid seed was found. From the 1st century AD 2 mineralised fig achenes and 2 fragments of carbonised fig mesocarp, 2.03 fragments of olive endocarp, 1 fragment of charred fruit exocarp, 3 charred fragments of almonds and 1 mineralised pinenut shell were retrieved.

Species	Bar of Pheobus		Preservation	1BC	1AD
Ficus carica L.	Fig	achene	Mineralised		X
Ficus carica L.	Fig mesocarp	fragment	Carbonised		X
Olea europea L.	Olive	endocarp	Carbonised		X
	Fruit exocarp	fragment	Carbonised		X
Nuts					
Amygdalus communis L.	Almond	shell fragment	Carbonised		X
Pinus pinea L.	Pinenut	shell fragment	Mineralised		X
Grasses					
Panicoid	Grass-like	whole	Carbonised	X	
Unknown					
	Indeterminate	fragment	Carbonised		X

TABLE 13: BAR OF *PHEOBUS* (VI.I.XVIII)

The Bar of Pheobus is dominated by both carbonised and mineralised fruit in the 1st century AD. Little to no archaeobotanical remains are recovered from the 1st century BC. This is likely due to its recent construction in the 1st century AD.

Bar of Acisculus (VI.I.xvii)

FIGURE 140: BAR OF ACISCULUS (PHOTOMOSAIC @ 2007 JENNIFER F. STEPHENS · ARTHUR E. STEPHENS)

The Bar of Acisculus was more profitable in terms of archaeobotanical remains recovered, both in variety of taxa [n=11] and quantity [n=149.57] when compared with the neighbouring Bar of Pheobus.

Species	Bar of Asiculus		Preservation	1BC	1AD
Cucumis L.	Melon	fragment	Mineralised	X	
Ficus carica L.	Fig	achene	Mineralised	X	X
Ficus carica L.	Fig mesocarp	fragment	Carbonised	X	X
Olea europea L.	Olive	endocarp	Carbonised	X	X
Vitis vinifera L.	Grape pip	pip	Carbonised	X	X
Vitis vinifera L.	Grape pip	pip	Mineralised	X	X
	Vasicular tissue indeterminate	fragment	Carbonised	X	X
	Fruit exocarp	fragment	Carbonised		X
Nuts					X
Pinus pinea L.	Pinenut	shell fragment	Carbonised	X	
	Nuts (generic)	shell fragment	Carbonised		X
Cereals					
T. cf. *aestivum/T.durum* L.	Free-threshing wheat	grain	Carbonised		X
Triticum sp	Wheat	fragment	Carbonised	X	X
Chenopodium/S.italica	Italian millet/chenopodium	whole	Carbonised		X
Panicum miliaceum L.	Common millet	whole	Mineralised	X	
Pulses					
Lens culinaris Medik.	Lentil	whole	Carbonised		X
Vicia faba L.	Broadbean	whole	Carbonised		X
Vicia sativa L.	Commmon vetch	whole	Carbonised		X
Vicia sp.	Bean	whole	Carbonised	X	X
Grasses					
Panicoid	Grass-like		Carbonised	X	X
Weeds					
Galium aparine L.	Goosegrass	whole	Carbonised	X	
Plant Parts					
	Thorn	fragment	Carbonised		X
Unknown					
	Indeterminate	fragment	Carbonised	X	X

TABLE 14: BAR OF ACISCULUS (VI.I.XVII)

No archaeobotanical remains, either mineralised or carbonised, were found in Room 3 from the 1st century BC. All the archaeobotanical assemblages from the three rooms from the Bar of Acisculus were dominated by fruit.

Soap Factory/Metal workshop (VI.I.xiv)

FIGURE 141: SOAP FACTORY/METAL WORKSHOP (PHOTOMOSAIC @ 2007 JENNIFER F. STEPHENS · ARTHUR E. STEPHENS)

There were 788 mineralised fig achenes recovered from the 1st century BC. A number of carbonised indeterminate vasicular tissue fragments [n=55] were recorded and likely relate to the large number of fragments of carbonised fig mesocarp recovered (although they could not be positively identified as fig mesocarp). A number of pulses were also recovered [n=35 carbonised common vetch]. Millet was also present [n=11]. The only cereals found were *Triticum* sp. and barley [n=14]. All of the above listed archaeobotanical remains could represent lower-status foods.

The Soap Factory was one of the few properties from Insula VI.I that had a higher ratio of archaeobotanical remains per context in the 1st century BC [n=23.27] or earlier period than in the later 1st century AD [n=2.62]. It may have been that this property saw heavier use in the 1st century BC than in the 1st century AD.

Species			Preservation	1BC	1AD
Ficus carica L.	Fig	achene	Mineralised	X	
Ficus carica L.	Fig	achene	Carbonised	X	
Ficus carica L.	Fig mesocarp	fragment	Carbonised	X	
Olea europaea L.	Olive	endocarp	Carbonised	X	X
Prunus persica (L.) Batsch.	Peach	whole	Carbonised		X
Vitis vinifera L.	Grape	fruit	Carbonised		X
Vitis vinifera L.	Grape	petiole	Carbonised	X	
Vitis vinifera L.	Grape pip	pip	Carbonised	X	X
Vitis vinifera L.	Grape	petiole	Mineralised		
Vitis vinifera L.	Grape	fruit	Mineralised	X	
Vitis vinifera L.	Grape pip	pip	Mineralised	X	
Vitis vinifera L.	Grape mesocarp	fragment	Carbonised		X
Vitis vinifera L.	Grape exocarp	fragment	Carbonised		X
	Vasicular tissue indeterminate	fragment	Carbonised	X	X
	Fruit exocarp	fragment	Carbonised	X	
	Fruit mesocarp	fragment	Carbonised	X	
Nuts					
Amygdalus communis L.	Almond	shell fragment	Carbonised		X
Corylus avellana L.	Hazelnut	shell fragment	Carbonised	X	X
cf. *Corylus avellana* L.	Hazelnut	shell fragment	Carbonised	X	
Juglans regia L.	Walnut	shell fragment	Carbonised	X	
Pinus pinea L.	Pinenut	shell fragment	Carbonised	X	X
Pinus pinea L.	Pinenut	shell fragment	Mineralised		X
	Nuts (generic)	shell fragment	Carbonised	X	X
Cereals					
Triticum sp	Wheat	grain	Carbonised	X	X
Hordeum vulgare L.	Barley	grain	Carbonised	X	
Millet					
Chenopodium/S.italica	Italian millet/chenopodium	whole	Carbonised	X	
Panicum cf. *miliaceum* L.	Panicum	whole	Carbonised		X
Panicum sp.	Panicum	whole	Carbonised	X	X
Setaria italica Beauv.	Italian millet	whole	Carbonised	X	
cf. *Setaria italica* Beauv.	Italian millet	whole	Carbonised	X	
Pulses					
Lens culinaris Medik.	Lentil	cotyledon	Carbonised	X	
cf. *Lens culinaris* Medik.	Lentil	fragment	Carbonised	X	
Pisum sativum L.	Pea	cotyledon	Carbonised	X	
Vicia ervilia (L.) Willd.	Bitter vetch	whole	Carbonised	X	
cf. *Vica faba* L.	Broadbean	cotyledon	Carbonised		X
Vicia cf. *sativa* L.	Common vetch	whole	Carbonised	X	
cf. *Vicia sativa* L.	Common vetch	cotyledon	Carbonised	X	
Vicia sp.	Bean	whole	Carbonised	X	
Grasses					
Panicoid	Grass-like	whole	Carbonised	X	X
Pinus pinea L.	Pinecones	whole	Carbonised		X
Weeds					
Trifolium L.	Clover	whole	Carbonised	X	
cf. *Echium vulgare*		whole	Mineralised		X
Arrhenatherum elatius ssp. *bulbosum*	Onion couch	fragment	Carbonised	X	X
Plant Parts					
	Thorn	fragment	Carbonised		X
Unknown					
	Indeterminate	fragment	Carbonised	X	X
	Indeterminate	fragment	Mineralised	X	

TABLE 15: SOAP FACTORY (VI.I.XIV)

Carbonised archaeobotanical remains were only recovered from Room 3 in the 1st century AD.

Mineralised remains were recovered from more rooms in the Soap Factory in the 1st century AD than in the 1st century BC. Fruit remains dominate the mineralised assemblage in both time periods. In the 1st century AD mineralised remains were found in the same rooms as the carbonised remains in the Soap Factory.

Discussion: Rubbish, Floors and Food

North

FIGURE 142: NORTH END OF INSULA VI.I (PHOTOMOSAIC @ 2007 JENNIFER F. STEPHENS · ARTHUR E. STEPHENS)

Triclinium

The Triclinium is located within the Inn. No contexts from the Triclinium could be dated to earlier than the 1st century AD. By far the most archaeobotanical remains [n= 15,402] from Insula VI.I came from this property due in large part to the exceptional preservation conditions created by the toilet feature.

Both carbonised and mineralised assemblages are dominated by fruit, although a greater diversity is present within the carbonised remains. The mineralised archaeobotanical assemblage largely consisted of mineralised fig achenes.

INN

FIGURE 143: INN (PHOTOMOSAIC @ 2007 JENNIFER F. STEPHENS · ARTHUR E. STEPHENS)

Species	Triclinium	Preservation	1AD
Cucumis L.	Melon	Mineralised	X
Ficus carica L.	Fig	Mineralised	X
Ficus carica L.	Fig	Carbonised	X
Ficus carica L.	Fig mesocarp	Carbonised	X
Malus domestica Borkh.	Apple	Mineralised	X
Olea europea L.	Olive	Carbonised	X
Punica granatum L.	Pomegranate	Carbonised	X
Punica granatum L.	Pomegranate	Mineralised	X
Prunus sp.		Mineralised	X
Vitis vinifera L.	Grape	Carbonised	X
Vitis vinifera L.	Grape pip	Carbonised	X
Vitis vinifera L.	Grape pip	Mineralised	X
	Vasicular tissue indeterminate	Carbonised	X
	Fruit exocarp	Carbonised	X
Nuts			
Amygdalus communis L.	Almond	Carbonised	X
Juglans regia L.	Walnut	Carbonised	X
Pinus pinea L.	Pinenut	Carbonised	X
Cereals			
Triticum sp	Wheat	Carbonised	X
cf. *Hordeum vulgare* L.	Barley	Carbonised	X
Chenopodium/S.italica	Italian millet/chenopodium	Carbonised	X
Panicum miliaceum L.	Common millet	Mineralised	X
Panicum sp	Panicum	Mineralised	X
cf. *Setaria italica* Beauv.	Italian millet	Carbonised	X
Pulses			
cf. *Lens culinaris* Medik.	Lentil	Mineralised	X
cf. *Vica faba* L.	Broadbean	Carbonised	X
Vicia sp.	Bean	Carbonised	X
Grasses			
Panicoid	Grass-like	Carbonised	X
Panicoid	Grass-like	Mineralised	X
Weeds			
	Cleavers	Carbonised	X
Illex aquifolium L.	Holly	Mineralised	X
	Mallow	Carbonised	X
Medicago	Medick	Mineralised	X
Ornithopus perpusillus/sativus	Bird's foot	Mineralised	X
Rubus	Blackberry	Mineralised	X
Cupressus sempervirens	Cypress	Mineralised	X
Plant Parts			
	Thorn	Mineralised	X
Unknown			
	Indeterminate	Carbonised	X
	Indeterminate	Mineralised	X

TABLE 16: TRICLINIUM

Species	Inn		Preservation	1BC	1AD
Cucumis L.	Melon	seed	Mineralised		X
Ficus carica L.	Fig	achene	Mineralised	X	X
Ficus carica L.	Fig	achene	Carbonised	X	
Ficus carica L.	Fig mesocarp	fragment	Carbonised		X
Olea europea L.	Olive	endocarp	Carbonised	X	X
Punica granatum L.	Pomegranate	seed	Carbonised	X	X
Prunus persica (L.) Batsch.	Peach	fragment	Carbonised		X
Vitis vinifera L.	Grape pip	pip	Carbonised	X	X
Vitis vinifera L.	Grape pip	pip	Mineralised	X	X
	Vasicular tissue indeterminate	fragment	Carbonised	X	X
	Fruit exocarp	fragment	Carbonised	X	
Nuts					
Corylus avellana L.	Hazelnut	shell fragment	Carbonised	X	
Juglans regia L.	Walnut	shell fragment	Carbonised	X	X
Pinus pinea L.	Pinenut	shell fragment	Mineralised		X
Pinus pinea L.	Pinenut	shell fragment	Carbonised	X	
	Nuts (generic)	shell fragment	Carbonised	X	
Cereals					
cf. *T.aestivum/durum* L.	Free-threshing wheat	grain	Carbonised	X	X
T. cf. *dicoccum* Schübl.	Emmer wheat	grain	Carbonised	X	
Triticum sp	Wheat	grain	Carbonised	X	X
Hordeum vulgare L.	Barley	grain	Carbonised		X
Secale cereale	Rye	grain	Carbonised		X
Chenopodium/S.italica	Italian millet/chenopodium	whole	Carbonised	X	
Panicum sp	Panicum	whole	Carbonised		X
Lens culinaris Medik.	Lentil	fragment	Carbonised	X	X
Pisum sativum L.	Pea	cotyledon	Carbonised		X
cf. *Vicia ervilia* (L.) Willd.	Bitter vetch	whole	Carbonised		X
Vicia sativa L.	Commom vetch	whole	Carbonised	X	X
Vicia cf. *sativa* L.	Commom vetch	cotyledon	Carbonised	X	
Vicia sp.	Bean	whole	Carbonised	X	X
Grasses					
Panicoid	Grass-like	whole	Carbonised	X	X
Panicoid	Grass-like	whole	Mineralised	X	
Weeds					
Agrostemma githago	Corncockle	whole	Mineralised	X	
Celtis L.	Hackberry	whole	Mineralised		X
Galium aparine L.	Goosegrass	whole	Carbonised		X
Illex aquifolium L.	Holly	whole	Mineralised		X
Lithospermum/Buglossoides		whole	Mineralised	X	X
Linum biene	Linum	whole	Mineralised		X
Plantago L.		whole	Carbonised	X	X
	Persicaria	whole	Carbonised	X	
	Polygonium	whole	Carbonised	X	
Sherardia arvensis L.	Field madder	whole	Carbonised		X
Sherardia arvensis L.	Field madder	whole	Mineralised	X	X
Trifolium L.	Clover	whole	Carbonised	X	
Viburnum lantana	Wayfaring tree	whole	Carbonised	X	
Arrhenatherum elatius var. *bulbosum*	Onion couch	fragment	Carbonised		X
Rubus	Blackberry	whole	Carbonised		X
Rubus	Blackberry	whole	Mineralised	X	X
Plant Parts					
	Thorn	fragment	Carbonised		X
Unknown					
	Indeterminate	fragment	Carbonised	X	X
	Indeterminate	fragment	Mineralised	X	X
	Bread/pastry	fragment	Carbonised	X	X

TABLE 17: INN PRESENCE AND ABSENCE

1578 mineralised fig achenes were recovered from the 1st century BC along with the equivalent of 119 carbonised olive endocarps. The 1st century AD assemblage consisted of carbonised wheat, barley, rye, millet, pea, bitter vetch, and common vetch; all of which could be classified as lower status foods in the Roman world. It is also the only property from Insula VI.I in which one specimen of rye was recovered.

There is a stronger presence of carbonised nuts in the 1st century BC and a visible increase in carbonised pulses in the 1st century AD from the Inn. Aside from Room 3 in the 1st century BC and Room 8 and Room 6 in the 1st century AD archaeobotanical remains were recovered from the same rooms, albeit in different proportions of types of archaeobotanical remains, in both the 1st century BC and AD.

Mineralised remains from the Inn were dominated by fruit from both the 1st century BC and the 1st century AD.

Inn Bar

FIGURE 144: INN BAR (PHOTOMOSAIC @ 2007 JENNIFER F. STEPHENS · ARTHUR E. STEPHENS)

Species	Inn Bar		Preservation	1BC	1AD
Ficus carica L.	Fig	achene	Mineralised	X	X
Ficus carica L.	Fig	achene	Carbonised	X	
Ficus carica L.	Fig mesocarp	fragment	Carbonised	X	X
Olea europea L.	Olive	endocarp	Carbonised	X	X
Prunus persica (L.) Batsch.	Peach	fragment	Carbonised		X
Vitis vinifera L.	Grape pip	pip	Carbonised		X
Vitis vinifera L.	Grape pip	pip	Mineralised	X	X
	Vasicular tissue indeterminate	fragment	Carbonised	X	X
Nuts					
Juglans regia L.	Walnut	shell fragment	Carbonised	X	
Pinus pinea L.	Pinenut	shell fragment	Carbonised	X	
	Nuts (generic)	shell fragment	Carbonised		X
Cereals					
Triticum sp	Wheat	fragment	Carbonised	X	X
Hordeum vulgare L.	Barley	fragment	Carbonised		X
Chenopodium/S.italica	Italian millet/chenopodium	whole	Carbonised	X	X
Panicum sp	Panicum	whole	Carbonised		X
Pulses					
Vicia sp.	Bean	whole	Carbonised		X
Weeds					
Echium vulgare	Viper's Bugloss	whole	Mineralised	X	
Lithospermum/Buglossoides		whole	Mineralised	X	
Cupressus sempervirens	Cypress	Seed	Mineralised	X	
Unknown					
	Indeterminate	fragment	Carbonised		x
	Indeterminate	fragment	Mineralised		x

TABLE 18: INN BAR PRESENCE AND ABSENCE

There is a visible decrease in the presence of nuts from the Inn Bar from the 1st century BC to the 1st century AD.

Mineralised fruit dominate the assemblage from the Inn Bar for both time periods. Mineralised fruit remains are recovered from Rooms 1B, 1B and 3B and 6B in the 1st century AD as opposed to only Room 6B from the 1st century BC.

Commercial properties

The commercial properties within Insula VI.I had an average of 7 archaeobotanical remains per context examined. The causal factors for this result may be manifold. In particular, the small physical size and limited excavation areas within the bar areas resulted in small volumes of residues available for examination (aside from the Inn Bar, which due to the large size of the Inn was more expansive). In addition, the conversion in the 1st century AD into what became defined as the bar areas was relatively recent. Hence, the archaeobotanical remains from the 1st century BC and earlier would most likely have been damaged or lost through numerous reconstruction/building and repair projects. Practically, in these small bar areas it may have been that food was carried away to be eaten elsewhere, perhaps by customers with small apartments with no cooking/heating facilities. Nothing was found or recorded from any of the *dolia*/ um from Insula VI.I. Potentially there could be a lack of archaeobotanical remains because these bar areas served drinks/wine rather than food.

Domestic properties

House of the Vestals (VI.I.vii)

FIGURE 145 HOUSE OF THE VESTALS (PHOTOMOSAIC @ 2007 JENNIFER F. STEPHENS · ARTHUR E. STEPHENS)

From the 1st century AD mineralised fig achenes [n=1703] were the most numerous archaeobotanical remains recovered from the House of the Vestals, as they were for the Triclinium. The archaeobotanical assemblage from the House of the Vestals (VI.I.vii) was the only one within Insula VI.I to include spices, as identified by Ciaraldi (2001). This would suggest that the House of the Vestals (VI.I.vii) had greater wealth to acquire such luxury items. It may also signify the overall increasing wealth of the city of Pompeii during the late 1st century BC to the 1st century AD (Ciaraldi and Richardson 2000, 75).

Carbonised remains were recovered from more rooms within the House of the Vestals from the 1st century AD than the 1st century BC. In the majority of rooms the archaeobotanical assemblage was dominated by carbonised fruit. There is a definite increase in the number of rooms in which carbonised archaeobotanical remains were recovered from the 1st century BC to the 1st century AD. The majority of archaeobotanical remains were recovered from rooms located around the bar areas and the front section of the House of the Vestals, what might be considered the publically accessible areas of the House of the Vestals.

remains were recovered from a number of rooms from this property in the 1st century AD. It is more likely that there was an absence of factors facilitating mineralisation.

Vestals bar

FIGURE 146: VESTALS BAR (PHOTOMOSAIC @ 2007 JENNIFER F. STEPHENS · ARTHUR E. STEPHENS)

Few archaeobotanical remains from the 1st century BC [n=7.6] aside from 2 mineralised fig achenes, 1 carbonised fig mesocarp fragment, 2.57 carbonised and 1.78 mineralised grape pip fragment, and 1 fragment of *Triticum* sp were recovered from this commercial property within the House of the Vestals (VI.I.vii). There is a visible shift from the 1st century BC [n=3] to the 1st century AD [n=16] towards a greater diversity of taxa at this property and archaeobotanical remains [n=84.2]. One of the few carbonised emmer spikelet forks recovered from Insula VI.I was found in the Vestals Bar dating to the 1st century AD.

Species	House of the Vestals		Preservation	1BC	1AD
Cerasus avium L.	Cherry	half	Mineralised		X
Cucumis melo L.	Melon	fragment	Carbonised		X
Cucumis L.	Melon	fragment	Mineralised	X	
Ficus carica L.	Fig	achene	Mineralised	X	X
Ficus carica L.	Fig	achene	Carbonised	X	X
Ficus carica L.	Fig mesocarp	fragment	Carbonised	X	X
Malus domestica Borkh.	Apple	seed	Carbonised		X
Olea europea L.	Olive	endocarp	Carbonised	X	X
Punica granatum L.	Pomegranate	seed	Carbonised	X	X
Prunus persica (L.) Batsch.	Peach	fragment	Carbonised		X
cf. *Prunus* sp.		fragment	Carbonised	X	
Vitis vinifera L.	Grape	petiole	Carbonised	X	
Vitis vinifera L.	Grape	pip	Carbonised	X	X
Vitis vinifera L.	Grape pip	pip	Mineralised	X	X
	Vasicular tissue indeterminate	fragment	Carbonised		X
	Fruit exocarp	fragment	Carbonised		X
Nuts					
Corylus avellana L.	Hazelnut	shell fragment	Carbonised		X
Juglans regia L.	Walnut	shell fragment	Carbonised	X	X
cf. *Juglans regia* L.	Walnut	shell fragment	Carbonised		X
Pinus pinea L.	Pinenut	shell fragment	Carbonised	X	X
	Nuts (generic)	shell fragment	Carbonised		X
Cereals					
T. aestivum/T.durum L.	Free-threshing wheat	grain	Carbonised		X
Triticum sp	Wheat	fragment	Carbonised	X	X
Hordeum vulgare L.	Barley	grain	Carbonised	X	X
cf. *Hordeum vulgare* L.	Barley	grain	Carbonised	X	
Panicum miliaceum L.	Common millet	whole	Carbonised	X	X
Panicum miliaceum L.	Common millet	whole	Mineralised		X
Panicum cf. *miliaceum* L.	Panicum	whole	Carbonised		X
Panicum sp	Panicum	fragment	Mineralised		X
Lens culinaris Medik.	Lentil	whole	Carbonised	X	X
cf. *Lens culinaris* Medik.	Lentil	whole	Carbonised		X
Pisum sativum L.	Pea	cotyledon	Carbonised		X
cf. *Pisum sativum* L.	Pea	fragment	Carbonised		X
cf. *Vica faba* L.	Broadbean	whole	Carbonised	X	
Vicia sativa L.	Common vetch	whole	Carbonised	X	X
Vicia sp.	Bean	whole	Carbonised		X
Grasses					
Panicoid	Grass-like		Carbonised	X	X
Weeds					
Agrostemma githago	Corncockle	whole	Mineralised	X	
Agrostemma githago	Corncockle	whole	Carbonised	X	
Echium vulgare	Viper's Bugloss	whole	Mineralised		X
Illex aquifolium L.	Holly	whole	Mineralised		X
Lithospermum/Buglossoides		whole	Mineralised	X	
Linum	Linum	whole	Mineralised		X
Ornithopus perpusillus/sativus	bird's foot	pod	Mineralised	X	
Plantago L.		whole	Carbonised		X
Trifolium L.	Clover	whole	Carbonised		X
Arrhenatherum elatius ssp. *bulbosum*	Onion couch	fragment	Carbonised		X
Viburnum lantana	Wayfaring Tree	whole	Mineralised		X
Plant Parts					
	Floret	fragment	Carbonised		X
Unkown					
	Indeterminate	fragment	Carbonised	X	X

TABLE 19: HOUSE OF THE VESTALS (VI.I.VII)

All the recovered mineralised assemblages from the rooms within the House of the Vestals were dominated by fruit. Only a few mineralised archaeobotanical remains were recovered from the 1st century AD from the House of the Vestals from Room 4. In the 1st century BC the mineralised remains appear to be cluster around the Vestals Bar area and adjacent rooms and small rooms in the centre of the house. This difference between the two time periods may be due to more favourable preservation conditions present in the 1st century BC, perhaps there was a higher household population. The paucity of mineralised remains from the 1st century AD cannot be attributed to rebuilding and construction events as carbonised archaeobotanical

Species	Vestals Bar			1BC	1AD
	Fruit		Preservation		
Ficus carica L.	Fig	achene	Mineralised	X	X
Ficus carica L.	Fig	achene	Carbonised		X
Ficus carica L.	Fig mesocarp	fragment	Carbonised	X	X
Olea europea L.	Olive	endocarp	Carbonised		X
Prunus sp.		fragment	Mineralised		X
Vitis vinifera L.	Grape pip	pip	Carbonised	X	X
Vitis vinifera L.	Grape pip	pip	Mineralised	X	X
	Vasicular tissue indeterminate	fragment	Carbonised		X
	Fruit exocarp	fragment	Carbonised	X	X
Nuts					
Corylus avellana L.	Hazelnut	shell fragment	Carbonised		X
Juglans regia L.	Walnut	shell fragment	Carbonised		X
Pinus pinea L.	Pinenut	shell fragment	Carbonised		X
	Nuts (generic)	shell fragment	Carbonised		X
Cereals					
T. dicoccum Schübl.	Emmer wheat	spikelet fork	Carbonised		X
Triticum sp	Wheat		Carbonised	X	X
Pulses					
cf. *Lens culinaris* Medik.	Lentil	cotyledon	Carbonised		X
cf. *Pisum sativum* L.	Pea	whole	Carbonised		X
Vicia sativa L.	Common vetch	whole	Carbonised		X
Vicia sp.	Bean	whole	Carbonised		X
Grasses					
Panicoid	Grass-like		Carbonised		X
Weeds					
Echium vulgare	Viper's Bugloss	whole	Mineralised		X
Lithospermum/Buglossoides		whole	Mineralised		X
	Mallow	whole	Mineralised		X
Ornithopus perpusillus/sativus	Bird's foot	pod	Mineralised		X
Ornithopus perpusillus/sativus	Bird's foot	pod	Carbonised		X
Rubus	Blackberry	whole	Carbonised		X
Rumex L.	Docks	whole	Carbonised		X
Plant Parts					
	Thorn	fragment	Carbonised		X
	Thorn	fragment	Mineralised		X
Unknown					
	Indeterminate	fragment	Carbonised	X	X
	Indeterminate	fragment	Mineralised		X

TABLE 20: VESTALS BAR

More carbonised archaeobotanical remains were recovered from the 1st century AD from the Vestals Bar in a greater number of rooms than in the 1st century BC. One possible explanation could be that there was an increase in the use of the Bar area in the 1st century AD.

There were no mineralised archaeobotanical remains recovered from Vestals Bar from the 1st century AD, whereas in the 1st century BC the mineralised assemblage is dominated by fruit, along with the presence of some weeds.

House of the Surgeon (VI.I.x)

FIGURE 147: HOUSE OF THE SURGEON (PHOTOMOSAIC @ 2007 JENNIFER F. STEPHENS · ARTHUR E. STEPHENS)

Despite its fame as one of the oldest houses in Pompeii there was a paucity [n=299.44] of archaeobotanical remains from this house. No evidence of crop-processing was recovered from this domestic residence and very little archaeobotanical remains in general were recovered from its front commercial rooms/shops (n=3.12 from Room 2; n=57.41 from Room 3 and 4).

Carbonised archaeobotanical remains are clustered around Rooms 22, 23, 10, 11 and the shop, Room 3 and 4, in the 1st century BC. In the 1st century AD no archaeobotanical remains are recovered from Rooms 22 and 7. These rooms, particular the Surgeon shop, rooms 3 and 4, are located in the more public areas of the House. In contrast, few archaeobotanical remains were recovered from the back of the House of the Surgeon, or what would be considered by Wallace-Hadrill (1994) as the 'private' area of the house as it was located deeper within the structure of the house, away from public view.

Nearly all the mineralised remains recovered from the House of the Surgeon were fruit. During the 1st century AD only the two front commercial rooms and Room 10 yielded archaeobotanical remains. This may suggest that the main domestic area of the house was left relatively unused during this final period.

Species	House of the Surgeon		Preservation	1BC	1AD
Ficus carica L.	Fig	achene	Mineralised	x	x
Ficus carica L.	Fig	achene	Carbonised	x	x
Olea europea L.	Olive	endocarp	Carbonised	x	x
Vitis vinifera L.	Grape	fruit	Carbonised	x	x
Vitis vinifera L.	Grape pip	pip	Carbonised	x	x
Vitis vinifera L.	Grape pip	pip	Mineralised	x	x
	Vasicular tissue indeterminate	fragment	Carbonised	x	
	Fruit exocarp	fragment	Carbonised	x	
Nuts					
Juglans regia L.	Walnut	shell fragment	Carbonised	x	x
Pinus pinea L.	Pinenut	shell fragment	Carbonised		x
Cereals					
T. aestivum/T.durum L.	Free-threshing wheat	grain	Carbonised	x	
T. cf. aestivum/T.durum L.	Free-threshing wheat	grain	Carbonised	x	
T. dicoccum Schübl.	Emmer wheat	grain	Carbonised	x	
T. cf. dicoccum Schübl.	Emmer wheat	grain	Carbonised	x	x
T cf. monococcum	Einkorn	grain	Carbonised		x
Triticum sp	Wheat	grain	Carbonised	x	x
Panicum miliaceum L.	Common millet	whole	Carbonised		x
Panicum cf. miliaceum L.	Panicum	whole	Carbonised	x	x
Panicum sp	Panicum	whole	Carbonised	x	x
Setaria italica Beauv.	Italian millet	whole	Carbonised	x	
cf. *Setaria italica* Beauv.	Italian millet	whole	Carbonised	x	
Pulses					
Lens culinaris Medik.	Lentil	whole	Carbonised	x	x
Lens culinaris Medik.	Lentil	whole	Mineralised	x	
cf. *Lens culinaris* Medik.	Lentil	cotyledon	Carbonised	x	
Vicia cf. ervilia (L.) Willd.	Bitter vetch	whole	Carbonised	x	
Vicia sativa L.	Common vetch	whole	Carbonised	x	x
Vicia cf. sativa L.	Common vetch	cotyledon	Carbonised	x	
cf. *Vicia sativa* L.	Common vetch	whole	Carbonised	x	
Vicia sp.	Bean	whole	Carbonised	x	x
Grasses					
Panicoid	Grass-like	whole	Carbonised	x	
Weeds					
Apiaceae			Carbonised	x	
Lamiaceae		whole	Carbonised	x	x
Ornithopus perpusillus/sativus	Bird's foot	pod	Carbonised	x	x
Polygonium		whole	Mineralised	x	
Solanum nigrum		whole	Carbonised		x
Solanum nigrum		whole	Mineralised	x	
Rumex	Docks	whole	Carbonised	x	
Plant Parts					
	Thorn	fragment	Carbonised	x	x
Unknown					
	Indeterminate	fragment	Carbonised	x	x

TABLE 21: HOUSE OF THE SURGEON

Shrine (VI.I.xiii)

FIGURE 148: SHRINE (PHOTOMOSAIC @ 2007 JENNIFER F. STEPHENS · ARTHUR E. STEPHENS)

It is believed that the area labelled the Shrine was in fact used as commercial space in the 1st century BC.

Five hundred and three mineralised fig achenes were recovered from the 1st century BC along with 69.22 carbonised olives, 25.83 mineralised and 22.14 fragments of carbonised grape pips and 123 carbonised fragments of indeterminate fruit vasicular tissue.

The Shrine was created and used as a roadside shrine for passers-by entering or leaving the city to make ritual burnt offerings in the 1st century AD. From the 1st century AD 1 *T. dicoccum* and 2 *T.*cf *dicoccum* spikelet forks were recovered as well as the largest number of carbonised *Triticum* sp. [n=49 fragments] from the entire Insula. As the cereal grains labelled *Triticum* sp. are difficult to identify to species level due to distortion and fragmentation it is likely that they were burnt quickly, under high heat conditions. Cereal grains were a common ritual offering and so may represent burnt offerings rather than evidence of food preparation and/or cooking. A number of charred pulses [n=64.67] were also recovered from the Shrine as well, which were known to have been an important component of ritual offerings in Roman times.

The Shrine had the highest count of carbonised archaeobotanical remains [n=1273.71] and the highest number of taxa [n=29 from the 1st century BC or earlier and n=37 from the 1st century AD] recovered in both time periods. The Shrine had a large number of indeterminates both carbonised [n=395] and mineralised [n=109]. A number of weedy taxa [n=9] were present. Perhaps these weedy species were burnt in a hearth and then added to the construction fill.

A greater range of carbonised archaeobotanical remains was recovered from room 1, 2 and 3 in the 1st century AD.

Like at the other properties with Insula VI.I mineralised remains from the Shrine were dominated by fruit.

Well

FIGURE 149: WELL AND FOUNTAIN AT SOUTHERN TIP OF INSULA VI.I (PHOTOMOSAIC @ 2007 JENNIFER F. STEPHENS · ARTHUR E. STEPHENS)

Few archaeobotanical remains [n=99.8] were recovered from this tiny structure at the southern tip of Insula VI.I.

Species	Well		Preservation	1BC	1AD
Ficus carica L.	Fig	achene	Mineralised		X
Ficus carica L.	Fig mesocarp	fragment	Carbonised		X
Olea europea L.	Olive	endocarp	Carbonised	X	
Olea europea L.	Olive	half of endocarp	Carbonised		X
Olea europea L.	Olive	fragment of endocarp	Carbonised		
Olea europea L.	Olive	Correction	Carbonised		
Vitis vinifera L.	Grape	petiole	Carbonised		X
Vitis vinifera L.	Grape pip	pip	Carbonised	X	
Vitis vinifera L.	Grape pip	fragment of pip	Carbonised		
Vitis vinifera L.	Grape pip	Correction	Carbonised		
Cereals					
Triticum sp	Wheat	fragment	Carbonised		X
cf. *Hordeum vulgare* L.	Barley	fragment	Carbonised		X
Grasses					
Panicoid	Grass-like		Carbonised		X
Unknown					
Indeterminate		fragment	Carbonised		X

TABLE 23: WELL

The carbonised assemblage from the Well was dominated by fruit in both time periods.

No mineralised remains were recovered from the 1st century BC and only mineralised fruit was recovered from the 1st century AD.

Species	Shrine		Preservation	1BC	1AD
Cucumis L.	Melon	seed	Carbonised		X
Cucumis L.	Melon	seed	Mineralised	X	
Ficus carica L.	Fig	achene	Mineralised	X	X
Ficus carica L.	Fig	achene	Carbonised	X	X
Ficus carica L.	Fig mesocarp	fragment	Carbonised	X	X
Malus domestica Borkh.	Apple	seed	Mineralised	X	
cf. *Malus domestica* Borkh	Apple	seed	Mineralised	X	
cf. *Malus domestica* Borkh	Apple	seed	Carbonised	X	
Olea europea L.	Olive	endocarp	Carbonised	X	X
cf. *Punica granatum* L.	Pomegranate	seed	Carbonised		X
Prunus persica (L.) Batsch.	Peach	whole	Carbonised		X
Prunus persica (L.) Batsch.	Peach	fragment	Carbonised		
Vitis vinifera L.	Grape	fruit	Carbonised	X	X
Vitis vinifera L.	Grape	Petiole	Carbonised		
Vitis vinifera L.	Grape	pip	Carbonised	X	X
Vitis vinifera L.	Grape	petiole	Mineralised		
Vitis vinifera L.	Grape	fruit	Mineralised	X	
Vitis vinifera L.	Grape pip	pip	Mineralised	X	X
Vitis vinifera L.	Grape mesocarp	fragment	Carbonised		
Vitis vinifera L.	Grape exocarp	fragment	Carbonised	X	X
	Vasicular tissue indeterminate	fragment	Carbonised	X	X
	Fruit exocarp	fragment	Carbonised		X
	Fruit mesocarp	fragment	Carbonised		X
Nuts					
Amygdalus communis L.	Almond	shell fragment	Carbonised		
Corylus avellana L.	Hazelnut	shell fragment	Carbonised		X
cf. *Corylus avellana* L.	Hazelnut	shell fragment	Carbonised		
Juglans regia L.	Walnut	shell fragment	Carbonised	X	X
cf. *Juglans regia* L.	Walnut	shell fragment	Carbonised	X	X
Pinus pinea L.	Pinenut	shell fragment	Carbonised	X	X
	Nuts (generic)	shell fragment	Carbonised	X	X
Cereals					
T. aestivum/T.durum L.	Free-threshing wheat	grain	Carbonised		X
T. cf. *aestivum/T.durum* L.	Free-threshing wheat	grain	Carbonised	X	
cf. *T.aestivum/durum* L.	Free-threshing wheat	grain	Carbonised	X	
T. dicoccum Schübl.	Emmer wheat	grain	Carbonised		X
T. dicoccum Schübl.	Emmer wheat	spikelet fork	Carbonised	X	X
T. cf. *dicoccum* Schübl.	Emmer wheat	grain	Carbonised	X	X
T. cf. *dicoccum* Schübl.	Emmer	spikelet fork	Carbonised	X	
cf. *T.dicoccum* Schübl.	Emmer	grain	Carbonised	X	
T. dicoccum (2-grained) Schübl.	two-grained Emmer wheat	grain	Carbonised	X	
T. durum	Bread wheat	spikelet fork	Carbonised	X	
Triticum sp	Wheat	grain	Carbonised	X	X
Triticum sp	Wheat	fragment	Carbonised	X	X
Triticum sp	Wheat	tip	Carbonised	X	X
Hordeum vulgare L.	Barley	grain	Carbonised	X	X
cf. *Hordeum vulgare* L.	Barley	grain	Carbonised	X	X
Chenopodium/S.italica	Italian millet/chenopodium	whole	Carbonised		X
Millet sp	Millet	whole	Carbonised		
Panicum miliaceum L.	Common millet	whole	Carbonised	X	X
Panicum miliaceum L.	Common millet	whole	Mineralised	X	X
Panicum cf. *miliaceum* L.	Panicum	whole	Carbonised	X	X
Panicum sp	Panicum	whole	Carbonised	X	X
Panicum sp	Panicum	whole	Mineralised	X	
Setaria italica Beauv.	Italian millet	whole	Carbonised		
Setaria cf. *italica* Beauv.	Italian millet	whole	Carbonised	X	X
cf. *Setaria italica* Beauv.	Italian millet	whole	Carbonised		
Pulses					
Lens culinaris Medik.	Lentil	whole	Carbonised	X	X

TABLE 22: SHRINE (VI.I.XIII)

	Well	Bar of Pheobus	Bar of Acisculus	Soap Factory	Shrine	House of the Surgeon	House of the Vestals	Vestals Bar	Inn	Inn Bar	Triclinium	Via Consolare	Vicolo di Narciso
Cereals													
Barley				x	x		x		x	x	x		x
Einkorn													
Einkorn/Emmer							x						
Emmer				x	x	x	x	x	x				x
Free-threshing wheat			x	x	x	x	x		x				
Oats							x						
Rye									x				
Millet													
Common millet			x	x	x	x	x	x	x	x	x	x	x
Italian millet				x	x	x	x				x		
Fruit													
Apple				x			x				x		x
Citron							x						
Date													
Fig	x	x		x	x	x	x		x	x	x		x
Grape	x		x	x	x	x	x	x	x	x	x	x	x
Peach				x	x		x		x				x
Pomegranate				x	x		x		x		x		x
Olive	x	x	x	x	x	x	x	x	x	x	x	x	x
Nuts													
Almond		x		x	x						x		
Hazelnut				x	x		x		x				x
Stone pine		x	x	x	x		x	x	x	x	x		x
Walnut				x		x		x	x	x	x		x
Pulses													
Bitter vetch				x	x	x			x				
Broadbean			x	x	x		x			x	x		
Chickpea													
Lentil			x	x	x	x	x	x	x		x		x
Pea				x	x	x		x	x				

TABLE 24: PRESENCE AND ABSENCE OF COMMON TAXA FROM PROPERTIES WITHIN INSULA VI.I
(INCLUDES DATA FROM CIARALDI (2001) PHD THESIS)

7.2 Sampling recovery

Few environmental studies address the impact of a comprehensive sampling scheme on archaeobotanical interpretation with empirical evidence (Lennstrom and Hastorf 1995, 701). This study hopes to redress this methodological issue by examining the results from the different properties within Insula VI.I regarding the different methods of recovery including the flot, residue, *in situ*/handpicked and dry sieving in order to ascertain the most effective method of recovery for the different taxa, preservation type and any biases present or created in this assemblage.

A classic archaeobotanical bias is present in the *in situ*/handpicked and 5 mm dry-screened category. The majority of the recovered botanical remains from these methods were large and readily identifiable fruit remains, mostly consisting of charred olive fragments. Olives are ubiquitous throughout Insula VI.I and are likely preserved and recovered due to their robust nature and the larger size of their endocarp. Recovery of archaeobotanical remains from this method is largely dependent upon existing soil types, excavation conditions and excavators' skill (Pearsall 2000, 11-12).

The majority of carbonised and mineralised remains from the properties examined in Insula VI.I were recovered from the flot samples. These results were largely expected as the majority of archaeobotanical remains from this assemblage were small and many were also fragmented. The majority of the mineralised remains were small and often honey-coloured-to-white and therefore difficult to pick out by excavators against the light-coloured archaeological matrix in the bright sunlight.

7.3 Indeterminates/unknowns

Veal (2009, 155) posits that there may be a correlation between areas not open to the public today such as the Inn and bar areas of Insula VI.I, which have lower rates of charcoal indeterminates, when compared with publicly accessible houses within Insula VI.I such as the House of the Surgeon (VI.I.x) and the House of the Vestals (VI.I.vii). She speculated that the effects of trampling on preservation would be more pronounced in the publicly accessible areas of Pompeii. These preliminary results on the charcoal may inform conservation issues to implement in the future at Pompeii and at other publicly accessible ancient sites.

However, this pattern was not replicated in the archaeobotanical record from Insula VI.I. Presumably, the same taphonomic factors which are affecting charcoal are also impacting upon archaeobotanical remains. However, there appears to be no correlation between the number of archaeobotanical indeterminates/unknowns and publically accessible and off-limit areas within Insula VI.I. This may be due to their different spatial distribution: charcoal was recovered from nearly every context excavated whereas archaeobotanical remains were sparsely distributed throughout the Insula. Archaeobotanical remains are also generally smaller and, particularly mineralised remains, can be more fragile than charcoal and therefore when partially damaged are often beyond recognition and often not identified and recorded as ecofacts.

7.4 Disposal

The disposal of food waste can be viewed as the final aspect of food consumption and reveals how a society views and defines space (Gumerman 1997). Ciaraldi and Richardson (2000, 81) noted that during the course of the early excavations of the House of the Vestals (VI.I.vii) and the House of the Surgeon (VI.I.x) both domestic properties were 'unusually clean' of ecofact material. One explanation for the paucity of cereal processing by-products, and archaeobotanical and faunal remains is that Insula VI is located next to the Herculaneum Gate. It is possible that household refuse was disposed of by simply tossing it over the city wall (Ciaraldi and Richardson 2000, 81; Richardson, Thompson and Genovese 1997, 94).

Both the back street, Vicolo di Narciso, located behind Insula VI.I and the busy front street, Via Consolare, onto which the front public entrances of the private residences, the House of the Surgeon (VI.I.x) and the House of the Vestals (VI.I.vii) faced, were examined in this study. Vicolo di Narciso proved to be more profitable in terms of archaeobotanical remains. This can largely be attributed to the drainage pipe which emptied, out of public view, onto the back street. It has also been postulated that household rubbish also left the house through the window at night and this could have added to the debris on the roads (Liebeschuetz 2000, 51). Rubbish in the back street was perhaps considered less objectionable, as this road was likely used more frequently by slaves and servants of the properties of Insula VI.I. Rainwater, running down the incline of Vicolo di Narciso towards the well, may have naturally swept away the debris periodically, particularly during the occasional torrential rainfall during the summer months.

It is known that in the 1st century AD the city of Rome had its own rubbish removal program. It is possible that a similar rubbish system may have also existed for the city of Pompeii. Ciaraldi and Richardson (2000, 81) argue that access to this municipal removal service would hint at the high status of the House of the Vestals (VI.I.vii) and the House of the Surgeon (VI.I.x), two elite residential properties within Pompeian society. However, the paucity of archaeobotanical remains from the bar properties suggests that this may not have been the case. However rubbish was being disposed of (or not) it is clear that in Pompeii dirt and offensive smells were taken as part of daily life and were not regarded as nearly as objectionable to the Romans as to modern sanitized standards; a striking example of this is the placement of the toilet within the kitchen area (Hobson 2009).

Although one would expect that stone floors and ornate mosaic pavements would be swept clean it has been suggested that they were simply covered over with fresh earth, straw or hay and some domestic refuse may have remained in the home (Liebschuetz 2000, 51). However, this does not appear to be the case as no carbonised remains of hay or straw were recovered from any of the properties from Insula VI.I or any other recorded archaeobotanical study to date from the Vesuvian area (aside from the carbonised hay from *Oplontis* found in the stables). As seen from the schematic diagram of the House of the Vestals (VI.I.vii) mosaics floors cover a large portion of the surface area within this house (Figure 175). The paucity of archaeobotanical remains from this property strongly suggests that these mosaic floors were most likely kept clean.

FIGURE 150: HOUSE OF THE VESTALS (VI.I.VII) PRESENCE OF MOSAIC FLOORS (IMAGE AFTER VEAL 2009)

The famous *trompe l'oeil* (trick-the-eye) asarotos oikos or 'unswept floor' comic floor mosaic from Roman Italy, now stored in the *Museo Gregoriano Profano* (Vatican Museum), illustrates this aspect of general housekeeping within the Roman house. The mosaic depicts, in a realistic fashion, the remains of a messy feast, composed of standard Roman fare, including seashells, fish bones, chicken legs, lobster claws, fruits and nuts (Dalby 2000, 256; Clarke 2007, 57). Servants were expected to clear the floor of food debris between courses. This is described in the passage in the section *Cena Trimalchionis* (The Banquet of Trimalchio) in the Roman work the *Satyricon* by the author Petronius. When a slave picks up discarded food on the floor he is told to put it back so that it may be swept up with the rest of the rubbish on the dining floor (Beard 2008a; Strong 2002, 4).

Clarke (2007, 57-59) interprets the *asarotos oikos* not as a warning to diners about slopping on the dining floor (as there were servants to clean the floors between courses) but as a humorous image for the guests to view during the meal. It may have even served as a game board in which diners might 'add to' or 'shoot for' the image with the real food item. A parallel example is found in the Greek world with *kottabos*, a Greek wine-drinking game. The mosaic floor may also have had a hint of a *memento mori*, reminding diners of the fleetingness of life and the eventual finality of all things.

This image of a mosaic floor 'in use' is particularly helpful as it permanently displays what has been literally swept away from archaeological view. This may be one factor in the paucity of archaeobotanical remains recovered from Insula VI.I, particularly from the domestic properties, the House of the Vestals (VI.I.vii) and the House of the Surgeon (VI.I.x), in which such dinner parties may have taken place. Hence, the presence of intact floors and mosaics and the lack of rubbish pits or middens provide evidence that rubbish was most likely removed from Insula VI.I.

7.5 Household changes

Due to Pompeii's unique preservation a preponderance of shrines, over 500, have been identified and recorded (in comparison with only two dozen from the rest of the entire Roman world). Many Pompeian properties possessed more than one shrine. The primary function of the household *lares* was cult protection. As the *lares* were often associated with the domestic family meal and it was the slaves who did the majority of the cooking they were naturally spatially coupled off together (Foss 1997, 217). Shrines were normally located in either kitchens or gardens as it was believed that burnt sacrificial offerings needed access to the open sky to reach the gods (Orr 1988, 293). A variety of offerings, normally composed of a portion of every meal, would be made to the *lares* and these could include incense, garlands of grain or individual grains such as spelt (presumably emmer wheat), grapes, honey cakes or honeycombs, the first fruits of the season, wine, and even blood sacrifices from animals (Clarke 1991, 9; Ciaraldi and Richardson 2000, 81). Cooking and food preparation and storage may have also taken place in the ambulatories or gardens. This would have been practical as dining rooms often opened up into the garden (Allison 2004, 89). Varro notes that meals were usually taken alfresco in the garden during the warmer months and by the hearth in the winter months (Jashemski 1979b, 89; Strong 2002, 28).

The general expansion of wealth and household size during the mid-to-late Republican period is accompanied by a more formal definition and use of space which can be witnessed in the layout of gardens (Jashemski 1979b), the shift from the atrium to the peristyle and the creation of separate domestic/cooking areas (Robinson et al. 2008). This compartmentalisation of domestic functions through the creation of separate service areas in the larger residences within Pompeii is seen in the House of the Vestals (VI.I.vii) and the House of the Surgeon (VI.I.x) (Jones and Robinson 2004). These changes in space may reflect a change in social attitudes with an increased concern for social distinction with the Augustan period (Perring 1992).

Kitchens were identified based upon the presence of cooking hearths but do not appear to have been restricted to any particular location in the house (Allison 2004, 99). Mau (1902, 266) dismissively states that kitchens within Pompeii 'had no fixed location. It was generally a small room, and was placed wherever it would least interfere with the arrangement of the rest of the house'. Hales (2003, 103) recently pointed out that the space designated for cooking was in close proximity to the front public entrance of the house. As kitchens and service areas were being confined to designed small areas within domestic properties daily tasks were removed from public view (Hales 2003). In fact, many of the identified Pompeian kitchens from the AD 79 layer were so strikingly small in size that Jashemski (1979) conjectured that Roman dishes may not have been the elaborate affair led to believe based upon Apicius's cookbook. Rather she speculates that it may have been the diversity and freshness of the produce from the local countryside which was used to impress the guests.

Within Roman households, based upon literary texts, it appears that rooms were largely labelled based upon the architecture of a room rather than their designated function. Therefore, activities may not have been confined to specific areas of the house. The majority of the so-called cubicula appear to have no definite function and could serve a variety of purposes including bedrooms, workrooms or rooms to entertain. All rooms within the Roman house could essentially be considered multifunctional and should not be restricted by modern definitions of domestic space (Hales 2003). Laurence (1995) argues for the temporal distribution of gendered space within the Roman household. After the men left the house the light-filled atrium could have been a space where the women and slaves of the household could work (Allison 2004, 155; Hales 2003). This could be one explanation for the presence of loam weights within atrium areas.

Allison's (2004, 6) study of artefactual assemblages from 30 Pompeian atrium houses used a quantified approach to examine the unquestioned assumptions regarding the spatial (re)-distribution of artefacts and the interpretations created by unsubstantiated cultural and social expectations, largely drawn from 19th century analogies from the middle and upper classes of Europe, during the rediscovery of Pompeii. From her analysis Allison (2004, 128) discovered evidence of food storage from 78 ground-floor rooms from both the front and rear of the house, while eight houses showed evidence of food storage on the upper floors, and three stairways showed evidence of temporary food storage. 'What is perhaps significant is that there was more evidence in the front area of the house for food storage than for food preparation. Food storage in the main courtyard areas of the house might conceivably be related to the role of certain foodstuffs as part of the household wealth' (Allison 2004, 130). It is possible that the earthquakes during the final fifteen years of Pompeii's existence may have affected where temporary food storage took place whilst restoration and repair were underway.

Allison's (2004) study reveals that food storage was not limited to out-of-view private areas of the house as previously believed based upon modern analogies. Indeed, food may have served as a signifier of the wealth of a household that was on display for clients and other members of the public to see. This also illustrates that environmental research strategies need to be carried out in more areas than what are simply regarded as 'kitchens' (based upon the presence of a cooking hearth) if scholars are to fully understand cultural attitudes towards food preparation, consumption and disposal (Gumerman 1997; Hastorf, 1991; Lennstrom and Hastorf 1995). The lack of conclusions regarding food storage is partly attributed by Allison (2004) to the lack of detailed studies on consumption patterns in Roman material culture.

Allison's (2004) analysis of household assemblage distribution is based upon artefactual usage in which the prevalent patterns are used to indicate habitual room use. In an attempt to 'see' the lower strata of Pompeian society service activities were identified by utilitarian assemblages related to food preparation and household industries were used to identify areas used by slaves or servants (Hastorf 1995). In a similar manner, this study assumes that the spatial distribution of preserved food remains can reveal some aspects of household use and change over time.

Hastorf (1995) comments that food remains scattered throughout a residence without a discernable pattern may imply that it was not important within that household to maintain food remains in one area, e.g. a hearth. In turn, she suggests that clustered remains may imply more regular and designated food processing, consumption or disposal areas within a residence. The very low recovery of archaeobotanical remains from Insula VI.I, with scattered remains found throughout each property, reinforces the theory of a general background 'archaeobotanical noise', present in the secondary/construction fill, used in levelling events and re-building materials throughout the Insula.

Artefactual assemblages in association with contextual information and the corresponding archaeobotanical assemblages could potentially provide insight into the more elusive symbolic meanings behind certain foodstuffs (Hodder 1987; Gumerman 1997). Unfortunately at this date, no artefactual data, e.g. pottery, metal, or glass, is available for comparison with the archaeobotanical data set from Insula VI.I. In the future, this would be a useful study to undertake alongside other lines of environmental data from Insula VI.I.

7.6 Changes over time in Insula VI.I

Pre-Roman to Roman

It has been proposed that Pompeii in the 3rd century BC was a network of scattered urban farms within the city walls. This idea is supported by the lack of archaeological evidence, aside from a few sherds of pottery. In addition, the results from the excavations of the House of Amarantus (I.IX.xii), lends credence to this theory with a lack of mineralised ecofacts which suggest that little latrine waste was present. This implies that a small population lived within a relatively open and undeveloped area. This unoccupied land could have been used for the cultivation of cereals and vegetables (Robinson 2002, 99).

Cereal by-products

Cereal and cereal processing by-products reveal at what stage archaeological sites are in terms of the spectrum from urban to rural. Sites with only cereal grains and no chaff or cereal by-products can be classified as urban consumer sites. Sites with evidence of cereal waste products such as barley, bread wheat or durum wheat rachis fragments, and culm nodes from straw imply that the inhabitants were undertaking early processing of cereals including threshing and winnowing and therefore can be classified as rural producer sites (Fuller and Stevens 2009, 37; Hillman 1973; Millet 1991). Sites with only evidence of emmer spikelet forks can be inferred to be practicing later stages of cereal processing including dehusking and winnowing. These differences in the preparation of cereal grains are important as they also hint at the power dynamics and control and trade issues present between producers and consumers and in turn the relationship between the city and countryside (Gumerman 1997).

Initially, the House of the Vestals (VI.I.vii) and House of the Surgeon (VI.I.x) were small properties situated around open courtyards with large areas of open space (see chapter 2). Arguably, based upon comparison with other elite houses in Pompeii, it is thought that the occupants of the early House of the Vestals (VI.I.vii) and the House of the Surgeon (VI.I.x) were engaging in crop-processing activities (Jones and Robinson 2004). There is little archaeobotanical or artefactual evidence to support this theory as there are few contexts which can be dated securely to the 2nd BC and earlier from the House of the Vestals (VI.I.vii), aside from the ritual foundation deposit (chapter 5). However, a few of these early deposits do contain cereal chaff providing evidence of cereal processing inside this domestic residence (Ciaraldi 2001, 2007). There is no evidence of any chaff or crop processing activity occurring within the House of the Surgeon (VI.I.x) from any time periods examined. However, different taphonomic and levelling and re-building events may have impacted upon the environmental results from this property.

Within Pompeian villas the colonnaded courtyard was thought to have been used for agricultural purposes including the threshing and winnowing of grain; a fact not mentioned in the ancient sources (Adams 2006, 27). Jones and Robinson (2004) argued that the House of the Vestals (VI.I.vii) maintained agricultural interests close to the city based upon its 'agricultural court', which had a rear entrance built around 100 BC onto Vicolo di Narciso, which could accommodate a horse and cart. They argue that this 'agricultural court' is similar in structure to the agricultural court from the House of the Menander (I.X.iv) (Jones and Robinson 2004, 2007).

Evidence of cereal by-products

The British School at Rome's excavations of the House of Amarantus (I.IX.xi-xii) revealed a lack of cereal by-products during the Roman period which represents a major shift in the agricultural dynamics of the city of Pompeii (Robinson 1999). Cereal processing may now have been confined to the many surrounding *villae rusticae* and shipped directly to the numerous bakeries in Pompeii (Ciaraldi and Richardson 2000, 76). At the time of its destruction in AD 79 Jashemski (1979b, 24) estimated that the 33 identified bakeries occupied 1.8% of the excavated area of Pompeii. These bakeries were also found to be distributed in a more regular fashion throughout the city than other shops (Pirson 2007). This would suggest that the majority of the population of the city had a bakery in or near their neighbourhood.

Insula VI.I

From the present study of Insula VI.I there is very little evidence for cereal processing occurring after the 1st century BC, aside from one charred spikelet fork from *T. dicoccum* (emmer wheat) from Vestals Bar dating to the 1st century AD (Table 25). The lack of cereal chaff suggests that cereal processing was no longer taking place within Insula VI.I and perhaps no longer within the city walls at all. This lack of evidence of cereal processing from Insula VI.I supports Robinson's (1999, 2002) argument that a fundamental shift in agricultural processing took place at Pompeii after the Sullan settlement in *circa* 80 BC.

Robinson (2002, 98) goes on to postulate that some of the differences observed between the pre-Roman and Roman burnt deposits from the House of Amarantus (I.IX. xii) may reflect 'local traditions becoming Romanised, following the establishment of Roman colonists in the town'. Knowledge of this specific historic event, the establishment of the Sullan colony *circa* 80 BC, adds weight to this argument. Cooley (2003, 114) states that '[c]hanges in what might be termed an 'archaeological culture' do not necessarily reflect the input of a particular ethnic group. Instead, it is more likely that what we are looking at is a complex cultural system derived from a mixture of influences, and it is often a thankless task to try to separate out different cultural trends and associate them with a particular group of inhabitants'. However, examining one aspect of an archaeological culture, such as crop-processing, with knowledge of a historically recorded event, may begin to reveal changes brought about by the presence of another culture. It may have been that as the population of the city grew, particularly with the sudden influx of Roman army veterans and their families, Pompeii became a more urbanized centre and there was no longer room in the city for crop processing. This shift resulted in the now visible urban 'Roman' food signature in the archaeobotanical assemblage in the 1st century AD, which no longer contained cereal processing by-products such as cereal rachis and spikelet forks.

7.7 Evidence of Urbanisation

Cleaned cereal grains are easier and less costly to transport and enable visual inspection of the quality of the naked grains (Fuller and Stevens 2009, 37). A number of Roman urban settlements and military sites show evidence of only clean grains, including *Durnovaria* (modern Dochester), Roman Colchester and Winchester and Roman London (Fuller and Stevens 2009, 57).

The Roman settlement of Silchester located in *Calleva Atrebatum* (Silchester, Hampshire) dating to the late 3rd to the 5th /7th century AD was never built-over after its abandonment. Archaeobotanical analysis has revealed a typical Roman diet consisting of cereals, dominated by spelt wheat and six-row barley, along with lentils, beans, and peas. Locally grown fruit recovered included apples, dominating the fruit assemblage, along with plum, cherry and perhaps wild/alpine strawberries. Nuts included hazelnuts and walnuts. Imported items included grape and fig. Herbs recovered 'included coriander, celery seed, summer savory, opium poppy seed, mustard (or at least hot-tasting Brassica or Sinapis seeds), and perhaps

Property	Time period	Type of Cereal	Quantity	Total
Vicolo di Narciso	2-1st century BC	*T. dicoccum* spikelet fork	1	1
Vestals Bar	1st century AD	*T. dicoccum* spikelet fork	1	1
Shrine	1st century BC	*T. dicoccum* spikelet fork	1	3
		T. cf. dicoccum spikelet fork	2	

TABLE 25: PRESENCE OF SPIKELET FORKS FROM INSULA VI.I

dill' (Robinson et al. 2006, 216-217). Interestingly, all the foodstuffs recovered from Insula IX from Silchester have been found in Roman London (J. Giorgi pers. comm.; Willcox 1977 cited in Robinson et al. 2006, 218). Significantly, at Silchester there is little evidence for the earlier stages of crop processing on the site (Robinson et al. 2006).

Robinson et al. (2006, 216-218) has put forward the hypothesis that Insula IX from late Roman Silchester was fully urbanised based upon the lack of crop-processing remains, evidence of trade by the presence of figs (exotics in Britain), and the recovery of mineralised plant remains due to the presence of cess from latrines. 'Urbanised' is here defined as a shift in the local economy in which the majority of the inhabitants are no longer directly participating in rural activities and were likely purchasing food goods, such as cereals, that had been cleaned and/or prepared for them.

Both Insula IX from late Roman Silchester and Insula VI.I from Pompeii had low concentrations of recovered carbonised archaeobotanical material (Robinson et al. 2006). At Insula VI.I the low concentration of carbonised archaeobotanical remains [17% of total archaeobotanical assemblage] is likely due to the lack of rubbish deposits and middens and perhaps even an organised waste removal system.

Also from Insula VI.I we see an increase in mineralised archaeobotanical remains, particularly in the House of the Vestals [n=23.8 to n=1802.4], Triclinium [n=15273], and to a lesser extent the Inn Bar [n=38.9 to n=58.2] from the 1st century BC to the 1st century AD, which may be attributed to an increase in the presence of cess in the form of informal cesspits or latrines within Insula VI.I. These results suggest the possibility of a growing population and increasing urbanism within the city of Pompeii in the 1st century AD.

The general results show that the range of foodstuffs recovered was slightly more diverse over time. There was an average of 10.85 taxa from 1st century BC and earlier and 15.92 taxa from the 1st century AD. In the present study of Insula VI.I we see that there is no evidence of exotics from Insula VI.I and therefore no statement can be made about long-distance trade.

7.8 City and country

Ciaraldi (2007) argues that an increasingly hierarchical relationship developed between the city and countryside as cereal dehusking was now taking place solely in the countryside. She further argues for the relocation of certain groups of people, including craftsmen and artisans, now based in the town, resulting in a greater compartmentalisation of knowledge; although Ciaraldi (2007) concedes that many of the economic and political structures within ancient Pompeian society are still poorly understood. Millet (1991, 185) suggests that '...sustained and systematic work on seeds from urban sites and their rural hinterlands could provide invaluable information about site interrelationships'. Further archaeobotanical investigations from the hinterland of Pompeii and within other areas of Pompeii with greater chronological depth could prove fruitful in examining this evolving relationship between the city and its surrounding agricultural land.

The anthracological study by Veal (2009) has shed insight into this relationship. Veal (2009) observed a noticeable trend of increasing diversity over time, specifically marked by an increase in orchard woods from the 1st century BC onwards. The finding of beech as the primary fuel wood is in direct contradiction with the advice of the ancient agricultural writers, who recommend the growing of riparian wood types. From the House of the Vestals (VI.I.vii) one taxon was predominant, the montane *Fagus sylvatica* (beech), with a marked increase in wood diversity after the Sullan colonisation. The results from the charcoal analysis, with the dominance of montane beech, which ecologically favours a high altitude, suggests that wealthy Pompeians had knowledge of and controlled tracts of land at higher altitudes within the nearby Lattari Mountains.

7.9 Preservation

A number of factors impeded archaeobotanical preservation within Insula VI.I, particularly the urban features of Pompeii, such as the presence of intact floors and mosaics. This is likely a huge contributing factor not simply because excavation below intact floors could not take place but they were also likely swept clean in the past and environmental material lost. The paucity of archaeobotanical remains from the majority of properties from Insula VI.I [Avg archaeobotanical remains per context =19.5] may also be attributed to the rebuilding, re-leveling and re-installation of mosaic floors throughout the occupation of the Insula and particularly the lack of specific features such as storage pits or middens, normally rich in ecofacts.

This study reveals the strong influence that preservation, particularly mineralisation, has upon an environmental assemblage. As van der Veen et al. (2008, 84) demonstrated in their comprehensive statistical analysis of archaeobotanical remains from Roman Britain there is a strong correlation between plant type and preservation. A similar pattern is seen in the archaeobotanical assemblages from Insula VI.I as carbonised archaeobotanical material is typically composed of olives and fruits and mineralised archaeobotanical material of small fruits seeds such as grape pips and fig achenes.

The taphonomic pressures present in different properties are illuminating in regards to the recovered archaeobotanical assemblage. The richest source of archaeobotanical material from Insula VI.I came from the Triclinium area due to the preservation conditions created by the toilet feature and the large number of small mineralised seeds, particularly fig seeds. Deposits within the House of the

Vestals (VI.I.vii) with exceptional preservation included the cesspit/drain, dating to the mid 2nd century BC to AD 79 and the foundation deposit analysed by Ciaraldi (2001, 2007). As both these deposits had the same preservation process, cess, Ciaraldi and Richardson (2000, 75) argue that these deposits offer the best reflection of the spectrum of food consumed within this house.

The majority of the archaeobotanical data from Insula VI.I came from secondary deposits from stratigraphic units within defined 'archaeological areas'. Admittedly, there are methodological issues with this system such as skill of the excavator in identifying new contexts, large variation in the size of the different archaeological areas from the different properties within Insula VI.I and the fact that secondary refuse deposits are often composed of a range of activities (Miksicek 1987). However, analysing aggregated archaeobotanical data not only provides larger and more reliable samples, an important aspect to consider when there are very low rates of recovery of carbonised and mineralised botanical remains as in this study, but it also allows broad consumption patterns to be visualised in the data (Gunerman 1997).

7.10 Lack of archaeobotanical remains

The House of the Vestals (VI.I.vii), with the inclusion of Ciaraldi's results from the foundation and drain deposits, was the only property within Insula VI.I which showed a similar pattern to the House of Amarantus (I.IX.xii). Both properties have evidence of crop-processing remains in the form of cereal chaff dating from the 2nd century BC which were absent in later deposits. Although there are a few spikelet forks presence in the 1st century BC and one in the 1st century AD there is no substantial evidence for cereal processing (including dehusking) in Insula VI.I. The presence of chaff in the 1st century BC is pretty rare for what one would expect at a rural site. It may have been that even in the 1st century BC few agricultural activities were still taking place within the city and they became rarer as the city became increasingly urbanised in the 1st century AD. It is suspected that the majority of contexts and deposits examined in this study from Insula VI.I were from secondary fill and therefore were too disturbed to show firm evidence of Romanisation in the archaeobotanical assemblage in the 1st century AD. However, the lack of evidence of crop processing in the 1st century AD in Insula VI.I would seem to support Robinson's (2002) theory that cereal processing was taking place outside of the city at this time.

All lines of environmental inquiry undertaken during the AAPP excavations have yet to be tallied and therefore it is too early to speculate upon whether the total environmental sampling strategy (see chapter 3) warranted the time and money invested. However, without this complete sampling strategy this study would not have obtained as many archaeobotanical remains as it did.

7.11 Summary

Like the majority of charred archaeobotanical material recovered via flotation, the archaeobotanical assemblage from Insula VI.I examined in this study proved rather unremarkable in terms of the standard Mediterranean taxa recovered (Fuller and Stevens 2009, 57). No spices or rare taxa were recovered. Firm identification of domestic apple seeds, both mineralised and carbonised, from Insula VI.I from both the 1st century BC and AD, were one of the few recorded to date from the known Vesuvian sites. Little evidence of early or later stages of crop processing were recovered from Insula VI.I and this may be taken as an index of Romanised urbanisation, with little to no cereal processing now taking place within the city.

This archaeobotanical assemblage will add to the growing body of environmental data collected from Pompeii and the Vesuvian region and provide a standard measure of typical botanical items against which perhaps richer contexts may be compared. Aside from the rich deposits recovered from the toilet feature from the Triclinium the majority of contexts examined from Insula VI.I were part of the general 'archaeobotanical background noise' normally found in secondary fill. This is one of the difficulties with urban archaeobotany: the lack of primary deposits containing archaeobotanical/environmental remains due to the presence of intact floors and multiple re-building and construction events.

Chapter 8

Conclusions

Pompeii, despite its fame as an archaeological site, still holds new problems and surprises for investigators, the general lack of archaeobotanical remains from Insula VI.I being one of them. Although quite a few properties from the Vesuvian area and Pompeii, in particular, have been uncovered, until recently, few were excavated with environmental priorities in mind. Although, admittedly, the results from Insula VI.I are rather limited, in terms of the quantity of archaeobotanical remains recovered, nevertheless, this study adds to the growing body of knowledge of paleoenvironmental research at Pompeii.

8.1 Integration of results

The archaeobotanical data retrieved and recorded from sites around Mount Vesuvius represents an opportunity to explore the range of preservation conditions, sites and taxa preserved (Meyers 1988). One of the ultimate goals of the present study was to integrate and situate the archaeobotanical results from Insula VI.I within this data set. The results from Insula VI.I are not unusual when compared with the archaeobotanical results from the surrounding Vesuvian sites. A standard Mediterranean archaeobotanical assemblage was recovered from Insula VI.I which included wheat, barley, legumes, olives, grapes and figs. The majority of contexts examined from the properties of Insula VI.I produced one to two taxa of preserved botanical remains. No spices or exotic items such as black pepper were recovered from Insula VI.I in the current study and hence no firm evidence for trade was present.

Whatever the type of property examined from Insula VI.I, whether domestic, commercial, and ritual, they all showed similar archaeobotanical scatter or 'background noise'. This reinforces the theory that the majority of contexts from Insula VI.I were composed of construction fill/levelling material and therefore composed of secondary fill. Although not spectacular in themselves, the archaeobotanical results from the majority of contexts from Insula VI.I present evidence of everyday 'background noise' at Pompeii as opposed to specific deposits of interest, e.g. foundation ritual deposit, drains, ritual offerings, or fodder. It is speculated that the results from Insula VI.I will be able to provide a comparison baseline from which both excavators and environmental specialists are able to distinguish archaeobotanical 'background noise' from secondary deposits from primary deposits and thereby assist in the interpretation of complex archaeology.

8.2 Preservation

No exceptional deposits, aside from the toilet feature from the Triclinium, were uncovered from Insula VI.I in the archaeobotanical analysis. Previous work by Ciaraldi (2001, 2007) did reveal several interesting deposits within Insula VI.I and these were included in some of the analyses. The lack of archaeobotanical results speak to the fact that the majority of archaeobotanical remains have been lost from the AD 62-79 destruction horizon when it was first exposed in the 18th century. Based upon Penelope Allison's work on artefact assemblages in Pompeian houses and recent paleoenvironmental research it is evident that the archaeological processes of deposition are as complicated as at any archaeological site. Beneath the AD 79 layer Pompeii, in regards to the archaeobotanical remains, is an archaeological site with no special preservation conditions or claims.

The majority of archaeobotanical remains from Insula VI.I were recovered via flotation. For the carbonised material there was a positive correlation between the volumes (L) of archaeological matrix floated and carbonised archaeobotanical material retrieved. The majority of the carbonised archaeobotanical assemblage from each property was dominated by fruit. This could in part be a preservational bias. Their seeds are often more numerous and disposed of more carelessly. Other taxa such as pulses, cereals, and millets have to be cooked before consumption and are often soaked or boiled and thus are less readily preserved via carbonisation. The majority of mineralised remains were composed of fruit, particularly small seeds, which pass through the human digestive system and are preferentially preserved.

8.3 Evidence of urbanisation

Urban archaeobotany presents a number of problems. The presence of intact floors, mosaics, re-building and construction events and lack of rubbish deposits or middens, makes it difficult to see specific activity areas and results in little preserved botanical material. The lack of archaeobotanical remains begs the question: where is the evidence of food remains? The mineralised archaeobotanical remains would suggest that it was either disposed of down cesspits or drains. The presence of cess hints at an increasing population living in a more confined space. Also, speculatively, food waste could have been carted away outside the city walls. There was also no archaeobotanical evidence of animal husbandry e.g. the presence of fodder or hay, in Insula VI.I. The absence of this type of archaeobotanical material again provides support for an urbanised Pompeii.

The general lack of evidence for crop-processing within Insula VI.I, from all properties examined, suggest that within this section of Pompeii, cereal processing was no longer occurring and was likely taking place in the nearby countryside outside the city gates. This new separation of what had previously been a routine domestic task, the final stages of cleaning/dehusking cereals for daily use within or near the household, can be read as an index of urbanisation in the Roman world. This developed urbanism, with its lack of crop-processing evidence, is not seen in the archaeobotanical record again until the modern industrial age, when agricultural activities were once again confined to the countryside and residents of urban cities became consumers, divorced from producing or processing their own food.

The few spikelet forks recovered from Vicolo di Narciso and the Shrine could have been either ritual offerings or remnants of waste. The one glume base of *T. monococcum /dicoccum* identified by Ciaraldi (2001) dating to the late 2nd century BC from the House of the Vestals (VI.I.vii) revealed a similar pattern to that seen from the House of Amarantus (I.IX.xii) and House of Hercules' Wedding (VII.IX.xlvii), in which evidence of crop-processing is present in the 2nd century BC and absent in later deposits. The archaeobotanical assemblage from Insula VI shows similarities to the 'urban' deposits from other Roman urban and military sites in which there is also a lack of crop-processing remains, recovery of mineralised plant remains (likely preserved through the presence of cess) and low concentration of carbonised archaeobotanical material.

8.4 Shifts in food preferences

No clear distinction between the archaeobotanical assemblages from the 1st century BC and earlier and those of the 1st century AD were seen. Similar Mediterranean taxa were recovered from both time periods from the majority of properties within Insula VI.I. Unlike the picture of Romanisation in the faunal record with the increase in the presence of pork, the archaeobotanical evidence is less clear.

Whole and fragmentary olive endocarps were present in every property from Insula VI.I, aside from the Via Consolare. The increase in the presence of olive remains in the 1st century AD from Insula VI.I coincides with an increase in arboreal olive pollen in southern Italy. This pattern could suggest that with an increasing Roman presence in the south of Italy intensification in olive cultivation and consumption was occurring at this time. It is well known that Romans ate olives at the table as well as consumed olive oil. This aspect of Roman cuisine, eating olives at the table, may account for the presence of olive endocarp throughout Insula VI.I as opposed to large deposits of olive lees created from olive oil pressing, often used as fertilizers in garden areas. However, olive pollen has been detected in low levels in southern Italy from the 7th/6th century BC onwards, so it is likely that the previous inhabitants of Pompeii were also familiar with and ate olives. Therefore it is nearly impossible to distinguish this Mediterranean trait of eating olives as a wholly Roman one.

The increase in orchard woods in the 1st century AD in the wood charcoal assemblage from Insula VI.I could correspond to the increase in fruits, particularly grape and fig in the same time period for the majority of properties from Insula VI.I. There is a general increase in the number of taxa over time in the charcoal and archaeobotanical record for nearly all properties within Insula VI.I. This may be due to the Roman presence and introduced food preferences in the 1st century AD. However, it is difficult to ascertain if these are solely Roman selections as all the archaeobotanical remains recovered were known Mediterranean taxa, which would have also been well-known to the Greek colonies, the Samnites and other native Italic people in southern Italy. Hence, the complex ethnic mix of the city of Pompeii, the majority of which were from Mediterranean area, complicates the issue of Roman cuisine.

Despite their exclusion from the traditional triad of Mediterranean food, consisting of wine, olive oil, and bread, pulses and other cereal grains were probably widely grown and eaten in Pompeii and were likely an important component of the diet. Only a handful of cereal grains and pulses were recovered from the majority of properties from Insula VI.I. This could be due several factors. Pulses are often soaked before cooking inhibiting preservation via carbonisation. Also, Pompeians were likely buying their bread from bakeries rather than purchasing cereal grains to make their own bread. Hence, bakeries were likely receiving the cereal grains already dehusked from the countryside. Therefore, there would have been less crop-processing waste and cereal grains expected to be found preserved in the archaeobotanical record throughout the city.

Shifts in grain preference from barley to glume wheats to free-threshing wheats are difficult to ascertain as relatively few cereal grains were recovered and no statistically valid conclusions could be reached. However, it does appear that emmer, barley, and millet, although according to ancient literary sources were regarded as unfashionable, persisted into the 1st century AD within most of the properties of Insula VI.I. No evidence of oats was discovered from this study. Oats were likely not cultivated as a distinct crop until the beginning of the 1st century AD, possibly as a replacement for millet. The lack of evidence from Insula VI.I could be regarded as additional support to this theory.

Based upon the lack of crop-processing remains Pompeii was an urban consumer society in the 1st century AD and likely into the late 1st century BC. Cereals, despite their relative absence in the archaeobotanical record from properties within the city, would have made an important component of the city's hinterland crop (Jongman 2007, 508; Moorman 2007, 435). Pompeii would have depended upon its fertile surrounding countryside for its food and in this manner is more like the recently modified consumer city model. Despite the recent charcoal evidence by Veal (2009) and Veal and Thompson (2008, 11) that shows that a sophisticated market system was in place for charcoal and wood selection at Pompeii in the 1st century AD the current archaeobotanical assemblage from Insula VI.I provides little evidence for trade or large-scale transport of crops or agro-business. All the botanical remains recovered from Insula VI.I were in small quantities and could have been grown in the open spaces or gardens within the city or surrounding hinterland.

8.5 Spatial distribution and patterns

No relationship was seen between archaeobotanical indeterminate/unknowns from specific areas within Insula VI.I and public versus private access to these areas. In regards to the spatial distribution of archaeobotanical remains throughout the Insula the general, very low scatter of a few fragments of standard 'Mediterranean' taxa e.g. mainly olive, grapes, and figs suggests that the majority of contexts were composed of secondary fill.

It also cannot be discounted that some of the charred archaeobotanical background scatter is composed of burnt offerings re-worked into the archaeological matrix (Robinson 2002). The recent archaeobotanical analysis by Matterne and Derreumaux (2008, 105) from the *Porta Nocera, Necropolis* revealed that the majority of burnt funerary offerings were composed of standard Mediterranean fare. Hence, it would be difficult to discern those food items created as burnt offerings and those that were disposed of culinary accidents if they were deposited in the same archaeological matrix.

In terms of the arrangement of domestic space it appears that rooms off to the side, *cubiculum*, within the domestic properties possessed little to no archaeobotanical remains, suggesting that their function or use did not involve food. Other rooms throughout the properties of Insula VI.I had low scatterings of archaeobotanical remains. Surprisingly, the garden area of the House of the Surgeon (VI.I.x) had few recovered archaeobotanical remains despite ancient literary sources stating that meals were taken alfresco, in the garden during the warm months, with daily offerings also made to the household gods. However, the very poor preservation conditions in the garden area certainly impacted upon the archaeobotanical results. Bar areas also yielded little archaeobotanical evidence and this may be due to the fact that they were recent additions to Insula VI.I in the 1st century AD and that food items or drinks were likely taken away by customers. However, both the Bar of Pheobus and the Bar of Acisculus did show the use of their back rooms in the 1st century AD as opposed to the 1st century BC.

8.6 Future research

As a finer chronology and stratigraphy become available for Insula VI.I it would be worthwhile to integrate all lines of artefactual and environmental evidence, e.g. pottery, charcoal, glass, metal and coins. Through this combined analysis it may be possible to see further details and draw new conclusions regarding the development and history of Insula VI.I.

Further archaeobotanical research is needed in other comparable Regiones and Insulae of Pompeii to further substantiate the theory of a fully urbanised city in the 1st century AD. It would be interesting to see if other areas within Pompeii produced similar archaeobotanical 'background noise' from secondary deposits, the lack of crop-processing remains from the 1st century AD and an absence of rubbish pits. In addition, excavation of rural properties dating to this time period outside of the city's walls in the surrounding hinterland could offer interesting insights into and allow further study of the interrelationship between the city and its hinterland.

8.7 Summary

The paucity of archaeobotanical remains, composed of standard Mediterranean taxa, from all properties within Insula VI.I reinforces the archaeological interpretation that the majority of contexts were composed of secondary fill. The archaeobotanical evidence from this study yielded no firm evidence of trade or conclusive information regarding the multiple cultural influences upon this city. Primarily, the archaeobotanical results have provided information on the urbanisation of Insula VI.I and the division of labour between the city and countryside. The general lack of crop-processing evidence from Insula VI.I reinforces the current understanding of an urbanised Pompeii in the 1st century AD.

8.8 Final remark

December 4, 2009 Pompeii was added to 'Google Street View' enabling anyone with access to the internet to 'travel' and 'explore' the ancient city of Pompeii. The Italian Ministry of Culture hopes this virtual 'experience' will increase tourism to the site. However, the real benefit to Pompeii may be the open access it allows 'visitors' whilst at the same time limiting the damaging effects that tourism has had upon this endangered site. It is clear that Pompeii still has the power to lure even virtual visitors to its dusty streets. It is hoped that renewed interest in the site will help generate new research on and funds for this unique archaeological site.

Bibilography

Adam, J. P., 2007. Building materials, construction techniques and chronologies. In: J. J. Dobbins and P. W. Foss (eds.), *The World of Pompeii*. London: Routledge, chap. 8, 98-116.

Adams, G. W., 2006. *The suburban Villas of Campania and their social function*. Oxford: Archaeopress.

Adams, K. R. and Gasser, R. E., 1980. Plant microfossils from archaeological sites: Research considerations and sampling techniques and approaches. *The Kiva* 45/4, 293-300.

Addley, Esther, November 11, 2010. *The Guardian*. Neglected ruins of Pompeii declared a 'disgrace to Italy'. Retrieved on 13 November 13, 2010 from World Wide Web: http://www.pasthorizons.com/index.php/archives/11/2010/neglected-ruins-of-pompeii-declared-a-disgrace-to-italy

Allen, J. R. M., Brandt, U., Brauer, A., Hubberten, H.-W., Huntly, B., Keller, J., Kraml, M., Mackensen, A., Mingram, J., Negendank, J. F. W., Nowaczyk, N. R., Oberhönsli, H., Watts, W. A., Wulf, S., and Zolitschka, B., 1999. Rapid environmental changes in southern Europe during the last glacial period. *Nature* 400/6746, 740-743.

Allen, J. R. M., Watts, W. A., and Huntly, B., 2000. Weichselian palynostratigraphy, palaeovegetation and palaeoenviroment: the record from Lago Grande di Monticchio, southern Italy. *Quaternary International* 73/74, 91-110.

Allison, P. M., 1991. Artefact Assemblages: not 'the Pompeii Premise'. In: E. Herring, R. Whitehouse, and J. Wilkins (eds.), *Papers of the fourth Conference of Italian Archaeology: New Developments in Italian Archaeology Part 1*. London: Accordia Research Centre, 49-56.

Allison, P. M., 1995a. House contents in Pompeii: data collection and interpretative procedures for a reappraisal of Roman domestic life and site formation processes. *Journal of European Archaeology* 3/1, 145-176.

Allison, P. M., 1995b. On-going seismic activity and its effects on the living conditions in Pompeii in the last decades. *Archäologie und Seismologie: la regione vesuviana dal 62 al 79 D.C. : problemi archaeologici e sismologici : Colloquium, Boscoreale, 26-27. November 1993*. München, Germany: Biering & Brinkmann, 183-189.

Allison, P. M., 1999. Labels for ladles: interpreting the material culture of Roman households. In: P. M. Allison (ed.), *The Archaeology of Household Activities*. London, New York: Routledge, 57-77.

Allison, P. M., 2004. *Pompeian households: an analysis of the material culture*. Los Angeles: Cotsen Institute of Archaeology at University of California.

Allison, P. M., 2006. The Insula of the Menander at Pompeii: *Volume III: The Finds, a Contextual Study*. Oxford: Clarendon Press.

Allison, P. M., 2007. Domestic spaces and activities. In: J. J. Dobbins and P. W. Foss (eds.), *The World of Pompei*. London: Routledge, chap. 17, 269-278.

Andreae, B., 2003. *Antike Bildmosaiken*. Mainz am Rhein: P. von Zabern.

Angiolillo, A., Mencuccini, M., and Baldoni, L., 1999. Olive genetic diversity assessed using amplified fragment length polymorphisms. *Theoretical Applied Genetics* 98, 411-421.

Appadurai, A., 1981. Gastro-politics in Hindu South Asia. *American Ethnological Society* 8/3, 494-511.

Arthur, P., 1986. Problems of the Urbanization of Pompeii: Excavations 1980-1981. *The Antiquaries Journal* 66, 29-44.

Ascher, R., 1961. Analogy in Archaeological Interpretaion. *Southwestern Journal of Anthropology* 17/4, 317-325.

Attema, P., Delvigne, J., and Haagsma, B.-J., 1999. Case studies from the Pontine Region in central Italy on settlement and environmental change in the first millennium BC. In: P. Leveau, F. Trement, K. Walsh, and G. Barker (eds.), *Environmental Reconstruction in Mediterranean Landscape Archaeology*. Oxford: Oxbow, 105-121.

Badal, E., Bernabeu, J., and Vernet, J.-L., 1994. Vegetation changes and human action from the Neolithic to the Bronze Age (7000- 4000 B.P.) in Alicante, Spain, based on charcoal analysis. *Vegetation History and Archaeobotany* 3, 155-166.

Bakels, C., 2002. Plant remains from Sardinia, Italy with notes on barley and grape. *Vegetation History and Archaeobotany* 11/1, 3-8.

Bakels, C. and Jacomet, S., 2003. Access to luxury foods in central Europe during the Roman period: the archaeobotanical evidence. *World Archaeology* 34/3, 542-557.

Bakker, J. C. (ed.), 1999. *The Mills-Bakeries of Ostia: description and interpretation*. Amsterdam: Gieben.

Banning, E. B., 2000. *The Archaeologist's Laboratory: the Analysis of Archaeological Data*. New York: Kluwer Academic/Plenum.

Barker, G., 1994. 'The busy countryside': Archaeological field survey and rual settlement in Roman Italy. In: B. Frenzel (ed.), *Evaluation of land surfaces cleared from forests in the Mediterranean region during the time of the Roman empire*. Strasbourg: European Science Foundation, 83-100.

Barker, G., 1995. *A Mediterranean Valley, Landscape Archaeology and Annales History in the Biferno Valley*. London, New York: Leicester University Press.

Baruch, U., 1990. Palynological evidence of human impact on the vegtation as recorded in late Holocene lake sediments in Israel. In: S. Bottema, G. Entjes-Nieborg and W. Van Zeist (eds.), *Man's role in the shaping of the eastern Mediterranean landscape*. Rotterdam, the Netherlands; Brookfield, VT: A.A. Balkema, 283-293.

Baxter, M. J. and Cool, H. E. M., 2006. *Notes on the Statistical Analysis of some Loom Weights from Pompeii*. Unpublished report.

BBC News, 2009. Ancient city of Pompeii added to Google Street View. Retrieved on 04 March 2010 from World Wide Web: http://news.bbc.co.uk/2/hi/8394384.stm

BBC News, 2009. Italy declares Pompeii emergency. Retrieved on 04 March 2010 from World Wide Web: http://news.bbc.co.uk/2/hi/7490735.stm

Beard, M., 2008a. *Pompeii, the Life of a Roman town*. London: Profile Books.

Beard, M., 2008b. *The Fires of Vesuvius: Pompeii lost and found*. Cambridge, Mass: Belknap Press of Harvard University Press.

Bergmann, B., 2008. Staging the supernatural: interior gardens of Pompeian houses. In: C. C. Mattusch (ed.), *Pompeii and the Roman villa: Art and Culture around the Bay of Naples*. London: Thames & Hudson, 53-69.

Berry, J., 1997. *The conditions of domestic life in Pompeii in AD 79: a case-study of houses 11 and 12, Insula 9, Region I*. Papers of the Brtish School at Rome 52, 103-125.

Berry, J., 2007. *The Complete Pompeii*. London: Thames & Hudson.

Bertacchi, A., Lombardi, T., Sani, A., and Tomei, P. E., 2008. Plant macroremains from the Roman harbour of Pisa (Italy). *Environmental Archaeology* 13/2, 181-188.

Besnard, G., De Casas, R. R., and Vargas, P., 2003. A set of primers for length and nucleotide-substitution polymorphism in chloroplastic DNA of Olea europaea L. (Oleaceae). *Molecular Ecology Notes* 3, 651-653.

Binford, L. R., 1981. Behavioural archaeology and the 'Pompeii premise'. *Journal of Anthropological Research* 37/3, 195-208.

Bird, J., 2007. Incense in Mthraic ritual: the evidence of the finds. In: D. Peacock and D. Williams (eds.), *Food for the Gods New Light on the Ancient Incense Trade*. Oxford: Oxbow, 122-134.

Bispham, E., 2007. *From Asculum to Actium: the municipalization of Italy from the Social War to Augustus*. Oxford: Oxford University Press.

Blondel, J., 2006. The 'design' of Mediterranean landscapes: a millennial story of humans and ecological systems during the historic period. *Human Ecology* 34, 713-729.

Boardman, S. and Jones, G., 1990. Experiments on the Effects of Charring on Cereal Plant Components. *Journal of Archaeological Science* 17/1, 1-11.

Bober, P. P., 1999. *Art, Culture, and Cuisine: Ancient and Medieval Gastronomy*. Chicago, London: The University of Chicago Press.

Bon, S. E., 1997. A city frozen in time or a site in perpetual motion? Formation processes at Pompeii. In: S. E. Bon and R. Jones (eds.), *Sequences and Space in Pompeii*. Oxford: Oxbow Books, 7-12.

Bon, S. E. and Jones, R. (eds.), 1997. *Sequence and Space in Pompeii*. Oxford: Oxbow Books.

Bon, S. E., Jones, R., Kurchin, B., and Robinson, D., 1995. Research in Insula VI, 1 by the Anglo-American Project in Pompeii, 1994-6. *Rivista di Studi Pompeiani* 7, 153-157.

Bon, S. E., Jones, R., Kurchin, B., and Robinson, D., 1997. The context of the House of the Surgeon: Investigations in Insula VI, 1 at Pompeii. In: S. E. Bon and R. Jones (eds.), *Sequences and Space in Pompeii*. Oxford: Oxbow Books, 32-49.

Borgongino, M., 1999. Suburban agriculture. In: A. Ciarallo and E. De Carolis (eds.), *Pompeii: Life in a Roman Town*. Milan: Electa, 89-91.

Borgongino, M., 2006. *Archeobotanica: reperti vegetali da Pompei e dal territorio vesuviano*, Studi della Soprintendenza Archeologica di Pompei, 16. Roma: L'Erma di Bretschneider.

Bouby, L. and Marinval, P., 2004. Fruits and seeds from Roman cremations in Limagne (Masstif Central) and the spatial variability of plant offerings in France. *Journal of Archaeological Science* 31/1, 77-86.

Braadbaart, F., 2008. Carbonisation and morphological changes in modern dehusked and husked *Triticum* dicoccum and *Triticum* aestivum grains. *Vegetation History and Archaeobotany* 17/1, 155-166.

Braadbaart, F., Boon, J. J., Veld, H., David, P., and van Bergen, P. F., 2004. Laboratory simulations of the transformation of peas as a result of heat treatment: changes of the physical and chemical properties. *Journal of Archaeological Science* 31/6, 821-833.

Braadbaart, F. and van Bergen, P. F., 2005. Digital imaging analysis of size and shape of wheat and pea upon heating under anoxic conditions as a function of the temperature. *Vegetation History and Archaeobotany* 14/1, 67-75.

Braadbaart, F., Bakels, C. C., Boon, J. J., and van Bergen, P. F., 2005. Heating experiments under anoxic conditions on varieties of wheat. *Archaeometry* 47/1, 103-114.

Bradley, K., 1998. The Roman family at dinner. In: I. Nielsen and H. S. Nielsen (eds.), *Meals in a social context: Aspects of the communal meal in the Hellenistic and Roman World*. Aarhus, Oxford: Aarhus University Press, 36-55.

Bray, T. L. (ed.), 2003. *The Archaeology and Politics of Food and Feasting in early States and Empires*. New York, London: Kluwer Academic/Plenum.

Briggs, D. E. G., and Wilby, P. R., 1996. The role of the calcium carbonate- calcium phosphate switch in the mineralisation of soft-bodied fossils. *Journal of the Geological Society of London*, 153, 665-8.

Brinkkemper, O., 2009. Review of People and Plants in Ancient Pompeii. A New Approach to Urbanism from the Microscope Room by M. Ciaraldi. *Environmental Archaeology* 14/2, 187-188.

Brion, M., 1973. Pompeii and Herculaneum: the Glory and the Grief. Trans.by John Rosenberg. London: Cardinal.

Brothers, A. J., 1996. Urban Housing. In: I. M. Barton (ed.), *Roman Domestic Buildings*. Exeter: University of Exeter Press, 33-64.

Brothwell, D. and Brothwell, P., 1998. *Food in Antiquity : a Survey of the Diet of early Peoples.* (Expanded edition). Baltimore, Md, London: Johns Hopkins University Press.

Bruan, T., 1995. Barley cakes and emmer bread. In: J. Wilkins, D. Harvey, and M. Dobson (eds.), *Food in Antiquity*. Exeter: University of Exeter Press, 25-37.

Brugiapaglia, V. E., Cheddadi, R., Ponel, P., Reille, M., De Beaulieu, J., and Barbero, M., 1999. A computerized data base for the palynological recording of human activity in the Mediterranean basin. In: G. Barker and D. J. Mattingly (eds.), *The Archaeology of Mediterranean Landscapes*. Oxford: Oxbow, chap. 2, 17-24.

Butterworth, A. and Laurence, R., 2005. *Pompeii: the Living City*. London: Weidenfeld & Nicolson.

Cappers, R. T. J., 2006. *Roman Foodprints at Berenike: Archaeobotanical Evidence of Subsistence and Trade in the eastern Desert of Egypt*, monograph 55, Berenike reports 6. Los Angeles: Cotsen Institute of Archaeology, University of California.

Cappers, R. T. J., Bekker, R. M., and Jans, J. E. A., 2006. *Digitale Zadenatlas van Nederland digital seed altas of the Netherlands*. Groningen, the Netherlands: Barkhuis.

Capps, E., Page, T. E., and Rouse, W. H. D, 1929. *The Geography of Strabo*. Trans. by H. L. Jones. London, New York: William Heinemann, Loeb Classical Library, vol. II.

Carafa, P., 2007. Recent work on early Pompeii. In: J. J. Dobbins and P. W. Foss (eds.), *The World of Pompeii*. London: Routledge, chap. 5, 63-72.

Carrington, R. C., 1932. The Etruscans and Pompeii. *Antiquity* 6/21, 5-23.

Carrington, R. C., 1933. The ancient Italian town-house. *Antiquity* 7/26, 133-152.

Carrington, R. C., 1936. *Pompeii*. Oxford: Clarendon Press.

Carrión, J. S., Dupré, M., Fumanal, M. P., and Montes, R., 1995. A palaeoenvironmental study in semi-arid southeastern Spain: the palynological and sedimentological sequence at Perneras Cave (Lorca, Murcia). *Journal of Archaeological Science* 22/3, 355-367.

Carroll, M. and Godden, D., 2000. The Sanctuary of Apollo at Pompeii: Reconsidering Chronologies and Excavation History. *American Journal of Archaeology* 104, 743-54.

Carruthers, W. J., 2000. The mineralised plant remains. In: A. J. Lawson (ed.), Potterne 1982-5. *Animal husbandry in later prehistoric Wiltshire*. Salisbury: Trust for Wessex, 47-70.

Carter, J. C. and Costantini, L., 1994. Settlement density, agriculture and the extent of productive land cleared from forest in the time of the Roman empire in *Magna Graecia*. In: B. Frenzel (ed.), *Evaluation of land surfaces cleared from forests in the Mediterranean region during the time of the Roman empire*. Strasbourg: European Science Foundation, 101-118.

Chow, D. S. L., 1984. *A study of ancient Roman dwellings Domus and Insulae, and their relationship to social classes, based on archaeological evidence in Pompeii and Ostia*. Unpublished MA thesis, University of London.

Ciaraldi, M., 2000. Drug preparation in evidence: an unusual plant and bone assemblage from the Pompeian countryside, Italy. *Vegetation History and Archaeobotany* 9, 91-98.

Ciaraldi, M. R. A., 2001. *Food and Fodder, Religion and Medicine at Pompeii*. Unpublished PhD thesis, Department of Archaeological Sciences, University of Bradford.

Ciaraldi, M., 2005. How many lives depended on plants? Specialisation and agricultural production at Pompeii. In: A. M. Mahon and J. Price (eds.), *Roman Working Lives and Urban Living*. Oxford: Oxbow Books, 191-201.

Ciaraldi, M., 2007. *People and plants in ancient Pompeii : a new approach to urbanism from the microscope room, the use of plant resources at Pompeii and in the Pompeian area from the 6th century BC to AD 79*. Accordia specialist studies on Italy vol. 12. London: Accordia Research Insititute, University of London.

Ciaraldi, M. R. A. and Richardson, J. 2000. Food, ritual and rubbish in the making of Pompeii. In: G. Fincham, G. Harrison, R. R. Holland and L. Revel (eds.), *TRAC 99, Proceedings of the Ninth annual Theoretical Roman Archaeology Conference Durham, April 1999*. Oxford: Oxbow Books, 74-82.

Ciarallo, A., 2000. *Gardens of Pompeii*. Roma: L'Erma di Bretschneider.

Ciarallo, A., 2002. About an ancient medical mixture found in Pompeii. In: J. Renn and G. Castagnetti (eds.), *Homo Faber: Studies on nature, technology and science at the time of Pompeii*. Studi della Soprintendenza archeologica di Pompei 6. Rome: L'Erma di Bretschneider, 152-167.

Ciarallo, A., 2005. *Le Stagioni nell'antica Pompei*. Napoli: Electa.

Ciarallo, A., 2007. *Flora Pompeiana Antica: Guida all'orto Botanico*. Napoli: Electa.

Ciarallo, A., 2009. Plants as a major element in the cultural framework of Pompeii. In: *Plants and Culture: seeds of the cultural heritage of Europe*. Soprintendenza Speciale ai Beni Archeologici di Napoli e Pompei. Retrieved on 13 April 2010 World Wide Web: http://www.plants-culture.unimore.it/book/06%20Ciarallo.pdf

Ciarallo, A. and De Carolis, E. 1998. *Lungo le mura di Pompei. L'antica città nel suo ambiente naturale*. Napoli: Electa.

Ciarallo, A. and Lippi, M. M., 1993. The garden of 'Casa dei Casti Amanti' (Pompeii, Italy). *Garden History* 21/1, 110-116.

Clarke, J. R., 1991. *The Houses of Roman Italy, 100 B.C.-A.D. 250: Ritual, Space, and Decoration*. Berkeley, London: University of California Press.

Clarke, J. R., 2007. *Looking at Laughter: Humour, Power and Transgression in Roman Visual Culture, 100 B.C-A.D 250*. Berkeley, California, London: University of California Press.

Cool, H. E. M., 2006. *Eating and Drinking in Roman Britain*. Cambridge: Cambridge University Press.

Cooley, A. E., 1997. Domestic space in the Roman world Pompeii and beyond. In: R. Laurence and A. Wallace-Hadrill (eds.), *Journal of Roman Archaeology*. Portsmouth, R.I: Journal of Roman Archaeology, supplementary series no. 22.

Cooley, A. E., 2003. *Pompeii*. London: Duckworth.

Cooley, A. E. and Cooley, M. G. L., 2004. *Pompeii: a Sourcebook*. London: Routledge.

Corbier, M., 2000. The broad bean and the moray: social hierarchies and food in Rome. In: J. L. Flandrin and M. Montanari (eds.), *Food, a culinary history from antiquity to the present*. New York: Penguin Books, 128-140.

Cornell, T.J. and Lomas, K. (eds.), 1995. *Urban Society in Roman Italy*. London: UCL Press.

Costantini, L. and Giorgi, J., 2001. Charred plant remains of the Archaic period from the Forum and Palatine. *Journal of Roman Archaeology* 14, 239-248.

Costantini, L. and Nencioni, L., 2001. Archaeobotanical evidence and biomolecular archaeology for the beginnings of agriculture in Italy. *Origini*. XXIII, 71-83.

Cremaschi, M., Marchetti, M., and Ravazzi, C., 1994. Geomorphological evidence for land surfaces cleared from forest in the central Po plain (northern Italy) during the Roman period. In: B. Frenzel (ed.), *Evaluation of land surfaces cleared from forests in the Mediterranean region during the time of the Roman empire*. Strasbourg: European Science Foundation, 119-132.

Curtis, R. I., 2009. Review of People and Plants in ancient Pompeii: a new approach to urbanism from the microscope room by M. Ciaraldi. *American Journal of Archaeology Online Book Review* 13/3.

Dalby, A., 2000. *Empires of pleasures, luxury and indulgence in the Roman world*. London: Routledge.

Day, J., 1932. *Agriculture in the life of Pompeii*. Yale Classical Studies. III, 165-208.

De Caro, S., 2007. The First Sanctuaries. In: J. J. Dobbins and P. W. Foss (eds.), *The World of Pompeii*. London: Routledge, chap.6, 73-81.

De Natale, A. and Pollio, A., 2007. Plants species in the folk medicine of Montecorvino Rovella (inland Campania, Italy). *Journal of Ethnopharmacology* 109, 295-303.

Dennell, R. W., 1976. The economic importance of plant resources represented on archaeological sites. *Journal of Archaeological Science* 3/3, 229-247.

De Sena, E. C., 2005. An assessment of wine and oil production in Rome's hinterland: ceramic, literary, art historical and modern evidence. In: A. Klynne and B. Santilli Frizell (eds.) *Roman Villas around the Urbs. Interaction with Landscape and Environment* Retrieved on 15 March 2010 from World Wide Web: *www.isvroma.org*.

De Sena, E. C. and Ikäheimo, J. P., 2003. The supply of amphora-borne commodities and domestic pottery in Pompeii 150 BC- AD79: preliminary evidence from the House of the Vestals. *European Journal of Archaeology* 6, 301-323.

Descœudres, J.P., 2007. History and historical sources. In: J. J. Dobbins and P. W. Foss, (eds.), *The World of Pompeii*. London: Routledge, chap. 2, 9-27.

Dickson, C., 1994. Macroscopic fossils of garden plants from British Roman and Medieval deposits. In: M. Dagfinn, J. H. Dickson and P. M. Jørgensen (eds.), *Garden history: garden plants, species, forms and varieties from Pompeii to 1800*: symposium held at the European University Centre for the Cultural Heritage, Ravello, June 1991. Rixensart, Belgium: Council of Europe, Division of Scientific Cooperation, PACT, 47-72.

Dickson, C. and Dickson, J., 2000. *Plants and People in Ancient Scotland*. Stroud, Gloucestershire, UK: Tempus.

Dietler, M., 1998. Feats and commensal politics in the political economy: food, power and status in prehistoric Europe. In: P. Wiessner and W. Schiefenhövel (eds.), *Food and the status quest an interdiscinplinary perspective*. Providence, R.I: Berghahn Books, 87-125.

Dietler, M. and Hayden, B., 2001. Digesting the feast: good to eat, good to drink, good to think: an introduction. In: M. Dietler and B. Hayden (eds.), Feasts :*Archaeological and Ethnographic Perspectives on Food, Politics, and Power*. Washington, D.C: Smithsonian Institution Press, 1-8.

Dimbleby, G. W. and Grüger, E., 2002. Pollen analysis of soil samples from the A.D 79 level Pompeii, *Oplontis* and Boscoreale. In: W. F. Jashemski and F. G. Meyer (eds.), *The Natural History of Pompeii*. Cambridge: Cambridge University Press, 181-216.

Dobbins, J. J. and Foss, P. W. (eds.), 2007. *The World of Pompeii*. London: Routledge.

Dormoy, I., Peyron, O., Combourieu-Neboutb, S., Goring, S., Kotthoff, U., Magny, M., and Pross, J., 2009. Terrestrial climate variability and seasonal changes in the Mediterranean region between 15 000 and 4000 years BP deduced from marine pollen records. *Climate of the Past Discussions* 5, 735-770.

Douglas, M., 1997. Deciphering a meal. In: C. Counihan and P. Van Esterik (eds.), *Food and Culture: a Reader*. New York, London: Routledge, 36-54.

Drescher-Schneider, R., 1994. Forest, forest clearance and open land during the time of the Roman empire in northern Italy (the botanical record). In: B. Frenzel (ed.), *Evaluation of land surfaces cleared from forests in the Mediterranean region during the time of the Roman empire*. Strasbourg: European Science Foundation, 45-58.

Drescher-Schneider, R., de Beaulieu, J., Magny, M., Walter-Simonnet, A. V., Bossuet, G., Millet, L., Brugiapaglia, E., and Drescher, A., 2007. Vegetation history, climate and human impact over the last 15,000 years at Lago dell'Accesa (Tuscany, Central Italy). *Vegetation History and Archaeobotany* 16/4, 279-299.

Dunbabin, K. M. D., 2003a. Houses and households of Pompeii. *Journal of Roman Archaeology*, 387-390.

Dunbabin, K. M. D., 2003b. *The Roman banquet: images of convivality*. Cambridge: Cambridge University Press.

Duncan-Jones, R., 1982. *The Economy of the Roman Empire : Quantitative studies*. (2nd edition). Cambridge: Cambridge University Press.

Dupont, F., 2000. The grammar of Roman dining. In: J. L. Flandrin and M. Montanari (eds.), *Food, a Culinary History from Antiquity to Present*. New York: Penguin Books, 113-127.

Dyson, S. L., 1997. Some random thoughts on a collection of papers on Roman archaeology. In: S. E. Bon and R. Jones (eds.), *Sequences and Space in Pompeii*. Oxford: Oxbow Books, 150-157.

Dyson, S. L., 2003. *The Roman Countryside*. London: Duckworth.

Edwards, J., 1985. *The Roman Cookery of Apicius*. Trans. and adapted by J. Edwards. London: Rider and Company.

Ellis, S. J. R., 2004. The distribution of bars at Pompeii: archaeological, spatial and viewshed analyses. *Journal of Roman Archaeology* 17, 371-384.

Ellis, S., 22/01/2005. Pompeii's bars not so seedy. *New Scientist* 2483.

Epstein, C., 1993. Oil production in the Golan Heights during the Chalcolithic Period. *Tel Aviv* 20, 133-146.

Erdkamp, P., 1998. *Hunger and the Sword: Warfare and Food Supply in Roman Republican Wars (264-30 B.C.)*. Amsterdam: Gieben.

Etienne, R., 1992. *Pompeii: the Day a City Died*. London: Thames and Hudson.

Farrar, L., 2000. *Ancient Roman Gardens*. (Revised paperback edition). Stroud, Gloucestershire, UK: Sutton.

Ferrio, J. P., Araus, J. L., Buxo, R., Voltas, J., and Bort, J., 2005. Water management practices and climate in ancient agriculture: inferences from the stable isotope composition of archaeobotanical remains. *Vegetation History and Archaeobotany* 14/4, 510-517.

Finley, M. I., 1973. *The Ancient Economy*. London: Chatto & Windus.

Fiori, A., 1969. *Nuova Flora Analitica D'Italia*. Bologna: Edagricole.

Follieri, M., Giardini, M., Magri, D., and Sadori, L., 1998. Palynostratigraphy of the last glacial period in the volcanic region of central Italy. *Quaternary International* 47/48, 3-20.

Ford, R. I., 1979. Paleoethnobotany in American Archaeology. In: M. Schiffer (ed.), *Advances in Archaeological Method and Theory*. New York: Academic Press, vol 2, 286-336.

Fortenberry, D. and Goalen, M., 2007. Report on the Conservation and Presentation of Regio VI Insula I, *Pompeii*. London: Academy Projects (Archaeology - Architecture) LLP.

Foss, J. E., 1988. Paleosols of Pompeii and *Oplontis*. In: R. I. Curtis (ed.), *Stvdia Pompeiana and Classica in honor of Wilhelmina F. Jashemski*. New Rochelle, New York: Orpheus, vol. 1, 127-148.

Foss, P. W., 1997. Watchful Lares: Roman household organization and the rituals of cooking and eating. In: Laurence, R. and Wallace-Hadrill, A. (eds.), *Journal of Roman Archaeology*, supplement 22, 197-218.

Foss, P. W., 2007. Rediscovery and Resurrection. In: J. J. Dobbins and P. W. Foss (eds.), *The World of Pompeii*. London: Routledge, chap. 3, 28-42.

Foss, J. E., Timpson, M. E., Ammons, J. T., Lee, S. Y., 2002. In: W. F Jashemski, and F. G. Meyer (eds.), *The Natural History of Pompeii*. Cambridge: Cambridge University Press, chap. 5, 65-79.

Francissen, F. P. M., 1987. A century of scientific research on plants in Roman mural paintings (1879-1979). *Rivista di Studi Pompeiani* 1, 111-122.

Frank, A. H. E., 1969. Pollen stratigraphy of the Lake of Vico (central Italy). *Palaeogeography, Palaeoclimatology, Palaeoecology* 6, 67-85.

Frankel, R., 1999. *Wine and Oil Production in Antiquity in Israel and other Mediterranean Countries*. Sheffield: Sheffield Academic Press.

Franklin, J. L., 2001. *Pompeis Difficile Est: Studies in the Political Life of Imperial Pompeii*. Ann Arbor: University of Michigan Press.

Frayn, J. M., 1979. *Subsistence Farming in Roman Italy*. Fontwell, UK: Centaur Press.

Frederiksen, M., 1984. *Campania*. London: British School at Rome.

Fulford, M. and Wallace-Hadrill, A., 1998. Unpeeling Pompeii. *Antiquity* 72/275, 128-145.

Fuller, D. Q. and Weber, S. A., 2005. Formation processes and paleoethnobotanical interpretation in south Asia. *Journal of Interdisciplinary Studies in History and Archaeology* 2/1, 93-115.

Fuller, D. Q. and Stevens, C. J., 2009. Agriculture and the development of complex societies: an archaeobotanical agenda. In: A. S. Fairbairn and E. Weiss (eds.), *From Foragers to Farmers, papers in honour of Gordon C. Hillman*. Oxford: Oxbow Books, chap. 6, 37-57.

Galili, E., Stanley, D. J., Sharvit, J., and Weinstein-Evron, M., 1997. Evidence for earliest olive-oil production in submerged settlements off the Carmel coast, Israel. *Journal of Archaeological Science* 24/12, 1141-1150.

Gardner, A. (2001). Identities in the late Roman army: material and textual perspectives. In G. Davies, A. Gardner and K. Lockyear (eds.) *TRAC 2000: Proceedings of the 10th Annual Theoretical Roman Archaeology Conference*, 35-47. Oxford: Oxbow Books.

Gardner, A. (2007). *An Archaeology of Identity, Soldiers and Society in late Roman Britain*. Walnut Creek, Calif: Left Coast Press, Publications of the Institute of Archaeology, University College London.

Gardner Coates, V. C. and Seydl, J.L., 2007. *Antiquity Recovered: the Legacy of Pompeii and Herculaneum*. Los Angeles: J. Paul Getty Museum.

Garnsey, P., 1999. *Food and Society in Classical Antiquity*. Cambridge, New York: Cambridge University Press.

Girling, M. A., 1979. Calcium Carbonate-replaced Arthropods from Archaeological Deposits. *Journal of Archaeological Science* 6/4, 309-320.

Goody, J., 1982. *Cooking, Cuisine, and Class: a Study in Comparative Sociology*. Cambridge: Cambridge University Press.

Gowers, E., 1992. *The Loaded Table: Representations of Food in Roman Literature*. Oxford: Clarendon Press.

Grahame, M., 2000. *Reading Space: Social Interaction and Identity in the Houses of Roman Pompeii: a syntactical approach to the analysis and interpretation of built space*. BAR international series 886. Oxford: Archaeopress.

Grant, M. and Forman, W., 1976. *Cities of Vesuvius: Pompeii and Herculaneum*. Harmondsworth: Penguin.

Grant, M., 2005. *Pompeii and Herculaneum: Cities of Vesuvius*. London: Folio Society.

Green, F. J., 1979. Phosphatic mineralization of seeds from archaeological sites. *Journal of Archaeological Science* 6/3, 297-284.

Greene, K., 1990. *The Archaeology of the Roman Economy*. Berkeley: University of California Press.

Greig, J., 1981. The investigation of a Medieval barrel-latrine from Worcester. *Journal of Archaeological Science* 8/2, 265-282.

Grove, A. T. and Rackman, O., 2001. *The Nature of Mediterranean Europe: an Ecological History*. New Haven, CT, London: Yale University Press.

Grüger, E., Thulin, B., Muller, J., Schneider, J., Alefs, J., and Welter-Schultes, F. W., 2002. Environmental changes in and around Lake Avernus in Greek and Roman times: A study of the plant and animal remains preserved in the Lake's sediments. In: W. F. Jashemski and F. G. Meyer (eds.), *The Natural History of Pompeii*. Cambridge: Cambridge University Press, 240-273.

Gumerman, G., 1997. Food and Complex Societies. *Journal of Archaeological Method and Theory* 4/2, 105-140.

Guzzo, G. P., 2007. City and Country: an Introduction. In: J. J. Dobbins and P. W. Foss (eds.), *The World of Pompeii*. London: Routledge, chap.1, 3-8.

Guzzo, P. G. and d'Ambrosio, A., 2002. *Pompeii: Guide to the Site*. Napoli: Electa.

Hales, S., 2003. *The Roman House and Social Identity*. Cambridge: Cambridge University Press.

Hall, A. R. and Kenward, H. K., 1990. *Environmental evidence from the Colonia: general accident and Rougier Street. The Archaeology of York* AY 14/6. London: Counc. Brit. Archaeology, 289-434.

Hally, D. J., 1981. Plant preservation and the content of paleobotanical samples: a case study. *Journal of Anthropological Research* 37, 195-208.

Hamilakis, Y., 2000. The anthropology of food and drink consumption and Aegean Archaeology. In: S. J. Vaughan and W. D. E. Coulson (eds.), *Palaeodiet in the Aegean: Papers from a Colloquium held at the 1993 meeting of the Archaeological Institute of American in Washington D.C.* Oxford: Oxbow Books for the Wiener Laboratory of the American School of Classical Studies at Athens, 55-63.

Hanf, M., 1983. *The arable weeds of Europe with their seedlings and seeds*. UK: BASF United Kingdom Limited.

Harris, W. V. (ed.), 2005. *Rethinking the Mediterranean*. Oxford: Oxford University Press.

Harris, W. V., 2007. The late Republic. In: W. Scheidel, I. Morris, and R. Saller (eds.), *The Cambridge Economic History of the Greco-Roman World*. Cambridge: Cambridge University Press, 511-542.

Harshberger, J. W., 1909. The plant remains of Pompeii. *Science*, New Series 30/773, 575-576.

Hastorf, C. A., 1993. *Agriculture and the onset of political inequality before the Inka*. Cambridge: Cambridge University Press.

Hastorf, C. A., 1995. Gender, space, and food in prehistory. In: J. M. Gero and M. W. Conkey (eds.), *Engendering Archaeology: Women and Prehistory*. Oxford: Blackwell, 132-159.

Hatcher, P. G., 2002. Wood associated with the 70 A.D. eruption: Its chemical characterization by solid state C as a guide to the degree of carbonization. In: W. F. Jashemski and F. G. Meyer (eds.), *The Natural History of Pompeii*. Cambridge: Cambridge University Press, 217-224.

Hayden, B., 1998. Feasting in prehistoric and traditional societies. In: P. Wiessner and W. Schiefenhövel (eds.), *Food and the Status Quest: an interdisciplinary perspective*. Providence, R.I.: Berghahan Books, 127-147.

Haynes, I., 1999. Introduction: The Roman army as a community. In: A. Goldsworthy (ed.), *The Roman Army as a Community*. Portsmouth, R.I.: Journal of Roman Archaeology Supplementary Series #34, 7-14.

Helbaek, H., 1956. Vegetables in the funeral meals of pre-urban Rome. *Early Rome* II 4/27/2, 287-294.

Hillman, G. C., 1973. Crop husbandry and food production: modern models for the interpretation of plant remains. *Anatolian Studies* 23, 241-244.

Hillman, G., Wales, S., McLaren, F., Evans, J., and Butler, A., 1993. Identifying problematic remains of ancient plant foods: a comparison of the role of chemical, histological and morphological criteria. *World Archaeology* 25/1, 94-121.

Hitchner, B. R., 2005. "The advantages of wealth and luxury": the case for economic growth in the Roman Empire. In: J. G. Manning and I. Morris (eds.), *The Ancient Economy: Evidence and Models*. Standford: Stanford University Press, 207-222.

Hobson, B., 2009. *Pompeii, Latrines and Down Pipes, A General Discussion and Photographic Record of Toilet facilities in Pompeii*. BAR International Series 2041. Oxford: John and Erica Hedges Ltd.

Hodder, I., 1987. The contextual analysis of symbolic meanings. In: I. Hodder (ed.), *The Archaeology of Contextual Meanings*. Cambridge: Cambridge University Press, 1-10.

Hodos, T., 2010. Local and Global Perspectives. In: S. Hales and T. Hodos (eds.), *Material Culture and Social Identities in the Ancient World*. New York, Cambridge: Cambridge University Press, chap. 1, 3-31.

Hopf, M., 1991. South and Southwest Europe. In: W. van Zeist, K. Wasylikowa, and K. E. Behre (eds.), *Progress in Old World Palaeoethnobotany: A retrospective view on the occasion of 20 years of the International Work Group for Palaeoethnobotany*. Rotterdam, the Netherlands: Balkema, 241-277.

Hubbard, R. N. L. B., 1992. Dichotomous keys for the identification of the major Old World crops. *Review of Palaeobotany and Palynology* 73, 105-115.

Hubbard, R. N. L. B. and al Azm, A., 1990. Quantifying preservation and distortion in carbonized seeds; and investigating the history of Friké production. *Journal of Archaeological Science* 17/1, 103-106.

Hubbard, R. N. L. B., and Clapham, A., 1992. Quantifying macroscopic plant remains. *Review of Palaeobotany and Palynology* 73/1-4, 117-132.

Hughes, D. J., 1994. *Pan's Travail: Environmental Problems of the Ancient Greeks and Romans*. Baltimore, Md., London: Johns Hopkins University Press.

Imazio, S., Labra, M., Grassi, A. S., and Failla, O., 2006. Chloroplast microsatellites to investigate the origin of grapevine. *Genetic Resources and Crop Evolution* 53, 1003-1011.

Jacobelli, L., 2003. *Gladiators at Pompeii*. Los Angeles: Getty Publications.

Jacobius, J. and Smolenaars, L., 2005. Earthquakes and volcanic eruptions in Latin Literature: reflections and emotional responses. In: M. S. Balmuth, D. K. Chester, and P. A. Johnston (eds.), *Cultural responses to the volcanic landscape: the Mediterranean and beyond*. Boston, Mass: Archaeological Institute of America, AIA colloquia and conference papers 8, 311-329.

Jacquat, C. and Martinoli, D., 1999. Vitis vinifera L.: wild or cultivated? Study of the grape pips found at Petra, Jordan, 150 B.C - A.D. 40. *Vegetation History and Archaeobotany* 8/1-2, 25-30.

James, S. (1999). The Community of Soldiers: a major identity and centre of power in the Roman empire. In P. Barker, C. Forcey, S. Jundy and R. Witcher (eds.) *TRAC 98: Proceedings of the 8th Annual Theoretical Roman Archaeology Conference, 14-25*. Oxford: Oxbow Books.

Janick, J., Paris, H. S., and Parrish, D. C., 2007. The cucurbits of Mediterranean antiquity: identification of taxa from ancient images and descriptions. *Annals of Botany* 100/7, 1441-1457.

Jansen, G. C. M., 2000. Systems for the disposal of waste and excreta in Roman cities: the situation in Pompeii, Herculaneum and Ostia. In: X. Dupré and J. A. Remolà (eds.), *Sordes urbis: la eliminación de residuos en la ciudad romana: actas de la Reunión de Roma (15-16 de noviembre de 1996)*. Roma: L'Erma di Bretschneider, 37-50.

Jansen, G. C. M., 2001. Water Pipe Systems in the Houses of Pompeii: Distribution and Use. In: A. O. Koloski-Ostrow (ed.), *Water Use and Hydraulics in the Roman City, Archaeological Institute of America Colloquia and Conference Papers*, Number 3. Dubuque, Iowa: Kendall/Hunt Publishing, 27-40.

Jashemski, W. F., 1963. The flower industry at Pompeii. *Archaeology* 16/2, 112-121.

Jashemski, W. F., 1968. Excavations in the 'Foro Boario' at Pompeii: a preliminary report. *American Journal of Archaeology* 72/1, 69-73.

Jashemski, W. F., 1970. University of Maryland Excavations at Pompeii, 1968. *American Journal of Archaeology* 74, 63-70.

Jashemski, W. F., 1973. The discovery of a large vineyard at Pompeii: University of Maryland excavtions 1970. *American Journal of Archaeology* 77/1, 27-41.

Jashemski, W. F., 1974. The Discovery of a Market-Garden Orchard at Pompeii: The Garden of the "House of the Ship Europa". *American Journal of Archaeology* 78/4, 391-404.

Jashemski, W. F., 1977. The Excavation of a shop-house garden at Pompeii (I.xx.5). *American Journal of Archaeology* 81/2, 217-227.

Jashemski, W. F., 1979a. Pompeii and Mount Vesuvius, A.D. 79. In: P. D. Sheets and D. K. Grayson (eds.), *Volcanic Activity and Human Ecology*. New York, London: Academic Press, 587-622.

Jashemski, W. F., 1979b. *The gardens of Pompeii, Herculaneum and the villas destroyed by Vesuvius*. New Rochelle, N.Y: Caratzas Brothers.

Jashemski, W. F., 1986. The garden of the house of the wedding of Alexander at Pompeii (VI-Ins-Occid-39-41). *American Journal of Archaeology* 90/2, 188.

Jashemski, W. F., 1987a. Introduction. In: E. B. Macdougall (ed.) *Ancient Roman villa gardens. Washington, D.C: Dumbarton Oaks Colloquium on the History of Landscape Architecture*, Dumbarton Oaks research library and collection, 3-5.

Jashemski, W. F., 1987b. Recently excavated gardens and cultivated land of the villas at Boscoreale and *Oplontis*. In: E. B. Macdougall (ed.) *Ancient Roman villa gardens. Washington, D.C: Dumbarton Oaks Colloquium on the History of Landscape Architecture*, Dumbarton Oaks research library and collection, 31-76.

Jashemski, W. F., 1992. The gardens of Pompeii, Herculaneum and the villas destroyed by Vesuvius. *Journal of Garden History* 12/2, 102-125.

Jashemski, W. F., 2002a. Introduction. In: W. F Jashemski, and F. G. Meyer (eds.), *The Natural History of Pompeii*. Cambridge: Cambridge University Press, chap. 1, 1-5.

Jashemski, W. F., 2002b. The Vesuvian Sites before A.D 79: The archaeological, literary, and epigraphical evidence. In: W. F Jashemski, and F. G. Meyer (eds.), *The Natural History of Pompeii*. Cambridge: Cambridge University Press, chap. 2, 6-28.

Jashemski, W. F., 2007. Gardens. In: J. J. Dobbins and P. W. Foss (eds.), *The World of Pompeii*. London: Routledge, chap. 31, 487-498.

Jashemski, S. A. and Jashemski, W. F., 1965. *Pompeii and the region destroyed by Vesuvius in A.D. 79*. Munich, Germany: Wilhelm Andermann.

Jashemski, W. F. and Meyer, F. G. (eds.), 2002. *The Natural History of Pompeii*. Cambridge: Cambridge University Press.

Jashemski, W. F., Meyer, F. G., and Ricciardi, M., 2002. Plants: evidence from wall paintings, mosaics, sculpture, plant remains, graffiti, inscriptions, and ancient authors, catalogue of plants. In: *The Natural History of Pompeii*. Cambridge: Cambridge University Press, chap. 6, 80-180.

Jasny, N., 1944. *The Wheats of Classical Antiquity*. Baltimore, M.D: Johns Hopkins Press.

Jensen, H. A., 1998. *Bibliography on Seed Morphology*. Rotterdam, Brookfield, VT: A. A. Balkema.

Jones, G. E. M and Rowley-Conwy, P., 1984. Plant remains from the North Italian lake- dwellings of Fiavé (1400-1200 BC). In: R. Perini (ed.), *Scavi archeologici nella zona palafitticole di Fiavé-Carera. Trento: Servizio beni culturali della provincia di Trento*, vol. 1, 323-355.

Jones, R., 2003a. Voices from the ashes (Pompeii archaeology). *Archaeology* 56/4, 28-31.

Jones, R., 2003b. The Urbanisation of Insula VI.I at Pompeii. *Rivista di Studi Pompeiani*, 139-146.

Jones, Rick. 2008. "The Urbanisation of Insula VI 1 at Pompeii." In *Nuove Ricerche Archeologiche nell'area Vesuviana (Scavi 2003-2006)*. Roma: L'Erma di Bretschneider, 139-146.

Jones, R. and Robinson, D., 2004. The making of an elite house: the House of the Vestals at Pompeii. *Journal of Roman Archaeology* 17, 107-130.

Jones, R. and Robinson, D., 2005a. *Anglo-American Project in Pompeii 2005 Resource Book*. Unpublished work.

Jones, R. and Robinson, D., 2005b. Water, wealth, and social status at Pompeii: The House of the Vestals in the first century. *American Journal of Archaeology* 109/4, 695-710.

Jones, R. and Robinson, D., 2006a. *Anglo-American Project in Pompeii 2006 Resource Book*. Unpublished work.

Jones, R. and Robinson, D. 2006b. The Development of Inequality in Pompeii: The Evidence from the Northern End of Insula VI.I. In: C. C. Mattusch, A. A. Donohue, and A. Brauer, (eds.), *Common Ground: Archaeology, Art, Science and the Humanities*. Proc. of the XVI International Congress of Classical Archaeology, Boston. Oxford: Oxbow Books, 498-502.

Jones, R. and Robinson, D., 2007. Intensification, heterogeneity and power in the development of insula VI.I. In: J. J. Dobbins and P. W. Foss (eds.), *The World of Pompeii*. London: Routledge, chap. 25, 389-406.

Jones, R., Robinson, D. and Stephens, J (eds.). 2006. *Anglo-American Project in Pompeii, The University of Bradford, 2006 Field School Handbook*. Unpublished work.

Jongman, W., 1988. *The Economy and Society of Pompeii*. Amsterdam: J.C. Gieben.

Jongman, W. M., 2007. The loss of innocence: Pompeian economy and society between past and present. In: J. J. Dobbins and P. W. Foss (eds.), *The World of Pompeii*. London: Routledge, chap. 32, 499-517.

Jürgen, R., and Castagnetti, G., (eds.), Divarci, L., and Rieger, S., (assist. eds.), 2002. *Homo Faber: Studies on Nature, Technology and Science at the time of Pompeii*. Presented at a conference at the Deutsches Museum, Munich, 21-22 March 2000. Studi della Suprintendenza Archeologica di Pompei, 6.Roma: L'Erma Bretschneider.

Keepax, C., 1977. Contamination of archaeological deposits by seeds of modern origin with particular reference to the use of flotation machines. *Journal of Archaeological Science* 4/3, 221-229.

Kenward, H. and Hall, A., 2000. Decay of delicate organic remains in shallow urban deposits: are we at a watershed? *Antiquity* 74, 519-525.

Kenward, H. and Hall, A., 2008. Urban organic archaeology: an irreplaceable palaeoecological archive at risk. *World Archaeology* 40/4, 584-596.

King, A., 1999. Diet in the Roman world: a regional inter-site comparison of the mammal bones. *Journal of Roman Archaeology* 12, 168-177.

Kislev, M. E., 1996. The Domestication of the Olive Tree. In: D. Eitam and M. Heltzer (eds.), *Olive Oil in Antiquity, Israel and Neighbouring Countries from the Neolithic to the Early Arab Period*. History of the Ancient Near East Studies, vol. VII. Padova: Sargon, 3-6.

Klynne, A. and Liljenstolpe, P., 2000. Investigating the gardens of the Villa of Livia. *Journal of Roman Archaeology* 13, 221-233.

Kooistra, L. I., 1996. *Borderland farming: possibilities and limitations of farming in the Roman period and early Middle Ages between the Rhine and Meuse*. Assen, The Netherlands: Van Gorcum & Comp.

Kron, G., 2000. Roman ley-farming. *Journal of Roman Archaeology* 13, 277-287.

Kron, G., 2005. Flora, fauna and more at Pompeii. *Journal of Roman Archaeology* 13, 607-612.

Krzyszowska, A., 2002. *Les cultes privés à Pompéi*. Wroclaw: Wydawnictwo Uniwersytetu Wroclawskiego.

Laidlaw, A., 2007. Mining the early published sources problems and pitfalls. In: J. J. Dobbins and P. W. Foss (eds.), *The World of Pompeii*. London: Routledge, chap. 39, 620-636.

Larew, H. G., 1988. Oak (Quercus) galls preserved at Herculaneum in A.D 79. In: R. I. Curtis (ed.), *Studia Pompeiana and classica in honor of Wilhelmina F. Jashemski*. New Rochelle, N.Y: Caratzas, vol 1, 145-148.

Laurence, R., 1994. *Roman Pompeii: Space and Society*. London: Routledge.

Laurence, R., 1995. The organization of space in Pompeii. In: T. J. Cornell and K. Lomas (eds.) *Urban Society in Roman Italy*. London: UCL Press.

Laurence, R., 2007. *Roman Pompeii: Space and Society*. (2nd edition). London, New York: Routledge.

Lazer, Estelle, 2009. *Resurrecting Pompeii*. London, New York: Routledge.

Lennstrom, H. A. and Hastorf, C. A., 1995. Interpretation in context: sampling and analysis in paleoethnobotany. *American Antiquity* 60/4, 701-721.

Lévi-Strauss, C., 1997. The culinary triangle. In: C. Counihan and P. Van Esterik (eds.), *Food and Culture: a Reader*. New York, London: Routledge, 28-35.

Liebeschuetz, W., 2000. Rubbish disposal in Greek and Roman cities. In: X. Dupré and J.A. Remolà (eds.), Sordes Urbis, *La Eliminaciódé residuos en la ciudad Romana, Actas de la Reunión de Roma (15-16 De Noviembre de 1996)*. Roma: L'Erma, 51-61.

Ling, R., 1995. Earthquake damage in Pompeii I., 10: one earthquake or two? In: *Deutsches Archäologisches Institut. Römische Abteilung, Italy, Soprintendenza archeologica di Pompei. Archäologie und Seismologie: la regione vesuviana dal 62 al 79 D.C.: problemi archaeologici e sismologici: Colloquium, Boscoreale, 26-27. November 1993*. Munich, Germany: Biering & Brinkmann, 201-209.

Ling, R., 1996. *Villae Rusticae* at Boscoreale. *Journal of Roman Archaeology* 9, 344-350.

Ling, R., 2005. *Pompeii: History, Life and Afterlife*. Stroud, Glouchestershire, UK: Tempus.

Lippi, M. M., 2000. The garden of the "Casa delle Nozze di Ercole ed Ebe" in Pompeii (Italy): palynological investigations. *Plant Biosystems* 134/2, 205-211.

Liphschitz, N., Gophna, R., Hartman, M., and Biger, G., 1991. The Beginning of Olive (*Olea europea*) cultivation in the Old World: a reassessment. *Journal of Archaeological Science* 18, 441-453.

Liphschitz, N., and Biger, G., 1990. Ancient dominance of the Quercus calliprinos- Pistacia palestina. *Journal of Vegetation Science* 1/1, 67-70.

Livadie, C. A., 2002. A first Pompeii: The early Bronze Age village of *Nola-Croce del Papa* (Palma Campania phase). *Antiquity* 76, 941-942.

Lomas, K., 1993. *Rome and the western Greeks, 350BC-AD200: conquest and acculturation in southern Italy*. London, New York: Routledge.

Lomas, K., 1995. Introduction. In: T. J. Cornell and K. Lomas (eds.), *Urban Society in Roman Italy*. London: UCL Press, 1-7.

Lomas, K., 2003. Public building, urban renewal and euergetism in early Imperial Italy. In: K. Lomas and T. Cornell (eds.), *'Bread and circuses' : euergetism and municipal patronage in Roman Italy*. London: Routledge.

Lomas, K. and Cornell, T. (eds.), 2003. *'Bread and circuses' : euergetism and municipal patronage in Roman Italy*. London: Routledge.

Lomas, K. and Cornell, T., 2003. Patronage and benefaction in ancient Italy. In: K. Lomas & T. Cornell (eds.), *'Bread and circuses': euergetism and municipal patronage in Roman Italy*. London: Routledge, 1-11.

Longo, O., 2000. The food of others. In: J. L. Flandrin and M. Montanari (eds.), *Food, a culinary history from antiquity to the present*. New York: Penguin Books, 153-162.

Lytton, E. B., 1834. *The Last Days of Pompeii*. London: Thomas Nelson and Sons.

McCobb, L. M. E., Briggs, D. E. G., Evershed, R. P., Hall, A., and Hall, R. A., 2001. Preservation of fossil seeds from a 10th century AD cess pit at Coppergate, York. *Journal of Archaeological Science* 28/9, 929-940.

MacKendrick, P., 1960. *The Mute Stones Speak: the Story of Archaeology in Italy*. London: Methuen.

MacKinnon, M., 2001. High on the hog: linking zooarchaeological, literary, and artistic data for pig breeds in Roman Italy. *American Journal of Archaeology* 105/4, 649-673.

MacKinnon, M., 2004. Production and consumption of animals in Roman Italy: integrating the zooarchaeological and textual evidence. Portsmouth, R.I: *Journal of Roman Archaeology*, supplementary series no. 54.

Magri, D., 1999. Late Quaternary vegetation history at Lagaccione near Lago di Bolsena (central Italy). *Review of Palaeobotany and Palynology* 106, 171-208.

Magri, D. and Sadori, L., 1999. Late Pleistocene and Holocene pollen stratigraphy at Lago di Vico, central Italy. *Vegetation History and Archaeobotany* 8/4, 247-260.

Manen, J. F., Bouby, L., Dalnoki, O., Marinval, P., Turgay, M., and Schlumbaum, A., 2003. Microsatellites from archaeological *Vitis vinifera* seeds allow a tentative assignment of the geographical origin of ancient cultivars. *Journal of Archaeological Science* 30/6, 721-729.

Manousis, T. and Moore, N. F., 1988. The Olive Tree. *Biologist* 35/1, 7-13.

Marchetti, M., 2002. Environmental changes in the central Po Plain (northern Italy) due to fluvial modifications and anthropogenic activities. *Geomorphology* 44/3-4, 361-373.

Margaritis, E. and Jones, M., 2006. Beyond cereals: crop processing and *Vitis vinifera* L. ethnography, experiment and charred grape remains from Hellenistic Greece. *Journal of Archaeological Science* 33/6, 784-805.

Margaritis, E. and Martin, J., 2008. Crop processing of Olea europaea L.: an experimental approach for the interpretation of archaeobotanical olive remains. *Vegetation History and Archaeobotany* 17/4, 381-392.

Marshall, L.J. R., Almond, M. J., Cook, S. R., Pantos, M., Tobin, M. J., and Thomas, L. A., 2008. Mineralised organic remains from cesspits at the Roman town of Silchester: Processes and preservation. *Spectrochimica Acta* 71, 854-861.

Martin, A. C. and Barkley, W. D., 2000. *Seed Identification Manual*. New Jersey: Blackburn Press.

Martinez, M. A., 2005. Agriculture and food from the Roman to the Islamic period in the north-east of the Iberian peninsula:Archaeobotanical studies in the city of Lleida (Catalonia, Spain). *Vegetation History and Archaeobotany* 14/4, 341-361.

Marturano, A. and Varone, A., 2005. The A.D 79 eruption: seismic activity and effects of the eruption on Pompeii. In: M. S. Balmuth, D. K. Chester, and P. A. Johnston (eds.), *Cultural Responses to the volcanic landscape: the Mediterranean and beyond*. Boston, Mass: Archaeological Institute of America, AIA colloquia and conference papers 8, 241-260.

Matterne, V. and Derreumaux, M., 2008. A Franco-Italian investigation of funerary rituals in the Roman world, "les rites et la mort a Pompei", the plant part: a preliminary report. *Vegetation History and Archaeobotany* 17/1, 105-112.

Mattingly, D. J., 1988. Oil for export? A comparison of Libyan, Spanish and Tunisian oil production in the Roman empire.. *Journal of Roman Archaeology* 1, 33-56.

Mattingly, D. J., 1996. First fruit? The olive in the Roman world. In: G. Shipley and J. Salmon (eds.), *Human Landscapes in Classical Antiquity: Environment and Culture*. London, New York: Routledge, 213-253.

Mattingly, D.J., 2004. Being Roman: expressing identity in a provincial setting. *Journal of Roman Archaeology* 17, 5-26.

Mattusch, C. C., 2008. *Pompeii and the Roman Villa, Art and Culture around the Bay of Naples*. London: Thames and Hudson.

Mau, A., 1902. *Pompeii: Its Life and Art*. Trans. into English by F. W. Kelsy. New York, London: MacMillan.

Mayeske, B.-J., 1979. Bakers, Bakeshops, and Bread: a Social and Economic Study. In: *Pompeii and the Vesuvian Landscape: papers of a symposium sponsored by the Archaeological Institute of America and the Smithsonian Institute*. Washington, D.C: Archaeological Institute of America.

Mazzini, I., 2000. Diet and medicine in the ancient world. In: J. L. Flandrin and M. Montanari (eds.), *Food, a culinary history from antiquity to the present*. New York: Penguin Books, 141-152.

Meadows, K., 1999. The appetites of households in early Roman Britain. In: P. M. Allison (ed.), *The Archaeology of Household Activities*. London, New York: Routledge, 101-120.

Megaloudi, F., 2005. Burnt sacrificial plant offerings in Hellenistic times: an archaeobotanical case study from Messene, Peloponnese, Greece. *Vegetation History and Archaeobotany* 14/4, 329-340.

Melillo, L., 1994. Diuretic plants in the paintings of Pompeii. *American Journal of Nephrology* 14, 423-425.

Mennell, S., Murcott, A., and van Otterloo, A. H., 1992. *The Sociology of Food: eating, diet and culture*. London: SAGE.

Mercuri, A. M., Accorsi, C. A., and Mazzanti, M. B., 2002. The long history of Cannabis and its cultivation by the Romans in central Italy, shown by pollen records from Lago Albano and Lago di Nemi. *Vegetation History and Archaeobotany* 11/4, 263-276.

Meyer, F. G., 1980. Carbonized food plants of Pompeii, Herculaneum, and the Villa at Torre Annunziata. *Economic Botany* 34/4, 401-437.

Meyer, F. G., 1988. Food plants identified from carbonized remains at Pompeii and other Vesuvian Sites. In: R. I. Curtis (ed.), *Studia Pompeiana and classica in honor of Wilhelmina F. Jashemski*. New Rochelle, N.Y: Caratzas, vol. 1, 183-230.

Meyer, F. G., 1994. Evidence of food plants of ancient Pompeii and other Vesuvian sites. In: D. Moe, J. H. Dickson, and P. M. Jorgensen (eds.), *Garden history: garden plants, species, forms and varieties from Pompeii to 1800: symposium held at the European Univeristy Centre for Cultural Heritage, Ravello, June, 1991*. Belgium: Council of Europe, 19-23.

Miksicek, C. H., 1987. Formation processes of the archaeobotanical record. *Advances in Archaeological Method and Theory* 10, 211-247.

Miller, J. I., 1998. *The Spice Trade of the Roman Empire, 29B.C. to A.D. 641*. London: Sandpiper Books Ltd.

Miller, N. F., 1991. The Near East. In: W. van Ziest, K. Wasylikowa and K.-E. Behre (eds.), *Progress in Old World palaeoethnobotany*. Rotterdam, The Netherlands: A.A. Balkema, 133-160.

Millett, M., 1991. Roman towns and their territories: an archaeological perspective. In: J. Rich and A. Wallace-Hadrill (eds.), *City and Country in the Ancient World*. New York, London: Routledge, vol. 2, 169-189.

Millet, M., 2002. Romanization: Historical issues and archaeological interpretation. In: T. Blagg and M. Millet (eds.), *The early Roman Empire in the West*. Oxford: Oxbow books, 35-41.

Millet, M., 2007. Urban Topography and social identity in the Tiber Valley. In: R. Roth and J. Killer (eds.), *Roman by Integration: dimensions of group identity in material culture and text*. Portsmouth, Rhode Island: Journal of Roman Archaeology, JRA supplementary Series # 66, 71-82.

Minnis, P. E., 1981. Seeds in archaeological sites: sources and some interpretive problems. *American Antiquity* 46, 143-152.

Montanari, M., 2000. Romans, Barbarians, Christians: The dawn of European food culture. In: J. L. Flandrin and M. Montanari (eds.), *Food, a culinary history from antiquity to the present*. New York: Penguin Books, 165-167.

Moorman, E. M., 2007. Villas surrounding Pompeii and Herculaneum. In: J. J. Dobbins and P. W. Foss (eds.), *The World of Pompeii*. London, New York: Routledge, chap. 28, 435-456.

Morel, J. P., 2007. Early Rome and Italy. In: W. Scheidel, I. Morris, and R. Saller (eds.), *The Cambridge Economic History of the Greco-Roman World*. Cambridge: Cambridge University Press, 487-510.

Morris, I., 2007. Introduction. In: W. Scheidel, I. Morris, and R. Saller (eds.), *The Cambridge Economic History of the Greco-Roman World*. Cambridge: Cambridge University Press, 1-14.

Motta, L., 2002. Planting the seed of Rome. *Vegetation History and Archaeobotany* 11/1, 71-77.

Murphy, C., Thompson, G., and Fuller, D.Q., 2013. Roman food refuse: urban archaeobotany in Pompeii, Regio VI, Insula 1. *Vegetation History and Archaeobotany* 22: 409-419.

Nappo, S., 1998. *Pompeii, Guide to the Lost City*. London: Weidenfeld & Nicolson.

Neef, R., 1990. Introduction, development and environmental implications of olive culture: the evidence from Jordan. In: S. Bottema, G. Entjes-Nieborg and W. Van Zeist (eds.), *Man's role in the shaping of the eastern Mediterranean landscape*. Rotterdam, The Netherlands; Brookfield, VT: A.A. Balkema, 295-306.

Newberry, P. E., 1937. *On some African Species of the Genus Olea and the original home of the cultivated olive-tree*. London: Proc. Linn. Soc. 150, 3-16.

Olmo, H. P., 1995. The origin and domestication of the Vinifera grape. In: P. E. Mc Govern, S. J. Fleming, and S. H. Katz (eds.), *The Origins and Ancient History of Wine*. New York: Gordon and Bread, 31-43.

Orr, D. G., 1988. Learning from Lararia: notes on the household shrines of Pompeii. In: R. I. Curtis (ed.), *Studia Pompeiana and Classica in honor of Wilhelmina F. Jashemski*. New Rochelle, N.Y: Caratzas, 293-304.

Palamarev, E., 1989. Paleobotanical evidences of the tertiary history and origin of the Mediterranean sclerophyll dendroflora. *Plant Systematics and Evolution* 162, 93-107.

Papi, E (ed.), 2007. Supplying Rome and the Empire: the proceedings of an international seminar held at Siena-Certosa di Pontignano on May 2-4, 2004, on Rome, the provinces, production and distribution. Portsmouth, R.I: *Journal of Roman Archaeology*, supplementary series no. 69.

Parker, A. J., 1990. The Wines of Roman Italy. *Journal of Roman Archaeology* 3, 325-331.

Peacock, D. P. S., 1989. The Mills of Pompeii. *Antiquity* 63/239, 205-214.

Pearsall, D. M., 2000. *Paleoethnobotany: a Handbook of Procedures*. (2nd edition). London: Academic Press.

Peña, J. T., 2007. *Roman pottery in the archaeological record*. Cambridge: Cambridge University Press.

Peña, J. T. and McCallum, M., 2009. The production and distribution of pottery at Pompeii: A review of the evidence; part 1, production. *American Journal of Archaeology* 113, 57-79.

Perring, D., 1992. Spatial organisation and social change in Roman towns. In: J. Rich and A. Wallace-Hadrill (eds.), *City and Country in the Ancient World*. London: Routledge, 273-294.

Pescatore, T., Senatore, M. R., Capretto, G., and Lerro, G., 2001. Holocene coastal environments near Pompeii before the AD 79 eruption of Mount Vesuvius, Italy. *Quaternary Research* 55/1, 77-85.

Petronius. *The Satyricon*. Trans., introduction and notes by P.G. Welsh, 1997. Oxford: Oxford University Press.

Pignatti, S., 1982. *Flora d'Italia*. Bologna: Edagricole, vols. I, II, and III.

Pirson, F., 2007. Shops and Industries. In: J. J. Dobbins and P. W. Foss (eds.), *The World of Pompeii*. London: Routledge, chap. 29, 457-473.

Poehler, E. E., 2006. The circulation of traffic in Pompeii's Regio VI. *Journal of Roman Archaeology* 19, 53-74.

Popper, V. S. and Hastorf, C. A., 1988. Introduction. In: C. A. Hastorf and V. S. Popper (eds.), *Current paleoethnobotany: analytical methods and cultural interpretations of archaeological plant remains*. Chicago, London: University of Chicago Press, 1-16.

Preiss, S., Matterne, V., and Latron, F., 2005. An approach to funerary rituals in the Roman provinces: plant remains from a Gallo-Roman cemetery at Faulquemont (Moselle, France). *Vegetation History and Archaeobotany* 14, 362-372.

Purcell, N., 1985. Wine and wealth in ancient Italy. *The Journal of Roman Studies* 75, 1-19.

Purcell, N., 1990. The Economy of an ancient Town. *Classical Review*, 111-116.

Purcell, N., 1995. The Roman villa and the landscape of production. In: T. J. Cornell and K. Lomas (eds.), *Urban Society in Roman Italy*. London: UCL Press, 151-179.

Reale, O. and Dirmeyer, P., 2000. Modeling the effects of vegetation on Mediterranean climate during the Roman Classical Period Part I: Climate history and model sensitivity. *Global and Planetary Change* 25/3-4, 163-184.

Reale, O. and Shukla, J., 2000. Modelling the effects of vegetation on Mediterranean climate during the Roman Classical period: Part II. Model simulation. *Global and Planetary Change* 25/3-4, 185-214.

Renfrew, J. M., 1973. *Palaeoethnobotany: The prehistory food plants of the Near East and Europe*. London: Methuen.

Renfrew, J. M., Monk, M., and Murphy, M., 1976. *First Aid for Seeds. UK*: Rescue Publication.

Ricciardi, M. and Aprile, G. G., 1988. Identification of some carbonized plant remains from the archaeological area of *Oplontis*. In: R. I. Curtis (ed.), *Studia Pompeiana and Classica in honor of Wilhelmina F. Jashemski*. New Rochelle, N.Y: Caratzas, vol. 1, 317-324.

Ricciardi, M., Aprile, G.G., La Valva, V., and Caputo, G., 1986. La Flora di Somma-Vesuvio. *Bollettino della Società dei Naturalisti in Napoli* 95, 3-121.

Richardson, J., 2006. *Food and fuel: Feeding the city. Preliminary findings Regio VI Insula I, 1994-2006*. Paper presented at the AAPP Pompeii Specialist Conference, Pompeii (NA), Italy, June 2006.

Richardson, J. E. and Ciarallo, A. M., 1996. Economy and environment at Pompeii. *American Journal of Archaeology* 100/2, 370.

Richardson, J., Thompson, G., and Genovese, A., 1997. New directions in economic and environmental research at Pompeii. In: S. E. Bon and R. Jones (eds.),

Sequences and Space in Pompeii. Oxford: Oxbow books, 88-101.

Riddle, J. M., Estes, J. W., and Russell, J. C., 1994. Birth control in the Ancient World. *Archaeology*, 27-33.

Robinson, D., 1996. The social texture of Pompeii. *American Journal of Archaeology* 100/2, 370.

Robinson, D., Anderson, M., and Jones, R., 2008. *New Light on the House of the Surgeon in Pompeii (VI.I.10)*. Unpublished report.

Robinson, M. A., 1999. The macroscopic plant remains. In: Fulford, M. and Wallace-Hadrill, A. (eds.), Towards a history of pre-Roman Pompeii: excavations beneath the House of Amarantus (I.9.11-12). *Papers of the British School at Rome* 67, 95-102, and 139-44.

Robinson, M. A., 2002. Domestic burnt offerings and sacrifices at Roman and pre-Roman Pompeii, Italy. *Vegetation History and Archaeobotany* 11/1-2, 93-99.

Robinson, M., Fulford, N., and Tootell, K., 2006. The macroscopic plant remains. In: M. Fulford, A. Clarke, and H. Eckardt (eds.), *Life and Labour in Late Roman Silchester, Excavations in insula IX since 1997*. London: Society for the Promotion of Roman Studies, Britannia monograph series no. 22, 206-216.

Rosenstein, N., 2008. Aristocrats and agriculture in the Middle and Late Republic. *Journal of Roman Studies* 98, 1-26.

Rossiter, J. J., 1981. Wine and Oil Processing at Roman Farms in Italy. *Phoenix* 35/4, 345-361.

Rossiter, J. J., 1998. Pressing Issues: wine- and oil-production. *Journal of Roman Archaeology* 11, 597-602.

Rostovtzeff, M.I., 1957. *The Social and Economic History of the Roman Empire*. (2nd edition revised by P.M. Fraser). Oxford: Oxford University Press.

Rovira, N. and Chabal, L., 2008. A foundation offering at the Roman part of Lattara (Lattes, France), the plant remains. *Vegetation History and Archaeobotany* 17/suppl.1, 191-200.

Runnels, C. N., and Hansen, J., 1986. The olive in the Prehistoric Aegean: The evidence for domestication in the early Bronze Age. *Oxford Journal of Archaeological Science* 5, 299-308.

Sadori, L. and Susanna, F., 2005. Hints of economic change during the late Roman Empire period in central Italy: a study of charred plant remains from "La Fontanaccia", near Rome. *Vegetation History and Archaeobotany* 14/4, 386-393.

Sallares, R., 2007. Ecology. In: W. Scheidel, I. Morris, and R. P. Saller (eds.), *The Cambridge Economic History of the Greco-Roman World*. Cambridge: Cambridge University Press, chap. 2, 15-37.

Sassatelli, G., 2000. The Diet of the Etruscans. In: J. L. Flandrin and M. Montanari (eds.), *Food, a culinary history from antiquity to the present*. New York: Penguin Books, 113-127.

Scheidel, W., Morris, I. and Saller, R. P. (eds.), 2007. *The Cambridge Economic History of the Greco-Roman World*. Cambridge: Cambridge University Press.

Schiffer, M. B., 1977. Towards a unified science of the cultural past. In: S. South (ed.), *Research Strategies in Historical Archaeology*. New York: Academic Press.

Schiffer, M. B., 1985. Is there a Pompeii premise in archaeology? *Journal of Anthropological Research* 41/1, 18-41.

Schiffer, M. B., 1987. *Formation processes of the archaeological record*. Albaquerque, N.M: University of New Mexico Press.

Shelton, J. A., 1998. *As the Romans did, a sourcebook in Roman social history*. (2nd edition). Oxford: Oxford University Press.

Sherrat, A. 1987. Cups that cheered. In: W. H. Waldren and R. C. Kennard (eds.), *Bell Beakers of the Western Mediterranean*. Oxford: British Archaeological Reports (SS331i), 81-114.

Sherrat, A., 1999. Cash-crops before cash: organic consumables and trade. In: C. Gosden and J. Hather (eds.), *The Prehistory of Food, Appetites for Change*. London: Routledge, 13-34.

Sigurdsson, H., 2007. The Environmental and Geomorphological context of the Volcano. In: J. J. Dobbins and P. W. Foss (eds.), *The World of Pompeii*. London: Routledge, chap. 4, 43-62.

Sigurdsson, H., Cashdollar, S., and Sparks, S. R. J., 1982. The eruption of Vesuvius in A.D. 79: reconstruction from historical and volcanological evidence. *American Journal of Archaeology* 86, 39-51.

Simoons, F. J., 1998. *Plants of life, plants of death*. Wisconsin: University of Wisconsin Press.

Sinopoli, C. M., 1994. The Archaeology of Empires. *Annual Review of Anthropology* 23, 159-180.

Small, A.M., 2007. Urban, suburban and rural religion in the Roman period. In: J. J. Dobbins and P.W. Foss (eds.), *The World of Pompeii*. London: Routledge, chap. 13, 184-211.

Smith, H. and Jones, G., 1990. Experiments on the Effects of Charring on Cultivated Grape Seeds. *Journal of Archaeological Science* 17/3, 317-327.

Smith, M. L., 2006. The Archaeology of Food Preference. *American Anthropologist* 108/3, 480-493.

Soprintendenza Archaeologica di Pompei, 2005. *Cibi e Sapori a Pompei e dintorni*. Pompeii (NA): Marius.

Šoštarić, R. and Küster, H., 2001. Roman plant remains from Veli Brijun (island of Brioni), Croatia. *Vegetation History and Archaeobotany* 10/4, 227-233.

Soyer, A., 1977. *The Pantropheon or a history of food and its preparation in ancient times*. New York, London: Paddington Press.

Spurr, M. S., 1983. The cultivation of millet in Roman Italy. *Papers of the British School at Rome*. London: British School at Rome, 1-15.

Spurr, M. S., 1986. Arable cultivation in Roman Italy c.200BC-c.AD 100. *Journal of Roman studies*, monograph no. 3. London: Society for the Promotion of Roman Studies.

Stallibrass, S. and Thomas, R. (eds.), 2008. *Feeding the Roman Army: the Archaeology of Production and Supply in NW Europe*. Oxford: Oxbow Books.

Stefani, G., 2003.*Uomo e ambiente nel territorio vesuviano*, Guida all'Antiquarium di Boscoreale. Pompei: Marius.

Stika, H.P., 2005. Early Neolithic agriculture in Ambrona, Provincia Soria, central Spain. *Vegetation History and Archaeobotany* 14/3, 189-197.

Strong, R., 2002. *Feast, a History of Grand Eating*. Orlando, Florida: Harcourt.

Tanno, K. and Willcox, G., 2006. The origins of cultivation of *Cicer arietinum* L. and *Vicia faba* L.: early finds from Tell el-Kerkh, north-west Syria, late 10th millennium B.P. *Vegetation History and Archaeobotany* 15/3, 197-204.

Tchernia, A., 1983. Italian wine in Gaul at the end of the Republic. In: P. Garnsey, K. Hopkins and C. R. Whittaker (eds.), *Trade in the Ancient Economy*. London: Chatto & Windus, 87-104.

Terral, J.-F., 1996. Wild and cultivated olive (*Olea europaea* L.): a new approach to an old problem using inorganic analyses of modern wood and archaeological charcoal. *Review of Palaeobotany and Palynology* 91, 383-397.

Terral, J.-F., 2000. Exploitation and management of the olive tree during prehistoric times in Mediterranean France and Spain. *Journal of Archaeological Science* 27/2, 127-133.

Terral, J.-F., Alonson, N., Capdevilla, B. I., Chatti, N., Fabre, L., Fiorentino, G., Marinval, P., Perez Jorda, G., Pradat, B., Rovira and Alibert, P., 2004. Historical biogeography of olive domestication (*Olea europaea* L.) as revealed by geometrical morphometry applied to biological and archaeological material. *Journal of Biogeography* 31, 63-77.

Terral, J.-F., Tabard, E., Bouby, L., Ivorra, S., Pastor, T. Figueiral, I., Picq, S., Chevance, J.-B., Jung, C., Fabre, L., Tardy, C., Compan, M., Bacilieri, R., Lacombe, T., and This, P., 2010. Evolution and history of grapevine (*Vitis vinifera*) under domestication: new morphometric perspectives to understand seed domestication syndrome and reveal origins of ancient European cultivars. *Annals of Botany* 105/3, 443-455.

The Local, Sweden's News in English, Anon, 2005. Pompeii Discovery for Swedish Archaeologists. Retrieved on 04 March 2010 from World Wide Web: *http://www.thelocal.se/1291/20050416/*

Turabian, K., Chicago, 1996. Revised by J. Grossman and A. Bennett. *A Manual for Writers of Term Papers, Theses, and Dissertations*. (6th edition). Chicago, London: the University of Chicago Press.

Turner, C., 1968. A note on the occurrence of Vitis and other new plant records from the Pleistocene deposits at Hoxne, Suffolk. *New Phytologist* 67, 333-334.

Tyree, E. L., and Stefanoudaki, E., 1996. The Olive Pit and Roman Oil Making. *The Biblical Archaeologist* 59/3, 171-178.

Urbanus, J., 2002. Re-analyzing the W half of Pompeii's Insula IX 1. *Journal of Roman Archaeology*, 591-592.

Valamoti, S. M., Mangafa, M., Koukouli-Chrysanthaki, C., and Malamidou, D., 2007. Grape-pressing from northern Greece: the earliest wine in the Aegean? *Antiquity* 81, 54-61.

van der Veen, M., 2003. When is food a luxury? *World Archaeology* 34/31, 405-427.

van der Veen, M., 2008. Food as embodied material culture: diversity and change in plant food consumption in Roman Britain. *Journal of Archaeological Science* 21, 83-110.

van der Veen, M. and Fieller, N., 1982. Sampling seeds. *Journal of Archaeological Science* 9/3, 287-298.

van der Veen, M., Livarda, A., and Hill, A., 2008. New plant foods in Roman Britain- dispersal and social access, archaeobotany of Roman Britain current state and identification of research priorities. *Environmental Archaeology* 13/1, 11-36.

van der Veen, M., Morales, J. and Cox, A., 2009. Food and culture: the plant foods from Roman and Islamic Quseir, Egypt. In: A. S. Fairbairn and E. Weiss (eds.), *From Foragers to Farmers, papers in honour of Gordon C. Hillman*. Oxford: Oxbow Books, chap. 28, 269-276.

van Zeist, W., 1983. Fruits in foundation deposits of two temples. *Journal of Archaeological Science* 10/4, 351-354.

van Zeist, W. and Woldring, H., 1980. Holocene vegetation and climate of northwestern Syria. *Palaeohistoria* 22, 111-125.

van Zeist, W., Woldring, H., and Neef, R., 1994. Plant husbandry and vegetation of early medieval Douai, northern France. *Vegetation History and Archaeobotany* 3/4 191-218.

Veal, R., 2014. Pompeii and its Hinterland Connection:The Fuel Consumption of the House of the Vestals (c. Third Century BC to AD 79). *European Journal of Archaeology* 17(1): 27-44.

Veal, R., 2009. *The Wood Fuel Supply to Pompeii third century BC to AD79: An environmental, historical and economic study based on charcoal analysis*. Unpublished PhD thesis, Department of Archaeology, University of Sydney.

Veal, R. and Thompson, G., 2006. *Fuel: Preliminary findings Regio VI Insula I, 1994- 2006*. Paper presented at the AAPP Pompeii Specialist Conference, Pompeii (NA), Italy, June 2006.

Veal, R. and Thompson, G., 2008. Fuel Supplies for Pompeii: Pre-Roman and Roman charcoals of the Casa delle Vestali. In: G. Fiorentino, G and D. Magri (eds.), *Charcoals from the past: Cultural and palaeoenvironmental implications*. Proceedings of the third International meeting of Anthracology, Cavallino- Leece (Italy), June 28-July 1st, 2004. British Archaeological Report S1807. Oxford: Archaeopress.

Wallace-Hadrill, A., 1990. The Social Spread of Roman Luxury: Sampling Pompeii and Herculaneum. *Papers of the British School at Rome* LVIII, 145-192.

Wallace-Hadrill, A., 1992. Elites and trade in the Roman town. In: J. Rich and A. Wallace-Hadrill (eds.), *City and Country in the Ancient World*. London: Routledge, 241-272.

Wallace-Hadrill, A., 1994. *Houses and Society in Pompeii and Herculaneum*. Princeton, N.J: Princeton University Press.

Wallace-Hadrill, A., 1995. *Horti* and Hellenization. In: M. Cima and E. La Rocca (eds.), *Horti romani: atti del convegno internazionale: Roma, 4-6 maggio 1995. Bullettino della Commissione archeologica comunale di Roma, Supplementi* 6, 1-13. Roma: L'Erma di Bretschneider.

Ward-Perkins, J. B., 1984. *From Classical antiquity to the Middle Ages: urban public building in Northern and Central Italy: AD 300-850*. Oxford: Oxford University Press.

Warnock, P., 2007. *Identification of ancient olive oil processing methods based on olive remains*. BAR international series 1635. Oxford: Archaeopress.

Watts, W. A., 1985. A long pollen record from Laghi di Monticchio, southern Italy: a preliminary account. *Journal of the Geological Society* 142/3, 491-499.

Watts, W. A., Allen, J. R. M., and Huntley, B., 1996. Vegetation history and palaeoclimate of the Last Glacial Period at Lago Grande di Monticchio, southern Italy. *Quaternary Science Reviews* 15, 113-132.

Weiss, C., 2007. *Determining Function of Pompeian Sidewalk Features through GIS Analysis*. Unpublished M.A. Thesis. Institute of Archaeology, University College London.

Welch, P. and Scarry, C. M. 1995. Status-Related variation in Foodways in the Moundville Chiefdom. *American Antiquity* 60/3, 397-419.

Westfall, C. W., 2007. Urban planning, roads, streets and neighborhoods. In: J. J. Dobbins and P. W. Foss (eds.), *The World of Pompeii*. London: Routledge, 129-139.

White, K. D., 1965. The Productivity of Labour in Roman Agriculture. *Antiquity* XXXIX, 102-107.

White, K. D., 1967. *Agricultural Implements of the Roman world. Cambridge*: Cambridge University Press.

White, K. D., 1970a. *A Bibliography of Roman Agriculture*. Reading: University of Reading.

White, K. D., 1970b. *Roman Farming: Aspects of Greek and Roman Life*. Great Britain: Camelot Press.

White, K. D., 1995. Cereals, bread and milling in the Roman world. In: J. Wilken, D. Harvey, and M. Dobson (eds.), *Food in Antiquity*. Exeter: Exeter University Press, 38-43.

Wiessner, P., and Schiefenhövel, W. (eds), 1998. *Food and the Status Quest: an interdiscinplinary perspective*. Providence, R.I: Berghahn Books.

Wilburn, D., 2000. The Geography and History of Campania and Pompeii. In: E. K. Gazda (ed.), *The Villa of the Mysteries in Pompeii: ancient ritual, modern muse*. Ann Arbor, Michigan: The Kelsey Musueum of Archaeology and the University of Michigan Museum of Art, 16-23.

Wilkinson, K. and Stevens, C., 2008. *Environmental Archaeology: approaches, techniques and applications*. (Revised edition). Stroud, Glouchestershire, UK: Tempus.

Wilson, A., 2002. Detritus, disease and death in the city. *Journal of Roman Archaeology* 15, 479-485.

Woolf, G. D., 2001. The Roman cultural revolution in Gaul. In: S. Keay and N. Terrenato (eds.), *Italy and the West Comparative Issues in Romanization*. Oxford: Oxbow, 173-186.

Wright, P. J., 2003. Preservation or destruction of plant remains by carbonization? *Journal of Archaeological Science* 30/5, 577-583.

Wright, P. J., 2005. Flotation samples and some paleoethnobotanical implications. *Journal of Archaeological Science* 32/1, 19-26.

Zanker, P., 1998. *Pompeii: Public and Private Life*. Trans. by L. Schneider. Cambridge, Mass., London: Harvard University Press.

Zohary, D., 1994. The wild genetic resources of the cultivated olive. *Acta Horticulturae* 356, 62-65.

Zohary, D. and Hopf, M., 2001. *Domestication of plants in the Old World: The origin and spread of cultivated plants in west Asia, Europe and the Nile Valley*. (3rd edition). Oxford: Oxford University Press.

Zohary, D. and Spiegel-Roy, P., 1975. Beginnings of Fruit Growing in the Old World. Science 31, 187/4174, 319-327.

			All	Vicolo di Narciso 1BC		Vicolo di Narciso 1AD		Via Consolare 1BC		Via Consolare 1AD		Triclinium 1AD		Shrine 1BC		Shrine 1AD		Soap Factory 1BC		Soap Factory 1AD		House of the Vestals 1BC		House of the Vestals 1AD	
				Backstreet		Backstreet		Frontstreet		Frontstreet		VI.I.i		VI.I.xiii		VI.I.xiii		VI.I.xiv		VI.I.xiv		VI.I.vii		VI.I.vii	
Preservation			Σ	Σ	Ub	Σ	Ub	Σ	Ub	Σ	Ub	Σ	Ub	Σ	Ub	Σ	Ub	Σ	Ub	Σ	Ub	Σ	Ub	Σ	Ub
Fruits/nuts:																									
Cerasus avium L.	stone	m	1	-	-	-	-	-	-	-	-	-	-	-	-	-	-	-	-	-	-	-	-	1	0.1
Cucumis melo L.	seed	c	2	-	-	-	-	-	-	-	-	-	-	-	-	-	-	-	-	-	-	-	-	1	0.1
Cucumis cf. melo L.	seed	c	4	-	-	1	0.4	-	-	-	-	-	-	-	-	1	0.2	-	-	-	-	1	1.0	-	-
Cucumis cf. melo L.	seed	m	9	-	-	-	-	-	-	-	-	5	0.03	3.00	0.302	-	-	-	-	-	-	-		-	-
Cucumis L.	seed	m	10	-	-	-	-	-	-	-	-	-	-	-	-	-	-	-	-	-	-	-	-	-	-
Ficus carica L.	achene	c	89	17	33.1	30	12.8	-	-	-	-	6	0.04	5.00	0.503	5	1.0	3	0.3	-	-	3	3.1	2	0.1
Ficus carica L.	achene	m	19983	5	9.7	18	7.7	-	-	-	-	15095	98.8	503.00	50.645	65	12.5	788	74.4	-	-	10	10.2	1703	88.0
Malus domestica Borkh	seed	c	1	-	-	-	-	-	-	-	-	-	-	-	-	-	-	-	-	-	-	-	-	1	0.1
Malus domestica Borkh.	seed	m	15	-	-	1	0.4	-	-	-	-	13	0.1	1.00	0.101	-	-	-	-	-	-	-	-	-	-
cf. Malus domestica Borkh	seed	m	1	-	-	-	-	-	-	-	-	-	-	1.00	0.101	-	-	-	-	-	-	-	-	-	-
cf. Malus domestica Borkh	seed	c	2	-	-	-	-	-	-	-	-	-	-	2.00	0.201	-	-	-	-	-	-	-	-	-	-
Olea europea L	endocarp	c	461	3	5.8	6	2.6	-	-	0.55	8.3	4	0.03	69.22	6.969	91	17.5	35	3.3	7	7.6	6	6.6	15	0.8
Punica granatum L	seed	c	12	2	3.9	-	-	-	-	-	-	2	0.01	-	-	-	-	-	-	-	-	1	1.0	5	0.3
Punica granatum L	seed	m	4	-	-	-	-	-	-	-	-	4	0.03	-	-	-	-	-	-	-	-	-	-	-	-
cf. Punica granatum L.	seed	c	2	-	-	-	-	-	-	-	-	-	-	-	-	1	0.2	-	-	-	-	1	1.0	-	-
Prunus persica (L.) Batsch.	endocarp	c	26	-	-	2	0.9	-	-	-	-	-	-	-	-	2	0.4	10	0.9	2	2.3	-	-	7	0.4
Prunus sp.	endocarp	m	3	-	-	-	-	-	-	-	-	2	0.01	-	-	-	-	-	-	-	-	-	-	-	-
Vitis vinifera L	seed	c	210	4	7.8	7	3.0	1	50.0	-	-	22	0.1	22.14	2.229	35	6.7	0.2	0.0	0.15	0.2	14	14.4	13	0.6
Vitis vinifera L	seed	m	466	-	-	134	57.0	-	-	2.06	31.2	73	0.5	25.83	2.601	16	3.1	37	3.5	37	42.5	8	8.0	46	2.4
Vasicular tissue indeterminate	fragment	c	424	7	13.6	4	1.7	-	-	-	-	3	0.02	123.00	12.384	48	9.2	55	5.2	10	11.5	11	11.2	12	0.6
Fruit exocarp	fragment	c	39	-	-	-	-	-	-	3.00	45.4	3	0.02	19.00	1.913	4	0.8	-	-	-	-	-	-	1	0.1
Amygdalus communis L.	shell fragment	c	5	-	-	-	-	-	-	-	-	1	0.01	-	-	-	-	-	-	1	1.1	-	-	-	-
Corylus avellana L	shell fragment	c	18	-	-	2	0.9	-	-	-	-	-	-	-	-	4	0.8	1	0.1	1	1.1	3	3.1	4	0.2
cf. Corylus avellana L.	shell fragment	c	3	-	-	-	-	-	-	-	-	-	-	-	-	-	-	3	0.3	-	-	-	-	-	-
Juglans regia L	seed	c	39	-	-	5	2.1	-	-	-	-	3	0.02	3.00	0.302	3	0.6	2	0.2	-	-	3	3.1	6	0.3
cf. Juglans regia L.	shell fragment	c	7	-	-	1	0.4	-	-	-	-	-	-	1.00	-	1	0.2	-	-	-	-	-	-	4	0.2
Pinus pinea L	shell fragment	c	60	1	1.9	6	2.6	-	-	-	-	3	0.02	1.00	0.101	4	0.8	5	0.5	3	3.4	1	1.0	25	1.3
Pinus pinea L.	shell fragment	m	3	-	-	-	-	-	-	-	-	-	-	-	-	-	-	-	-	1	1.1	-	-	-	-
Nuts (generic)	shell fragment	c	49	4	7.8	2	0.9	-	-	-	-	-	-	2.00	0.201	1	0.2	4	0.4	14	16.1	4	4.1	10	0.5
Cereals																									
T. aestivum/T.durum L.	caryopsis	c	4	-	-	-	-	-	-	-	-	-	-	-	-	2	0.4	-	-	-	-	-	-	1	0.1
T. cf. aestivum/T.durum L.	caryopsis	c	28	-	-	1	0.4	-	-	-	-	-	-	24.00	2.416	-	-	-	-	-	-	-	-	-	-
cf. T.aestivum/durum L	caryopsis	c	6	-	-	-	-	-	-	-	-	-	-	1.00	0.101	-	-	-	-	-	-	-	-	-	-
T. dicoccum Schübl.	caryopsis	c	5	1	1.9	-	-	-	-	-	-	-	-	1.00	0.101	1	0.2	-	-	-	-	-	-	-	-
T. cf. dicoccum Schübl.	caryopsis	c	14	-	-	-	-	-	-	-	-	-	-	8.00	0.805	1	0.2	-	-	-	-	-	-	-	-
cf. T.dicoccum Schübl.	caryopsis	c	1	-	-	-	-	-	-	-	-	-	-	1.00	0.101	-	-	-	-	-	-	-	-	-	-
T cf. monoccocum	caryopsis		1	-	-	-	-	-	-	-	-	-	-	-	-	-	-	-	-	-	-	-	-	-	-
Triticum sp	caryopsis	c	197	2	3.9	3	1.3	-	-	-	-	5	0.03	53.00	5.336	66	12.7	9	0.8	4	4.6	4	4.1	7	0.4
Hordeum vulgare L	caryopsis	c	20	-	-	-	-	-	-	-	-	-	-	3.00	0.302	2	0.4	1	0.1	-	-	4	4.1	4	0.2
cf. Hordeum vulgare L.	caryopsis	c	9	1	1.9	2	0.9	-	-	-	-	1	0.01	1.00	0.101	1	0.2	-	-	-	-	2	2.0	-	-
Secale cereale	caryopsis	c	1	-	-	-	-	-	-	-	-	-	-	-	-	-	-	-	-	-	-	-	-	-	-
Setaria italica Beauv	caryopsis	c	2	-	-	-	-	-	-	-	-	-	-	-	-	-	-	1	0.1	-	-	-	-	-	-
Setaria cf. italica Beauv.	caryopsis	c	3	-	-	-	-	-	-	-	-	-	-	1.00	0.101	2	0.4	-	-	-	-	-	-	-	-
cf. Setaria italica Beauv	caryopsis	c	5	-	-	-	-	-	-	-	-	1	0.01	-	-	-	-	2	0.2	-	-	-	-	-	-
cf. Setaria	caryopsis	c	7	-	-	-	-	-	-	-	-	1	0.01	-	-	1	0.2	1	0.1	-	-	-	-	-	-
Panicum miliaceum L	caryopsis	c	18	-	-	-	-	1	50.0	-	-	-	-	4.00	0.403	9	1.7	-	-	-	-	2	2.0	1	0.1
Panicum miliaceum L	caryopsis	m	58	-	-	1	0.4	-	-	-	-	3	0.02	30.00	3.021	12	2.3	-	-	-	-	-	-	2	0.1
Panicum cf. miliaceum L.	caryopsis	c	12	-	-	-	-	-	-	-	-	-	-	5.00	0.503	2	0.4	-	-	1	1.1	-	-	1	0.1
Panicum sp	caryopsis	c	35	-	-	-	-	-	-	-	-	-	-	3.00	0.302	5	1.0	4	0.4	2	2.3	-	-	-	-
Panicum sp	caryopsis	m	6	-	-	-	-	-	-	-	-	3	0.02	2.00	0.201	-	-	-	-	-	-	-	-	1	0.1
Pulses																									
Lens culinaris Medik	seed	c	23	1	1.9	-	-	-	-	-	-	-	-	2.33	0.235	3	0.5	2	0.2	-	-	3	2.7	3	0.2
cf. Lens culinaris Medik.	seed	c	5	-	-	1	0.4	-	-	-	-	-	-	-	-	-	-	0.3	0.0	-	-	-	-	3	0.1
Lens culinaris Medik	seed	m	18	-	-	1	0.4	-	-	-	-	-	-	16.00	1.611	-	-	-	-	-	-	-	-	-	-
cf. Lens culinaris Medik.	seed	m	1	-	-	-	-	-	-	-	-	0	0.00	-	-	-	-	-	-	-	-	-	-	-	-
Pisum sativum L	seed	c	3	-	-	-	-	-	-	-	-	-	-	-	-	1	0.1	1	0.1	-	-	-	-	0.5	0.03
cf. Pisum sativum L.	seed	c	1	-	-	-	-	-	-	-	-	-	-	-	-	-	-	-	-	-	-	-	-	0.3	0.02
Vicia ervilia (L.) Willd	seed	c	1	-	-	-	-	-	-	-	-	-	-	-	-	-	-	1	0.1	-	-	-	-	-	-
Vicia cf. ervilia (L.) Willd	seed	c	41	-	-	-	-	-	-	-	-	-	-	-	-	-	-	40	3.8	-	-	-	-	-	-
cf. Vicia ervilia (L.) Willd	seed	c	2	-	-	-	-	-	-	-	-	-	-	-	-	-	-	1	0.1	-	-	-	-	-	-
Vicia faba L.	seed	c	6	-	-	-	-	-	-	-	-	-	-	2.33	0.235	2	0.4	-	-	-	-	-	-	-	-
cf. Vica faba L.	seed	c	7	-	-	0.3	0.1	-	-	-	-	0	0.00	1.17	0.118	3	0.5	-	-	1	0.6	2	2.0	-	-
Vicia sativa L	seed	c	42	-	-	1	0.4	-	-	-	-	-	-	7.17	0.722	15	2.8	-	-	-	-	1	1.0	5	0.3
Vicia cf. sativa L	seed	c	3	-	-	-	-	-	-	-	-	-	-	-	-	-	-	-	-	-	-	-	-	-	-
cf. Vicia sativa L	seed	c	5	-	-	-	-	-	-	-	-	-	-	-	-	-	-	-	-	-	-	-	-	-	-
Vicia sp.	seed	c	121	0.3	0.6	2	0.9	-	-	-	-	4.5	0.03	10.00	1.007	19	3.6	12	1.1	1	1.0	6	6.3	11	0.6
Weedy Species																									
Panicoid	seed	c	66	-	-	1	0.4	-	-	1.00	15.1	-	-	10.00	1.007	7	1.3	3	0.3	1	1.1	3	3.1	4	0.2
Panicoid	seed	m	2	-	-	-	-	-	-	-	-	-	-	1.00	0.101	-	-	-	-	-	-	-	-	-	-
cf. Panicoid	seed	c	1	1	1.9	-	-	-	-	-	-	-	-	-	-	-	-	-	-	-	-	-	-	-	-
Agrostemma githago	seed	c	1	-	-	-	-	-	-	-	-	-	-	-	-	-	-	-	-	-	-	1	1.0	-	-
Agrostemma githago	seed	m	3	-	-	-	-	-	-	-	-	-	-	1.00	0.101	-	-	-	-	-	-	1	1.0	-	-
Apiaceae	seed	c	3	-	-	-	-	-	-	-	-	-	-	1.00	0.101	-	-	-	-	-	-	-	-	-	-
Arrhenatherum elatius ssp. bulbosum	seed	c	94	-	-	-	-	-	-	-	-	-	-	3.00	0.302	50	9.6	37	3.5	1	1.1	-	-	2	0.1
Celtis L.	seed	m	1	-	-	-	-	-	-	-	-	-	-	-	-	-	-	-	-	-	-	-	-	-	-
Cupressus sempervirens	seed	c	6	-	-	-	-	-	-	-	-	-	-	1.00	0.101	5	1.0	-	-	-	-	-	-	-	-
Cupressus sempervirens	seed	m	5	-	-	-	-	-	-	-	-	3	0.02	-	-	1	0.2	-	-	-	-	-	-	-	-
Echium vulgare	seed	c	3	-	-	-	-	-	-	-	-	-	-	1.00	0.101	1	0.2	-	-	-	-	-	-	1	0.1
Echium vulgare	seed	m	5	-	-	-	-	-	-	-	-	-	-	1.00	0.101	2	0.4	-	-	-	-	-	-	-	-
cf. Echium vulgare	seed	m	1	-	-	-	-	-	-	-	-	-	-	-	-	-	-	-	-	1	1.1	-	-	-	-
Galium aparine L	seed	c	3	-	-	-	-	-	-	-	-	-	-	-	-	-	-	-	-	-	-	-	-	-	-
Galium cf. aparine L	seed		1	-	-	-	-	-	-	-	-	-	-	-	-	-	-	-	-	-	-	-	-	-	-
Illex aquifolium L	seed	m	5	-	-	-	-	-	-	-	-	1	0.01	-	-	-	-	-	-	-	-	-	-	2	0.1
Lamiaceae	seed	c	2	-	-	-	-	-	-	-	-	-	-	-	-	-	-	-	-	-	-	-	-	-	-
Linum biene	seed	m	2	-	-	-	-	-	-	-	-	-	-	-	-	-	-	-	-	-	-	-	-	1	0.1
Lithospermum/Buglossoides	seed	m	20	-	-	-	-	-	-	-	-	-	-	1.00	0.101	1	0.2	-	-	-	-	1	1.0	7	0.4
Medicago	seed	m	3	-	-	-	-	-	-	-	-	1	0.01	2.00	0.201	-	-	-	-	-	-	-	-	-	-
Ornithopus perpusillus/sativus	seed	c	9	2	3.9	-	-	-	-	-	-	-	-	2.00	0.201	-	-	-	-	-		-		-	
Ornithopus perpusillus/sativus	seed	m	42	-	-	-	-	-	-	-	-	18	0.12	-	-	2	0.4	-	-	-	-	2	2.0	19	1.0
Persicaria	seed	c	1	-	-	-	-	-	-	-	-	-	-	-	-	-	-	-	-	-	-	-	-	-	-
Plantago L.	seed	c	22	-	-	-	-	-	-	-	-	-	-	2.00	0.201	8	1.5	-	-	-	-	-	-	1	0.1
Polygonium	seed	c	2	-	-	-	-	-	-	-	-	-	-	-	-	-	-	-	-	-	-	-	-	-	-
Polygonium	seed	m	1	-	-	-	-	-	-	-	-	-	-	-	-	-	-	-	-	-	-	-	-	-	-
Sheradia arvensis	seed	c	3	-	-	-	-	-	-	-	-	-	-	-	-	2	0.4	-	-	-	-	-	-	-	-
Sheradia arvensis	seed	m	2	-	-	-	-	-	-	-	-	-	-	-	-	-	-	-	-	-	-	-	-	-	-
Solanum nigrum	seed	c	1	-	-	-	-	-	-	-	-	-	-	-	-	-	-	-	-	-	-	-	-	-	-
Solanum nigrum	seed	m	1	-	-	-	-	-	-	-	-	-	-	-	-	-	-	-	-	-	-	-	-	-	-
Trifolium L.	seed	c	11	-	-	1	0.4	-	-	-	-	-	-	3.00	0.302	3	0.6	1	0.1	-	-	-	-	1	0.1
Trifolium L.	seed	m	3	-	-	-	-	-	-	-	-	-	-	3.00	0.302	-	-	-	-	-	-	-	-	-	-
Rubus sp.	seed	c	16	-	-	1	0.4	-	-	-	-	-	-	2.00	0.201	11	2.1	-	-	-	-	-	-	-	-
Rubus sp.	seed	m	9	-	-	-	-	-	-	-	-	1	0.01	-	-	-	-	-	-	-	-	-	-	-	-
cf. Rubus sp.	seed	c	1	-	-	-	-	-	-	-	-	-	-	1.00	0.101	-	-	-	-	-	-	-	-	-	-
Rumex	nutlet	c	3	-	-	-	-	-	-	-	-	-	-	-	-	-	-	-	-	-	-	-	-	-	-
Rumex	nutlet	m	1	-	-	-	-	-	-	-	-	-	-	1.00	0.101	-	-	-	-	-	-	-	-	-	-
Umbelliferae	seed	m	1	-	-	1	0.4	-	-	-	-	-	-	-	-	-	-	-	-	-	-	-	-	-	-
Viburnum lantana	seed	c	1	-	-	-	-	-	-	-	-	-	-	-	-	-	-	-	-	-	-	-	-	-	-
Viburnum lantana	seed	m	1	-	-	-	-	-	-	-	-	-	-	-	-	-	-	-	-	-	-	-	-	1	0.1
Total			23018	51	100	235	100	2	100	7	100	15278	100	993	100	520	100	1059	100	87	100	98	100	1935	100

APPENDIX

		House of the Surgeon 1BC		House of the Surgeon 1AD		Inn 1BC		Inn 1AD		Vestals Bar 1BC		Vestals Bar 1AD		Inn Bar 1BC		Inn Bar 1AD		Well 1BC		Well 1AD		Bar of Acisculus 1BC		Bar of Acisculus 1AD		Bar of Pheobus 1BC		Bar of Pheobus 1AD	
		VI.I.x		VI.I.x		VI.I.iv		VI.I.iv		VI.I.v		VI.I.v		VI.I.iii		VI.I.iii		VI.I.xix		VI.I.xix		VI.I.xvii		VI.I.xvii		VI.I.xviii		VI.I.xviii	
Preservation		∑	Ub	∑	Ub	∑	Ub	∑	Ub	∑	Ub	∑	Ub	∑	Ub	∑	Ub	∑	Ub	∑	Ub	∑	Ub	∑	Ub	∑	Ub	∑	Ub
Fruits/nuts:																													
Cerasus avium L.	m	-	-	-	-	-	-	-	-	-	-	-	-	-	-	-	-	-	-	-	-	-	-	-	-	-	-	-	-
Cucumis melo L.	c	-	-	1	0.5	-	-	-	-	-	-	-	-	-	-	-	-	-	-	-	-	-	-	-	-	-	-	-	-
Cucumis cf. melo L.	c	-	-	1	0.5	-	-	-	-	-	-	-	-	-	-	-	-	-	-	-	-	-	-	-	-	-	-	-	-
Cucumis cf. melo L.	m	-	-	-	-	-	-	1	0.5	-	-	-	-	-	-	-	-	-	-	-	-	-	-	-	-	-	-	-	-
Cucumis L.	m	-	-	-	-	-	-	-	-	-	-	-	-	-	-	-	-	-	-	-	-	10	12.9	-	-	-	-	-	-
Ficus carica L.	c	1	2.5	-	-	4	0.2	-	-	-	-	1	1.1	2	4.0	-	-	-	-	10	12.0	-		-	-	-	-	-	-
Ficus carica L.	m	1	2.5	24	12.5	1578	84.2	3	1.6	2	23.8	19	20.3	30	59.9	56	69.1	-	-	65	78.3	15	19.4	1	2.6	-	-	2	22.2
Malus domestica Borkh	c	-	-	-	-	-	-	-	-	-	-	-	-	-	-	-	-	-	-	-	-	-	-	-	-	-	-	-	-
Malus domestica Borkh.	m	-	-	-	-	-	-	-	-	-	-	-	-	-	-	-	-	-	-	-	-	-	-	-	-	-	-	-	-
cf. Malus domestica Borkh	m	-	-	-	-	-	-	-	-	-	-	-	-	-	-	-	-	-	-	-	-	-	-	-	-	-	-	-	-
cf. Malus domestica Borkh	c	-	-	-	-	-	-	-	-	-	-	-	-	-	-	-	-	-	-	-	-	-	-	-	-	-	-	-	-
Olea europea L.	c	8	18.7	48	25.0	120	6.4	24	12.7	-	-	2	2.6	1	2.3	2	2.5	1	9.9	5	6.0	1	0.7	9	24.5	-	-	2	22.6
Punica granatum L	c	-	-	-	-	1	0.1	1	0.5	-	-	-	-	-	-	-	-	-	-	-	-	-	-	-	-	-	-	-	-
Punica granatum L	m	-	-	-	-	-	-	-	-	-	-	-	-	-	-	-	-	-	-	-	-	-	-	-	-	-	-	-	-
cf. Punica granatum L.	c	-	-	-	-	-	-	-	-	-	-	-	-	-	-	-	-	-	-	-	-	-	-	-	-	-	-	-	-
Prunus persica (L.) Batsch	c	-	-	-	-	-	-	1	0.5	-	-	-	-	-	-	2	2.5	-	-	-	-	-	-	-	-	-	-	-	-
Prunus sp.	m	-	-	-	-	-	-	-	-	-	-	1	1.1	-	-	-	-	-	-	-	-	-	-	-	-	-	-	-	-
Vitis vinifera L.	c	3	7.4	33	17.3	29	1.5	3	1.4	3	30.6	7	7.1	-	-	0.1	0.2	12	90.1	-	-	0.4	0.5	3	7.6	-	-	-	-
Vitis vinifera L.	m	3	7.4	2	1.1	19	1.0	21	11.1	2	21.2	14	15.4	0.9	1.8	1	1.5	-	-	-	-	24	30.5	0	0.2	-	-	-	-
Vasicular tissue indeterminate	c	-	-	6	3.1	68	3.6	37	19.5	-	-	4	4.3	3	6.0	12	14.8	-	-	-	-	7	9.1	14	36.5	-	-	-	-
Fruit exocarp	c	-	-	-	-	2	0.1	-	-	1	11.9	3	3.2	-	-	-	-	-	-	-	-	-	-	2	5.2	-	-	1	11.1
Amygdalus communis L.	c	-	-	-	-	-	-	-	-	-	-	-	-	-	-	-	-	-	-	-	-	-	-	-	-	-	-	3	33.3
Corylus avellana L	c	-	-	-	-	2	0.1	-	-	-	-	1	1.1	-	-	-	-	-	-	-	-	-	-	-	-	-	-	-	-
cf. Corylus avellana L.	c	-	-	-	-	-	-	-	-	-	-	-	-	-	-	-	-	-	-	-	-	-	-	-	-	-	-	-	-
Juglans regia L	c	1	2.5	4	2.1	4	0.2	1	0.5	-	-	3	3.2	1	2.0	-	-	-	-	-	-	-	-	-	-	-	-	-	-
cf. Juglans regia L.	c	-	-	-	-	-	-	-	-	-	-	-	-	-	-	-	-	-	-	-	-	-	-	-	-	-	-	-	-
Pinus pinea L	c	1	2.5	-	-	2	0.1	-	-	-	-	2	2.1	2	4.0	-	-	-	-	-	-	4	5.2	-	-	-	-	-	-
Pinus pinea L	m	-	-	-	-	-	-	1	0.5	-	-	-	-	-	-	-	-	-	-	-	-	-	-	-	-	-	-	1	11.1
Nuts (generic)	c	-	-	-	-	1	0.1	-	-	-	-	1	1.1	3	6.0	1	1.2	-	-	-	-	1	1.3	1	2.6	-	-	-	-
Cereals																													
T. aestivum/T.durum L.	c	-	-	1	0.5	-	-	-	-	-	-	-	-	-	-	-	-	-	-	-	-	-	-	-	-	-	-	-	-
T. cf. aestivum/T.durum L.	c	-	-	2	1.0	-	-	-	-	-	-	-	-	-	-	-	-	-	-	-	-	-	-	1	2.6	-	-	-	-
cf. T.aestivum/durum L	c	-	-	-	-	3	0.2	2	1.1	-	-	-	-	-	-	-	-	-	-	-	-	-	-	-	-	-	-	-	-
T. dicoccum Schübl.	c	-	-	1	0.5	-	-	-	-	-	-	1	1.1	-	-	-	-	-	-	-	-	-	-	-	-	-	-	-	-
T. cf. dicoccum Schübl.	c	1	2.5	3	1.6	1	0.1	-	-	-	-	-	-	-	-	-	-	-	-	-	-	-	-	-	-	-	-	-	-
cf. T.dicoccum Schübl.	c	-	-	-	-	-	-	-	-	-	-	-	-	-	-	-	-	-	-	-	-	-	-	-	-	-	-	-	-
T cf. monoccocum		1	2.5	-	-	-	-	-	-	-	-	-	-	-	-	-	-	-	-	-	-	-	-	-	-	-	-	-	-
Triticum sp	c	4	9.9	21	10.9	4	0.2	5	2.6	1	11.9	3	3.2	1	2.0	1	1.2	-	-	1	1.2	2	2.6	1	2.6	-	-	-	-
Hordeum vulgare L	c	-	-	-	-	-	-	5	2.6	-	-	-	-	-	-	1	1.2	-	-	-	-	-	-	-	-	-	-	-	-
cf. Hordeum vulgare L.	c	-	-	-	-	-	-	-	-	-	-	-	-	-	-	-	-	-	-	1	1.2	-	-	-	-	-	-	-	-
Secale cereale	c	-	-	-	-	-	-	1	0.5	-	-	-	-	-	-	-	-	-	-	-	-	-	-	-	-	-	-	-	-
Setaria italica Beauv	c	-	-	1	0.5	-	-	-	-	-	-	-	-	-	-	-	-	-	-	-	-	-	-	-	-	-	-	-	-
Setaria cf. italica Beauv.	c	-	-	-	-	-	-	-	-	-	-	-	-	-	-	-	-	-	-	-	-	-	-	-	-	-	-	-	-
cf. Setaria italica Beauv	c	-	-	2	1.0	-	-	-	-	-	-	-	-	-	-	-	-	-	-	-	-	-	-	-	-	-	-	-	-
cf. Setaria	c	-	-	-	-	1	0.1	-	-	-	-	-	-	1	2.0	1	1.2	-	-	-	-	-	-	1	2.6	-	-	-	-
Panicum miliaceum L	c	1	2.5	-	-	-	-	-	-	-	-	-	-	-	-	-	-	-	-	-	-	-	-	-	-	-	-	-	-
Panicum miliaceum L	m	-	-	-	-	-	-	-	-	-	-	-	-	-	-	-	-	-	-	-	-	10	12.9	-	-	-	-	-	-
Panicum cf. miliaceum L.	c	1	2.5	2	1.0	-	-	-	-	-	-	-	-	-	-	-	-	-	-	-	-	-	-	-	-	-	-	-	-
Panicum sp	c	4	9.9	15	7.8	-	-	1	0.5	-	-	-	-	-	-	1	1.2	-	-	-	-	-	-	-	-	-	-	-	-
Panicum sp	m	-	-	-	-	-	-	-	-	-	-	-	-	-	-	-	-	-	-	-	-	-	-	-	-	-	-	-	-
Pulses																													
Lens culinaris Medik	c	2	4.9	5	2.3	0.50	0.03	0.3	0.2	-	-	0.8	0.9	-	-	-	-	-	-	-	-	2	2.6	0.3	0.9	-	-	-	-
cf. Lens culinaris Medik.	c	-	-	0.5	0.3	-	-	-	-	-	-	-	-	-	-	-	-	-	-	-	-	-	-	-	-	-	-	-	-
Lens culinaris Medik	m	-	-	1	0.5	-	-	-	-	-	-	-	-	-	-	-	-	-	-	-	-	-	-	-	-	-	-	-	-
cf. Lens culinaris Medik.	m	-	-	0.5	0.3	-	-	-	-	-	-	-	-	-	-	-	-	-	-	-	-	-	-	-	-	-	-	-	-
Pisum sativum L	c	-	-	-	-	-	-	0.5	0.3	-	-	-	-	-	-	-	-	-	-	-	-	-	-	-	-	-	-	-	-
cf. Pisum sativum L.	c	-	-	-	-	-	-	-	-	-	-	1	1.1	-	-	-	-	-	-	-	-	-	-	-	-	-	-	-	-
Vicia ervilia (L.) Willd	c	-	-	-	-	-	-	-	-	-	-	-	-	-	-	-	-	-	-	-	-	-	-	-	-	-	-	-	-
Vicia cf. ervilia (L.) Willd	c	-	-	1	0.5	-	-	-	-	-	-	-	-	-	-	-	-	-	-	-	-	-	-	-	-	-	-	-	-
cf. Vicia ervilia (L.) Willd	c	-	-	-	-	-	-	1	0.5	-	-	-	-	-	-	-	-	-	-	-	-	-	-	-	-	-	-	-	-
Vicia faba L.	c	-	-	-	-	-	-	-	-	-	-	-	-	-	-	-	-	-	-	-	-	-	-	1	2.6	-	-	-	-
cf. Vica faba L.	c	-	-	-	-	-	-	-	-	-	-	-	-	-	-	-	-	-	-	-	-	-	-	-	-	-	-	-	-
Vicia sativa L	c	1	2.5	0.5	0.3	1	0.1	7	3.7	-	-	3	3.2	-	-	-	-	-	-	-	-	-	-	1	2.6	-	-	-	-
Vicia cf. sativa L	c	-	-	0.5	0.3	3	0.1	-	-	-	-	-	-	-	-	-	-	-	-	-	-	-	-	-	-	-	-	-	-
cf. Vicia sativa L	c	-	-	1	0.7	4	0.2	-	-	-	-	-	-	-	-	-	-	-	-	-	-	-	-	-	-	-	-	-	-
Vicia sp.	c	5	12.3	7	3.6	-		38	19.8	-	-	6	6.8	-	-	3	3.3	-	-	-	-	0.3	0.4	2	4.3	-	-	-	-
Weedy Species																													
Panicoid	c	-	-	1	0.5	4	0.2	16	8.4	-	-	11	11.8	-	-	-	-	-	-	1	1.2	1	1.3	1	2.6	1	100.0	-	-
Panicoid	m	-	-	-	-	1	0.1	-	-	-	-	-	-	-	-	-	-	-	-	-	-	-	-	-	-	-	-	-	-
cf. Panicoid	c	-	-	-	-	-	-	-	-	-	-	-	-	-	-	-	-	-	-	-	-	-	-	-	-	-	-	-	-
Agrostemma githago	c	-	-	-	-	-	0.0	-	-	-	-	-	-	-	-	-	-	-	-	-	-	-	-	-	-	-	-	-	-
Agrostemma githago	m	-	-	-	-	1	0.1	-	-	-	-	-	-	-	-	-	-	-	-	-	-	-	-	-	-	-	-	-	-
Apiaceae	c	-	-	2	1.0	-	-	-	-	-	-	-	-	-	-	-	-	-	-	-	-	-	-	-	-	-	-	-	-
Arrhenatherum elatius ssp. bulbosum	c	-	-	-	-	-	-	1	0.5	-	-	-	-	-	-	-	-	-	-	-	-	-	-	-	-	-	-	-	-
Celtis L.	m	-	-	-	-	-	-	1	0.5	-	-	-	-	-	-	-	-	-	-	-	-	-	-	-	-	-	-	-	-
Cupressus sempervirens	c	-	-	-	-	-	-	-	-	-	-	-	-	-	-	-	-	-	-	-	-	-	-	-	-	-	-	-	-
Cupressus sempervirens	m	-	-	-	-	-	-	-	-	-	-	-	-	1	2.0	-	-	-	-	-	-	-	-	-	-	-	-	-	-
Echium vulgare	c	-	-	-	-	-	-	-	-	-	-	-	-	-	-	-	-	-	-	-	-	-	-	-	-	-	-	-	-
Echium vulgare	m	-	-	-	-	-	-	-	-	-	-	1	1.1	1	2.0	-	-	-	-	-	-	-	-	-	-	-	-	-	-
cf. Echium vulgare	m	-	-	-	-	-	-	-	-	-	-	-	-	-	-	-	-	-	-	-	-	-	-	-	-	-	-	-	-
Galium aparine L	c	-	-	-	-	-	-	3	1.6	-	-	-	-	-	-	-	-	-	-	-	-	-	-	-	-	-	-	-	-
Galium cf. aparine L		-	-	-	-	-	-	-	-	-	-	-	-	-	-	-	-	-	-	-	-	1	1.3	-	-	-	-	-	-
Illex aquifolium L	m	-	-	-	-	-	-	2	1.1	-	-	-	-	-	-	-	-	-	-	-	-	-	-	-	-	-	-	-	-
Lamiaceae	c	1	2.5	1	0.5	-	-	-	-	-	-	-	-	-	-	-	-	-	-	-	-	-	-	-	-	-	-	-	-
Linum biene	m	-	-	-	-	1	0.1	-	-	-	-	-	-	-	-	-	-	-	-	-	-	-	-	-	-	-	-	-	-
Lithospermum/Buglossoides	m	-	-	-	-	4	0.2	2	1.1	-	-	1	1.1	3	6.0	-	-	-	-	-	-	-	-	-	-	-	-	-	-
Medicago	m	-	-	-	-	-	-	-	-	-	-	-	-	-	-	-	-	-	-	-	-	-	-	-	-	-	-	-	-
Ornithopus perpusillus/sativus	c	1	2.5	1	0.5	-	-	-	-	-	-	3	3.2	-	-	-	-	-	-	-	-	-	-	-	-	-	-	-	-
Ornithopus perpusillus/sativus	m	-	-	-	-	-	-	-	-	-	-	1	1.1	-	-	-	-	-	-	-	-	-	-	-	-	-	-	-	-
Persicaria	c	-	-	-	-	1	0.1	-	-	-	-	-	-	-	-	-	-	-	-	-	-	-	-	-	-	-	-	-	-
Plantago L.	c	-	-	-	-	3	0.2	8	4.2	-	-	-	-	-	-	-	-	-	-	-	-	-	-	-	-	-	-	-	-
Polygonium	c	-	-	-	-	2	0.1	-	-	-	-	-	-	-	-	-	-	-	-	-	-	-	-	-	-	-	-	-	-
Polygonium	m	-	-	1	0.5	-	-	-	-	-	-	-	-	-	-	-	-	-	-	-	-	-	-	-	-	-	-	-	-
Sheradia arvensis	c	-	-	-	-	-	-	1	0.5	-	-	-	-	-	-	-	-	-	-	-	-	-	-	-	-	-	-	-	-
Sheradia arvensis	m	-	-	-	-	1	0.1	1	0.5	-	-	-	-	-	-	-	-	-	-	-	-	-	-	-	-	-	-	-	-
Solanum nigrum	c	1	2.5	-	-	-	-	-	-	-	-	-	-	-	-	-	-	-	-	-	-	-	-	-	-	-	-	-	-
Solanum nigrum	m	-	-	1	0.5	-	-	-	-	-	-	-	-	-	-	-	-	-	-	-	-	-	-	-	-	-	-	-	-
Trifolium L.	c	-	-	-	-	2	0.1	-	-	-	-	-	-	-	-	-	-	-	-	-	-	-	-	-	-	-	-	-	-
Trifolium L.	m	-	-	-	-	-	-	-	-	-	-	-	-	-	-	-	-	-	-	-	-	-	-	-	-	-	-	-	-
Rubus sp.	c	-	-	-	-	-	-	1	0.5	-	-	1	1.1	-	-	-	-	-	-	-	-	-	-	-	-	-	-	-	-
Rubus sp.	m	-	-	-	-	7	0.4	1	0.5	-	-	-	-	-	-	-	-	-	-	-	-	-	-	-	-	-	-	-	-
cf. Rubus sp.	c	-	-	-	-	-	-	-	-	-	-	-	-	-	-	-	-	-	-	-	-	-	-	-	-	-	-	-	-
Rumex	c	-	-	1	0.5	-	-	-	-	-	-	2	2.1	-	-	-	-	-	-	-	-	-	-	-	-	-	-	-	-
Rumex	m	-	-	-	-	-	-	-	-	-	-	-	-	-	-	-	-	-	-	-	-	-	-	-	-	-	-	-	-
Umbelliferae	m	-	-	-	-	-	-	-	-	-	-	-	-	-	-	-	-	-	-	-	-	-	-	-	-	-	-	-	-
Viburnum lantana	c	-	-	-	-	1	0.1	-	-	-	-	-	-	-	-	-	-	-	-	-	-	-	-	-	-	-	-	-	-
Viburnum lantana	m	-	-	-	-	-	-	-	-	-	-	-	-	-	-	-	-	-	-	-	-	-	-	-	-	-	-	-	-
Total		41	100	192	100	1875	100	190	100	8	99	94	100	50	100	81	100	14	100	83	100	77	100	38	100	1	100	9	100